Simon Hornig

State Space Theory of Discrete Linear Control

State Space Theory of Discrete Linear Control

VLADIMÍR STREJC

Czechoslovak Academy of Sciences
Institute of Information Theory and Automation
Prague, Czechoslovakia

A Wiley-Interscience Publication

JOHN WILEY & SONS

Chichester • New York • Brisbane • Toronto

Scientific Editor
Professor Ing. Stanislav Kubík, DrSc.
Corresponding Member of the Czechoslovak Academy of Sciences

Reviewer
Professor Ing. Zdeněk Kotek, DrSc.
Corresponding Member of the Czechoslovak Academy of Sciences

Published in co-edition with ACADEMIA, Publishing House of the Czechoslovak Academy of Sciences, Prague

Copyright © Vladimír Strejc, Prague, 1981

All rights reserved

No part of this publication may be reproduced by any means, nor transmitted, nor translated into a machine language without the written permission of the publisher

Library of Congress Cataloging in Publication Data

Strejc, Vladimír.
   State space theory of discrete linear control.
   'A Wiley-Interscience publication.'
   Bibliography: p
   1. Control theory. I. Title.
QA402.3.S819   1980   629.8'312   79-991
ISBN 0 471 27594 8

Printed in Czechoslovakia

Contents

PREFACE 9

1. INTRODUCTION 13

2. SYSTEM STATE 15

   2.1 Input, state and output 15
   2.2 Zadeh's definition of system state 18

3. EQUATIONS OF CONTINUOUSLY WORKING SYSTEMS 21

   3.1 Mathematical model of systems in state space 21
   3.2 Transcription of an ordinary differential equation to state equations . . . 26
      3.2.1 Procedure I — the state variables are defined by the relation $x_{i+1} = x'_i$ . . 27
      3.2.2 Procedure II — the state variables are a linear combination of the derivatives of the plant input and output variables 30
      3.2.3 State equation under the action of a controlling and a disturbing variables . 32
   3.3 Continuous multi-input/multi-output controlled plant 34

4. OPERATORS 38

   4.1 Basic properties 38
   4.2 Linearity 39
   4.3 Stationarity in time 40
   4.4 System equivalence 40

5. SOLUTION OF THE CONTINUOUS-TIME STATE EQUATION 42

   5.1 Solution in the time domain 42
   5.2 Decomposition into partial fractions — Jordan form 44
   5.3 Relationship to Laplace transformation 47

6. TRANSCRIPTION OF AN ORDINARY DIFFERENCE EQUATION INTO STATE EQUATIONS 48

   6.1 Single-input/single-output plant 48
      6.1.1 Procedure I — the state variables are defined by the relation $x_i(k+1) = x_{i+1}(k)$ 48

  6.1.2 Procedure II — the state variables are a linear combination of the values of the plant input and output variables 51
  6.1.3 Discrete-plant state equation under action of the controlling and disturbing variables 54
 6.2 Multi-input/multi-output plant 55

7. SOLUTION OF THE DISCRETE-TIME STATE EQUATION 59

 7.1 Solution in the time domain 59
 7.2 Weighting matrix 60
 7.3 Decomposition into partial fractions — Jordan form 61
 7.4 Connection of the solution with the Z-transformation 65
 7.5 Continuously working controlled plant — impulse sequence input signal 67
 7.6 Continuously working controlled plant — staircase function input signal 68
 7.7 Relationships between the coefficients of continuous and discrete system models . 74
  7.7.1 Calculation according to the definition of exp $[FT]$ 74
  7.7.2 Use of the relationship to Laplace transformation 75
  7.7.3 Faddejev's algorithm 76
  7.7.4 Ježek's method for the calculation of discrete and continuous transfer functions 79
 7.8 Special types of sampling 86
  7.8.1 Asynchronous sampling 87
  7.8.2 Finite sampling time 90

8. MUTUAL CONFIGURATION OF DISCRETE SYSTEMS 92

 8.1 Parallel configuration 92
 8.2 Series configuration 93
 8.3 Feedback configuration 94

9. SYSTEM PROPERTIES 97

 9.1 Reachability, controllability and stabilizability 97
 9.2 Observability, reconstructability and detectability 102
 9.3 Identifiability 105
 9.4 Canonical decomposition 106

10. TRANSFORMATIONS 110

 10.1 Linear transformation 110
 10.2 Jordan matrix 113
  10.2.1 Distinct eigenvalues 113
  10.2.2 Multiple eigenvalues 117
  10.2.3 Invariant factors 129
  10.2.4 Using the Jordan matrix for numerical calculations 134
  10.2.5 Calculations involving matrices with complex elements 137
 10.3 Canonical form of reachability 139
 10.4 Canonical form of controllability 141

CONTENTS

    10.5 Canonical form of observability 143
    10.6 Canonical form of reconstructability 145
    10.7 Relationship of the characteristic equation to impulse response 147

11. MULTI-INPUT/MULTI-OUTPUT SYSTEMS 149

    11.1 Asynchronous sampling 150
    11.2 Index of reachability and observability of multi-input/multi-output systems . . . 154
    11.3 Canonical forms of multi-input/multi-output systems 162
        11.3.1 Jordan's canonical form 162
        11.3.2 Canonical forms of reachability and observability with one-way coupling . . 165
        11.3.3 Canonical forms with two-way coupling 170
    11.4 Minimum description 181

12. STABILITY 186

    12.1 Basic definitions 186
    12.2 Routh-Shur stability test 195
    12.3 Lyapunov's stability theorem 197

13. IDENTIFICATION OF MODEL PARAMETERS 205

    13.1 Deterministic methods 207
        13.1.1 Minimum realization 207
        13.1.2 Determination of the difference equation coefficients 212
    13.2 The least squares method 214
        13.2.1 Mathematical controlled plant models 215
        13.2.2 Least sum of squares 220
        13.2.3 Geometrical interpretation 221
        13.2.4 Increasing the number of parameters 222
        13.2.5 Increasing the number of samples 223
        13.2.6 Properties of least squares estimates 226
    13.3 Method of maximum likelihood 228
        13.3.1 Calculation of maximum likelihood estimates 229
        13.3.2 Properties of maximum likelihood estimates 233
        13.3.3 Determination of state model parameters 235

14. SYSTEM STATE VECTOR ESTIMATION 237

    14.1 Deterministic state estimator 237
        14.1.1 Estimator of the order n 237
        14.1.2 Estimator of the reduced order 239
    14.2 Statistical and probabilistic state estimation 241
        14.2.1 State estimation in the Kalman sense 241
        14.2.2 State estimate by the maximum likelihood method 252
        14.2.3 State estimate in the Bayes' sense 253
        14.2.4 Parameter estimate using the Kalman filter 257

15. PRINCIPLES OF DETERMINISTIC SYNTHESIS 259

    15.1 Pole assignment problem 263
        15.1.1 Use of the canonical form of controllability 263
        15.1.2 Use of the general form of the matrix of dynamics 266
    15.2 Finite number of control steps 270
        15.2.1 Transition to equilibrium state. 270
        15.2.2 Transition to steady state 279
        15.2.3 Weak version of the finite number of control steps 290
    15.3 Determination of the characteristic polynomial of an estimator 298
    15.4 Principle of separability 309
    15.5 Invariants 313
    15.6 Command control 317
    15.7 Disturbance-variable compensation 323

16. DETERMINISTIC CONTROLLER DESIGN BASED ON QUADRATIC COST FUNCTIONS 332

    16.1 Optimum control based on the second method of Lyapunov 334
        16.1.1 General procedure 334
        16.1.2 Simplified modifications 338
    16.2 The discrete maximum principle 339
        16.2.1 General procedure 339
        16.2.2 Simplified modifications 343
    16.3 Dynamic programming 348
        16.3.1 General procedure 348
        16.3.2 Infinite number of control steps 350
    16.4 Numerical solution 350
    16.5 Analytical solution 355
        16.5.1 Positive definite matrix R 355
        16.5.2 Plant output optimization 362
    16.6 Minimum sum of squares of the control error 368
    16.7 Application of the estimator of non-measurable state variables 378

17. SYNTHESIS OF STOCHASTIC SYSTEMS 382

    17.1 Synthesis in the Kalman sense 382
    17.2 The separation principle 390
    17.3 Principle of duality 393

APPENDICES 396

    Appendix A: Matrix inversion lemma 396
    Appendix B: Test of the linear dependence of vectors 397
    Appendix C: Definiteness and semidefiniteness 399
    Appendix D: Numerical solution of an overdetermined set of linear algebraic equations . 402

REFERENCES 407

SUBJECT INDEX 422

Preface

When perusing the journals and books dealing with theoretical problems of automatic control we find that in most of the literature of the last decade the description of dynamic behaviour of systems, the system analysis and the design of optimum control is based on the state space representation. The formerly used methods, designated today as classical, based e.g. on frequency analysis, on root locus investigating the roots of the characteristic equations or on the compensation of unsuitable root factors in the polynomials of the transfer functions of linear systems, already receded into the background. Their place was taken up by a new, considerably more abstract system theory and by methods of analysis and synthesis which enable us to solve, besides a deeper theoretical analysis, much more complex problems than the classical methods do and, in addition, facilitate the formalization of results for a computer-aided numerical solution. These advantages were given utter preference over the simplicity and clarity of the classical methods. It may also be noted that if a different system theory should be taken up in the future it would have to bring at least one additional advantage besides the two mentioned above.

Although the book is a monothematically written work devoted exclusively to the state space and its use for the solution of automatic control problems, the treatment of the subject was intentionally selected so as to relate the current concepts of the classical theory with the theory of state space in order that the connection with the previously acquired methods of the classical solution could as far as possible be preserved.

The subject matter selected relates to continuously working technological plants controlled by means of a digital computer. In this case the continuous plant variables are given only by their discrete values at selected instants of time, while closed loops, control systems and their parts are described by discrete state equations with discrete and discretized variables. Therefore, the book is devoted exclusively to the methods of analysis and synthesis of discrete systems in the state space, although the aim is in fact the control of continuously working plants. The description in the state space with continuous variables in Chap. 3 and 5 serves only to show the relationships between the two versions of the representation of a continuously working plant, i.e. the continuous and the discrete version of representation.

On this occasion it should be pointed out that the direct digital control of complex systems corresponds to the real technical possibilities which have already been successfully verified for these purposes. On the other hand, the continuously working control technique is restricted, for economical reasons, merely to simple plants. Therefore, the analysis and synthesis of continuous systems in the state space has only a negligible importance for the continuous control of technological systems and was for this reason intentionally not included in this publication.

As can be seen from the contents of the book, the text proceeds from the simpler problems to the more complex ones. It proceeds, for instance, from single-input/single-output to multi-input/multi-output systems, from deterministic to stochastic problems, etc. This procedure enables the beginner to become acquainted first with the simpler problem solutions and, if need be, to extend his knowledge to more complex problems.

The description of systems in the state space as well as the methods of analysis and synthesis using the state space are based on the representation in terms of matrices and vectors. Although this publication includes all special matrix and vector properties, as far as they are closely related with control theory, the basic concepts and rules, particularly of matrix calculus, will have to be looked up in the appropriate literature.

It certainly need not be emphasized that anybody knowing the concepts and methods of the classical control theory will find the mathematical means of the state space theory easier to acquire. In this case, the understanding of the relationships with the classical theory and physical interpretation is also substantially facilitated. Nevertheless, the author took every effort to treat the subject matter in such a way as to enable the reader to understand it without the knowledge of the classical theory. Practically, this means that the reader not familiar with the classical theory may simply omit the text portions dealing with the relationships to this theory. This concerns particularly Sections 5.3, 7.4, 10.7 and Par. 7.7.2.

The book is intended for those who need to get acquainted with the state space theory of control and who wish to acquire a deeper knowledge of system theory. It is also intended for those who design and realize control systems of technological processes and are engaged in writing control algorithms. Besides, the book may serve as a superior course for selected students and as an aid for a post-graduate specialization in state space theory.

In conclusion I take this opportunity to extend my sincere appreciation to all my collaborators who helped me in whatever way in the preparation of the manuscript, drawing the illustrations, performing certain test calculations on the computer, etc. or assisted indirectly by solving problems not related to this manuscript thus enabling me in turn to work on the manuscript more intensively. I wish to express my thanks particularly to Ing. V. Peterka, DrSc., Ing. V. Kučera, DrSc., RNDr. A. Tuzar, CSc.

for their numerous comments and suggestions, to Mrs. J. Novovičová, grad. math., for the careful proof-reading of the entire manuscript and to Ing. R. Major for his unusual conscientiousness in providing the English translation.

Professor Ing. V. Strejc, DrSc.

Corresponding Member of the Czechoslovak Academy of Sciences

1

Introduction

In the period of recent years a tremendous development of state space methods in the field of automatic control has been recorded. A similar general trend may also be observed in other fields of application of system theory. The control methods based on frequency analysis, transfer function algebra, Laplace transformation and Z-transformation, which may now be regarded as classical, played a significant role in the development and application of control theory and the related fields of automation. Owing to their simplicity and clear connection to physical reality they will apparently keep their position among the more modern methods even in the future. However, they cannot hold their position in the solution of multivariable systems and complex control loops, where the classical methods frequently fail for purely computational reasons, while the state space methods permit a clean-cut formalization and mechanization of the calculation procedures.

The term "state space methods" is in fact only a new name for all kinds of methodical procedures that have already been used for a long time, e.g. in analytical dynamics, quantum mechanics, theory of stability, in the solution of ordinary differential equations and in other fields. The application of these methods to automatic control was stimulated in the second half of the fifties mainly by the work of L. S. Pontryagin et al. [61.3], by the method of dynamic programming suggested by R. E. Bellman [57.1] and by the general theory of filtering and the control theory elaborated by R. E. Kalman [59.2]. The present works taking advantage of state space are very numerous and their list would be very extensive.

Of the substantial state space advantages we mention, for instance, the uniform formulation of different problems, the possibility of an easy solution of control problems with a larger number of controlling and controlled variables, the solution of asynchronous sampling problems, finite sampling time problems and nonperiodic sampling problems, the investigation of time-variant and nonlinear control systems.

The description of systems in state space allows us to reveal and investigate such properties which, when described by the classical input/output methods and by frequency analysis, would remain concealed. The matrix form of writing has indisputable advantages in the computer-aided numerical solution, while the clearness of mathematical formulation as well as the solution itself remain unimpaired even for multivariable and complex systems.

However, the state space control theory has also some disadvantages. If the state variables are directly measurable or at least simply definable physical quantities, the relationship between the mathematical description and physical reality remains preserved and also the system simulation on analogue and digital computers directly follows from the mathematical description. If, however, the choice of the state variables is contingent on special circumstances, e.g. on the necessity to simplify the matrices of the state description so as to facilitate as much as possible the numerical calculation, then the state variables will often be defined in a rather complex form as linear combinations of measurable or definable variables of the system. The connection with physical reality is lost and the calculations cannot be checked by a physical consideration.

2

System State

2.1 Input, state and output

The dynamics of a system is described by its mathematical model. Such a model represents the mathematical relations between three sets of variables: input, output and state variables.

The system input expressed by a set of time functions or by a set of time sequences of input values represents the external variables acting on the system. The output, expressed similarly to the input, represents the description of a directly observable behaviour of the system.

The basic property of any dynamic system is that its behaviour at any instant depends not only on the variables acting on it at that very instant, but also on the variables having acted on it in the past. We may consider the system as having a "memory" storing the contribution of the variable that acted on it in the past to its behaviour at the instant of observation. The system state defined as a set of values of the so-called state variables represents the instantaneous state of the "cells" of this memory. If the state at an arbitrary instant t_0 and the input segment $u\langle t_0, t\rangle$ are known, the output and state at any instant within the interval $t \geqq t_0$ may be determined.

The afore-said definition of state is only approximate, giving merely a preliminary information on the concepts "state" and "state variables". The exact definition is given in Sec. 2.2.

The concept "dynamic system" is also only approximate. The conventional meaning of the term "dynamic" is almost the same as of the term "causal": past events influence future events but not vice versa. A mathematical notion of the dynamic system leads to emphasizing and formalizing the flow of causation from the past to the future. From the viewpoint of mathematics the dynamic system is an axiomatic concept. We accept here the definition formulated by R.E. Kalman [69.4].

DEFINITION 2.1: *A dynamic system Σ is a complex mathematical concept defined by the following axioms:*

(a) *The following sets are given: the set of instants T, the set of states X, the set of input values U, the set of acceptable input functions $\Omega = \{\omega: T \to U\}$, the set of output values Y and the set of output functions $\Gamma = \{\gamma: T \to Y\}$.*

(b) *(Time orientation). T is an ordered subset of real numbers.*

(c) *The input space Ω satisfies the conditions:*

   (i) *(Non-triviality). Ω is not an empty set.*

   (ii) *(Input concatenation). The input segment $\omega(t_1, t_2\rangle$ is a function $\omega \in \Omega$ bounded to the time interval $(t_1, t_2\rangle \cap T$. If $\omega, \omega' \in \Omega$ and $t_1 < t_2 < t_3$, then a function $\omega'' \in \Omega$ will exist for which $\omega''(t_1, t_2\rangle = \omega(t_1, t_2\rangle$ and $\omega''(t_2, t_3\rangle = \omega'(t_2, t_3\rangle$.*

(d) *A state-transition function φ is given representing the state $x(t) = \varphi(t; \tau, x, \omega) \in X$ reached at the instant $t \in T$ from the initial state $x = x(\tau) \in X$ at the initial instant $\tau \in T$ under the action of the input $\omega \in \Omega$. The function φ has the following properties:*

   (i) *(Time orientation). The functon φ is defined for all values of $t \geq \tau$ but may not necessarily be defined for all values of $t < \tau$.*

   (ii) *(Consistency). It holds that $\varphi(t; t, x, \omega) = x$ for all $t \in T$, $x \in X$ and $\omega \in \Omega$.*

   (iii) *(Composition property). For any value of $t_1 < t_2 < t_3$ it holds that $\varphi(t_3; t_1, x, \omega) = \varphi(t_3; t_2, \varphi(t_2; t_1, x, \omega), \omega)$ for all states $x \in X$ and all inputs $\omega \in \Omega$.*

   (iv) *(Causality). If $\omega, \omega' \in \Omega$ and $\omega(t, \tau\rangle = \omega'(t, \tau\rangle$, then $\varphi(t; \tau, x, \omega) = \varphi(t; \tau, x, \omega')$.*

(e) *The output representation $\eta: T \times X \to Y$ defining the output $y(t) = \eta(t, x(t))$ exists. The representation $\eta(\sigma, \varphi(\sigma; \tau, x, \omega))$ for $\sigma \in (\tau, t\rangle$ is the output segment, i.e. the segment $\gamma(\tau, t\rangle$ of some output function $\gamma \in \Gamma$ restricted to the interval $(\tau, t\rangle$.*

It can be stated that the pair (τ, x) with $\tau \in T$ and $x \in X$ represents an event in the dynamic system Σ. The set $T \times X$ determines the space of events in the dynamic system Σ.

The system is *physically realizable* if its output and state at an arbitrary instant t_0 can be a function of only such inputs which had been acting before the instant t_0.

Sometimes it is useful to consider as physically realizable such a system whose output and state is affected by the input $u(t, t_0\rangle$, i.e. up to and inclusive of the instant t_0. In such a case there exists an output representation $\eta: T \times X \times U \times \to Y$ defining the output $y(t) = \eta(t, x(t), u(t))$ or $y(\sigma) = \eta(\sigma, \varphi(\sigma; \tau, x, \omega), u(\sigma))$ for $\sigma \in (\tau, t\rangle$.

SYSTEM STATE

A system is said to be *deterministic* if its output and state at any instant t can be determined with certainty by means of its state at some instant $t_0 < t$ and from a known input within the semiclosed interval $\langle t_0, t)$.

Conversely, a system is *stochastic* if the knowledge of its state at some instant t_0 and of its input within the interval $\langle t_0, t)$ allows to determine the system output and state at the instant $t > t_0$ with only a certain probability or by other statistical means.

Since input, state and output are constituted by finite sets of variables we express them as vectors. Thus, for instance, the input of m variables of a controlled plant is denoted by the vector

$$u^T = [u_1, u_2, ..., u_m] = [u_i], \quad i = 1, 2, ..., m$$

Similarly, state is expressed by a vector of state variables. For instance, in a linear controlled plant of the n-th order, formed by a chain of n series-connected members of the 1-st order, the state may be represented by the variables measured between the individual members of the chain and at the end of the chain. Let us denote them by $x_1, x_2, ..., x_n$.

State variables may also be selected, say, as the output variable derivatives $x_1(t)$, $dx_1(t)/dt, ..., d^{n-1}x_1(t)/dt^{n-1}$ or any other function, usually a linear combination of such system variables that the vector of the state variables fully determines the state of the system. Hence it follows that the choice of the state vector variables is unlimited, provided that all alternatives fully describe the state of the system. When choosing the state vector variables we prefer such state variables for which the calculations to be performed are simplified or which can easily be measured, etc.

The output too, is expressed by the vector of output variables. This vector is a unique function of the state vector. The output vector may even be identical with the state vector, or the output variables may be directly certain variables of the state vector. For instance, in a controlled plant constituted by a chain of series-connected members of the 1-st order, the chain output variable x_1 is one of the state vector variables

$$x^T = [x_1, x_2, ..., x_n]$$

as well as of the state vector

$$x^T = [x_1(t), dx_1(t) dt, ..., d^{n-1}x(t)/dt^{n-1}]$$

The set of all possible system inputs u will be designated as the space of inputs, U. Analogously, the set of all possible system states x will be referred to as the state space X and the set of all possible system outputs y as the space of outputs, Y.

If the input, state and output vectors are defined for each t of some interval (continuous time) we speak of a *continuous system, continuous controlled plant,* etc. If, however, the input and state vectors are defined only in the discrete instants t_k, where k is a sequence of numbers usually integers of some interval, we speak of *discrete systems.*

2.2 Zadeh's definition of system state

Although state space is very often used in the literature for the description of dynamic systems, it is quite difficult to formulate a satisfactory definition of state and state variables. Perhaps the most satisfactory definition is that published by L. A. Zadeh [63.9] introducing the state properties of the system and its state variables.

Consider the system Σ, i.e. the mathematical model of a real system with the inputs u_i, $i = 1, 2, \ldots, r$, and the outputs y_j, $j = 1, 2, \ldots, p$. It is assumed that u_i and y_j are functions of time and $u_i(t)$, $y_j(t)$ denote the values of these quantities at the instant t. The set of inputs u_i, $i = 1, 2, \ldots, r$, is expressed by the input vector u and similarly the set of outputs by the output vector y.

We denote the set of all possible input vectors, i.e. the vector functions of time, $u(t)$, which may appear at the input of a system Σ, by the symbol Ω. This is the space of input functions of the system Σ. The set of all values assumed by the vector u at the instant t is denoted as the input space U of the system Σ at the instant t. If the input space U is a finite set, it is usually called the input alphabet. In most of the cases which we shall discuss here the input space will be the Euclidean space E^n.

Consider now the following experiment: At the initial instant t_0 let the system Σ be acted upon by the input variables u up to the instant t_1, $t_1 \geq t_0$. We know by experience that the output vector $y_0(t_0, t_1\rangle$ depends, in general, not only on the input vector $u(t_0, t_1\rangle$ but also on the initial conditions of the system Σ at the beginning of the experiment. Roughly speaking, these initial conditions form the initial state of the system Σ. Therefore, having the initial state $x(t_0)$ of the system Σ we can state that the output $y(t_0, t_1)$ depends on $x(t_0)$ and $u(t_0, t_1)$ but not on the values of u before the instant t_0. In other words, the initial state and the following input values u is all we need to know for the determination of the output y after application of the input u.

The initial conditions of the system Σ may, in general be expressed in various ways. Therefore, the question arises whether a certain description of initial conditions may be designated as the initial state of Σ. Assume that $x(t_0)$ is a vector variable which can vary within the space X. Denote this space as the state space of Σ and assume, for simplicity, that X is independent of t_0. Furthermore, let $x(t) = \alpha$ be the state of Σ, where α is an element of the space X. On these assumptions we can now formulate the following definition:

SYSTEM STATE

DEFINITION 2.2: *The variable $x(t)$ may be designated as the state of the system Σ at the instant t if it satisfies the following three conditions of consistency:*

(i) *The vectors $x(t_0)$ and $u(t_0, t\rangle$, where $u(t_0, t\rangle$ denotes the input segment within the semiclosed interval $(t_0, t\rangle$, uniquely determine the output segment $y(t_0, t\rangle$ for all initial states $x(t_0) \in X$, for all values of $t \geq t_0$ and all input functions $u \in \Omega$, i.e. from the input functions space of Σ. We have*

$$y(t_0, t\rangle = \bar{F}[x(t_0); u(t_0, t\rangle] \qquad (2.1)$$

or

$$y(t) = F[x(t_0); u(t_0, t\rangle] \qquad (2.2)$$

Note 2.1: From the foregoing it follows that Σ is a deterministic system (has no random elements) and is not anticipative, which means that the present output values do not depend on future output values.

(ii) Let t_1 be any instant between t_0 and t. Then for each input vector $u(t_0, t\rangle$ and the observed output vector $y(t_0, t\rangle$, considered from the instant t_1 so that they give the segments $u(t_1, t\rangle$ and $y(t_1, t\rangle$, a non-empty subset of elements from the space X exists, denoted as $S[x(t_0); u(t_0, t\rangle]$, whose each element α satisfies the relation

$$y(t_1, t\rangle = F[\alpha, u(t_0, t\rangle] \qquad (2.3)$$

Note 2.2: This condition ensures that to each pair $u(t_1, t\rangle$, $y(t_1, t\rangle$ corresponds an initial state $x(t_1)$ in X.

(iii) If $u(t_0, t_1\rangle$ is invariable and $u(t_1, t)$ varies over all inputs of the input functions space of Σ, then the intersection of the sets $S[x(t_0); u(t_0, t\rangle]$, considered over all values of $u(t_1, t)$ is a non-empty set.

Note 2.3: This condition ensures the existence of at least one state in the space X, which refers to all possible pairs of inputs and outputs $u(t_1, t\rangle$ and $y(t_1, t\rangle$, respectively.

From conditions 2 and 3 it follows that the state of Σ at the instant t is determined by the initial state $x(t_0)$ and the input function vector $u(t_0, t)$:

$$x(t) = G[x(t_0); u(t_0, t\rangle] \qquad (2.4)$$

where G is a unique function of its own arguments. It should be noted that any two states α' and α'' are equivalent if for all values of u of the input functions space of Σ the response of the system corresponding to the initial state α' is identical with that corresponding to the initial state α''.

In the above definition and notes it makes no difference whether the independent variable of time, t, is continuously varying or only its selected discrete values are considered, as is usual with discrete computer control.

At the instants of process sampling the physical system may at any sampling instant be described by the values of the finite set of variables $x_1(k), x_2(k), \ldots, x_n(k)$ with a minimum number of elements of this set. The time variable k assumes integer values for a constant sampling period. The values $x_i(k)$, $i = 1, 2, \ldots, n$, are the values of the state variables. They constitute the components of the state vector $x(k)$. The dynamic behaviour of such systems may be described by difference equations or by recurrent relations of the form

$$x(k+1) = f[x(k)], \quad x(0) = x_0 \tag{2.5}$$

where $k = 0, 1, 2, \ldots$. For an increasing k, eqn. (2.5) may easily be solved by solving it first for $k = 0$:

$$x(1) = \Phi(1, x_0) = f(x_0) = F(x_0) \tag{2.6}$$

From eqn. (2.6) it is apparent that the values of the state vector x at the instant $k = 1$ may be considered as a transformation $F(x_0)$ of the initial state x_0 for $F(x) = f(x)$. Then

$$x(2) = \Phi[2, x(1)] = \Phi[2, \Phi(1, x_0)] = \Phi[2, F(x_0)] = f[x(1)] =$$
$$= f[f(x_0)] = F^2(x_0)$$

and, in the general form,

$$x(v) = \Phi(v, x_0) = F^v(x_0) \tag{2.7}$$

The calculation of the vector $x(k)$, $k = v$, indicated above represents the successive k-th iteration of the function of the vector x_0.

3

Equations of Continuously Working Systems

3.1 Mathematical model of systems in state space

The mathematical model may be set up for controlled plants, controllers or complete control loops. If we know the physical description of the system and can write the equations describing the behaviour of its individual elementary parts, the system state equations are usually very easily established. We shall show the procedure by a few examples.

Example 3.1: Set up the state equation of the mechanical system shown in Fig. 3.1.

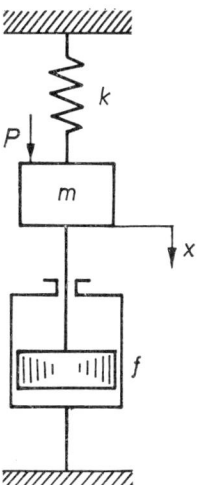

Fig. 3.1. Mechanical plant.

This is a simple system for which the ordinary differential equation may directly be written as follows:

$$m\,x''(t) + f\,x'(t) + k\,x(t) = P(t) \tag{3.1}$$

The state equation may, for instance, be determined by choosing the state vector as

$$\begin{bmatrix} x(t) \\ x'(t) \end{bmatrix} = \begin{bmatrix} x(t) \\ v(t) \end{bmatrix} \tag{3.2}$$

The choice of the state variables is not unique, however. Using the state variables selected, eqn. (3.1) assumes the form

$$m\,v'(t) + f\,v(t) + k\,x(t) = P(t) \tag{3.3}$$

and the given system may now be described in state space by the mathematical model

$$\begin{bmatrix} x'(t) \\ v'(t) \end{bmatrix} = \begin{bmatrix} 0, & 1 \\ -\dfrac{k}{m}, & -\dfrac{f}{m} \end{bmatrix} \begin{bmatrix} x(t) \\ v(t) \end{bmatrix} + \begin{bmatrix} 0 \\ \dfrac{1}{m} \end{bmatrix} P(t) \tag{3.4}$$

$$x'(t) = F\,x(t) + g\,u(t) \tag{3.5}$$

where the force $P(t)$ was replaced by $u(t)$, the general input variable notation. The output equation is

$$y(t) = h^T\,x(t) \tag{3.6}$$

where, in this example, $h^T = [1, 0]$. Equation (3.4) may be solved, i.e. the values of $x(t)$ and $v(t)$ for $t \geq t_0$ calculated, if we know the initial values $x(t_0)$ and $v(t_0)$ and the force $P(t)$ acting on the system at $t \geq t_0$.

Example 3.2: Set up the state equation of the *RLC* circuit shown in Fig. 3.2.

The solution is very much like that in the preceding example. The dynamic behaviour of the system is fully defined for $t \geq t_0$ if the initial values $i(t_0)$ and $e_c(t_0)$ and

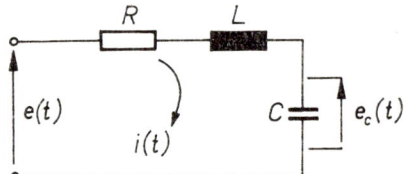

Fig. 3.2. Electric circuit.

the input voltage $e(t)$ for $t \geq t_0$ are known. Therefore, the system is completely determined by the state variables $i(t)$ and $e_c(t)$. The state variables could, however, be also chosen differently.

For the state variables $i(t)$ and $e_c(t)$ we have

$$L\,i'(t) + R\,i(t) + e_c(t) = e(t) \tag{3.7}$$

$$C\,e'_c(t) = i(t) \tag{3.8}$$

EQUATIONS OF CONTINUOUSLY WORKING SYSTEMS

In the vector-matrix form we obtain

$$\begin{bmatrix} i'(t) \\ e'_c(t) \end{bmatrix} = \begin{bmatrix} -\dfrac{R}{L}, & -\dfrac{1}{L} \\ \dfrac{1}{C}, & 0 \end{bmatrix} \begin{bmatrix} i(t) \\ e_c(t) \end{bmatrix} + \begin{bmatrix} \dfrac{1}{L} \\ 0 \end{bmatrix} e(t) \qquad (3.9)$$

which is the mathematical model of the given system in state space.

Example 3.3: Consider a stick balancer (inverted pendulum mounted with a ball-bearing pivot on a motor driven cart) as shown in Fig. 3.3. The external force $u(t)$ is changing both the cart position $x(t)$ and the balancer position. Provided we measured the angle $\varphi(t)$ and $d\varphi(t)/dt$ we could, by a suitable controller, develop such a force $u(t)$ that the balancer would remain in the vertical position except for small deviations. Such small deviations are necessary for ensuring the controller operation. This plant represents a simplified plane model of an equipment for the control of the vertical position of a rocket after starting. Set up the state equation of this model.

Solution: The plant is described by two equations, i.e. by the equation of force equilibrium

$$ml\,\varphi''(t) + (M + m)\,x''(t) = u(t) \qquad (3.10)$$

and by the equation of force moment equilibrium

$$(J + ml^2)\,\varphi''(t) + ml\,x''(t)\cos\varphi(t) - mgl\sin\varphi(t) \qquad (3.11)$$

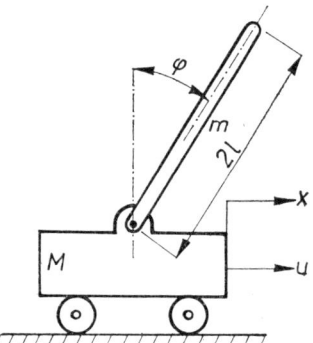

Fig. 3.3. Inverted pendulum on a cart.

where we may put $\sin\varphi(t) \doteq \varphi(t)$ and $\cos\varphi(t) \doteq 1$ if the angle $\varphi(t)$ is small. The symbol J denotes the moment of inertia of the balancer with respect to the mass centre, g is the acceleration of gravity and the meaning of the rest of the quantities follows from Fig. 3.3.

For a balancer of the length $2l$ the moment of inertia

$$J = \frac{ml^2}{3}$$

and eqn. (3.11) may be simplified in the form

$$\tfrac{4}{3}ml^2\, \varphi''(t) + ml\, x''(t) - mgl\, \varphi(t) = 0$$

From this equation

$$x''(t) = -\tfrac{4}{3}l\, \varphi''(t) - g\, \varphi(t) \tag{3.12}$$

Substituting eqn. (3.12) into eqn. (3.10) we obtain

$$[ml - \tfrac{4}{3}(M + m)\, l]\, \varphi''(t) + (M + m)\, g\, \varphi(t) = u(t) \tag{3.13}$$

Introducing now the angular velocity

$$\omega(t) = \varphi'(t)$$

we may write the state equation

$$\begin{bmatrix} \varphi'(t) \\ \omega'(t) \end{bmatrix} = \begin{bmatrix} 0, & 1 \\ -\dfrac{(M+m)g}{A}, & 0 \end{bmatrix} \begin{bmatrix} \varphi(t) \\ \omega(t) \end{bmatrix} + \begin{bmatrix} 0 \\ \dfrac{1}{A} \end{bmatrix} u(t) \tag{3.14}$$

where $A = [ml - \tfrac{4}{3}(M + m)\, l]$. Again, if we know the vector of initial conditions and the force $u(t)$ for $t \geq t_0$, eqn. (3.14) can be solved.

Example 3.4: Figure 3.4 shows a d.c. motor with a constant excitation Φ and an armature voltage $u(t)$. The armature current is $i(t)$, the angular displacement $\alpha(t)$ and the angular velocity $\omega(t)$.

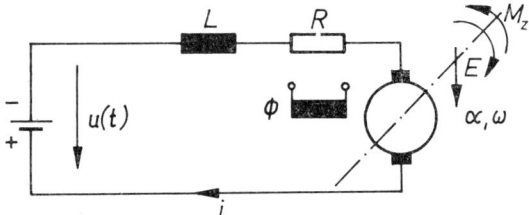

Fig. 3.4. D. C. motor with constant excitation.

The electromotive force

$$E = k_1 \Phi \omega$$

The motor torque

$$M = k_2 \Phi i$$

The moment of load

$$M_z = k_3 \omega$$

EQUATIONS OF CONTINUOUSLY WORKING SYSTEMS

The differential equations of the electric circuit

$$L\frac{di}{dt} + Ri = u - E$$

$$L\frac{di}{dt} + Ri = u - k_1\Phi\omega$$

The differential equations of the rotating part

$$\frac{d\alpha}{dt} = \omega$$

$$J\frac{d\omega}{dt} = M - M_z = k_2\Phi i - k_3\omega \tag{3.15}$$

where J is the moment of inertia of the rotor.

The state equation has, therefore, the form

$$\begin{bmatrix} \alpha' \\ \omega' \\ i' \end{bmatrix} = \begin{bmatrix} 0, & 1, & 0 \\ 0, & -\dfrac{k_3}{J}, & \dfrac{k_2\Phi}{J} \\ 0, & -\dfrac{k_1\Phi}{L}, & -\dfrac{R}{L} \end{bmatrix} \begin{bmatrix} \alpha \\ \omega \\ i \end{bmatrix} + \begin{bmatrix} 0 \\ 0 \\ \dfrac{1}{L} \end{bmatrix} u \tag{3.16}$$

In many cases not all of the controlled plant variables are measurable but only those which are elements of the output vector $y(t)$. Among the measurable variables usually are also the variables to be controlled.

The number of mutually independent controlled variables of a system defines the *system dimension*. If each of N controlled variables of a multivariable system is to be controlled independently of the other controlled variables, then the number of input variables (i.e. controlling variables) of the controlled plant, r, must be $r \geq N$ with all the input variables being also mutually independent. They are usually acting at different input points of the controlled plant.

The *order of a system* depends on the number of state variables. This order is equal to the number of state variables. For instance, the controlled plant of Example 3.4 is of the third order.

For the control of the motor speed measured by a tachometer the output variable (i.e. controlled variable) is

$$y = h^T x = [0, 1, 0]\, x \tag{3.17}$$

but for the control of the angular displacement α it is

$$y = [1, 0, 0]\, x \tag{3.18}$$

Although eqns. (3.17) and (3.18) can be used to determine two different controlled variables of the controlled plant of Example 3.4, this is a single-input/single-output plant since the angular displacement α and the angular velocity ω are mutually dependent variables. This dependence is expressed by eqn. (3.15).

The general expressions for the dynamics of the plant of Example 3.4 in state space are again given by eqs. (3.5) and (3.6).

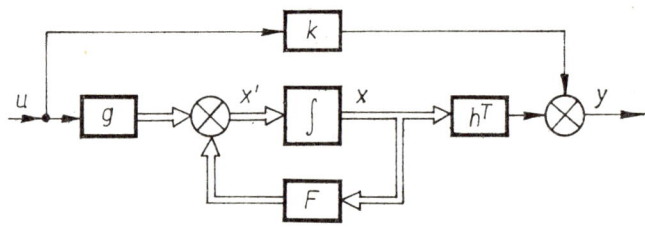

Fig. 3.5. Block diagram relating to the state description of a continuously working single-input/single-output plant.

If a single-input/single-output plant is capable to follow an input step-change, the description of such a plant in state space will have the form

$$x' = Fx + gu \qquad (3.19)$$

$$y = h^T x + ku \qquad (3.20)$$

The output equation has changed by the term ku. If the plant order is n, the individual matrices in eqns. (3.19) and (3.20) have the dimensions $F(n; n)$, $g(n; 1)$, $h^T(1; n)$, $k(1; 1)$ and hence follow the dimensions of the corresponding vectors. In the subsequent discussions we shall use the following nomenclature: Set of state equations consisting of the *equation of dynamics* (3.19) and of the *output equation* (3.20). In the equation of dynamics let F be the *matrix of dynamics* and g the *input column matrix* (*input matrix*). In the output equation let h^T be the *output row matrix* (*output matrix*) and k the *input factor* (*input matrix factor*). The terms in brackets refer to multivariable systems.

The block diagram in Fig. 3.5 relates to eqns. (3.19) and (3.20). The integration sign denotes n parallel integrators which integrate the components of the vector x'.

3.2 Transcription of an ordinary differential equation to state equations

The examples of the determination of a mathematical controlled plant model, given in Sec. 3.1, were all based on the plant design data and on the description of its elementary processes. Sometimes, however, it is desirable to set up the mathematical model in state space from a known differential equation describing the plant's dynamics. Since, as already mentioned, the choice of the state variables is not unique, it would not suffice to show merely one method of transforming the differential equation to state space. We shall, therefore, give some examples of frequently occurring transformations. The examples given in this section will be referred to single-input/single-output plants.

3.2.1 Procedure I — the state variables are defined by the relation $x_{i+1} = x_i'$

Let us consider a plant, e.g. a time invariant controlled plant, with the input variable u and the output variable y (Fig. 3.6), whose dynamics is described by the linear differential equation with constant coefficients

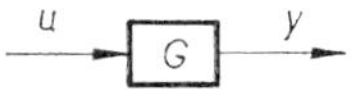

Fig. 3.6. Continuously working single-input/single-output plant.

$$\sum_{i=0}^{n} a_i\, y^{(i)}(t) = \sum_{j=0}^{m} b_j\, u^{(j)}(t)\,, \quad n \geqq m \tag{3.21}$$

Introducing the operator

$$D = \frac{\mathrm{d}}{\mathrm{d}t} \tag{3.22}$$

eqn. (3.21) may formally be rewritten in the form

$$\sum_{i=0}^{n} a_i D^i\, y(t) = \sum_{j=0}^{m} b_{(j)} D^j\, u(t)\,, \quad n \geqq m \tag{3.23}$$

or

$$\tilde{A}\, y(t) = \tilde{B}\, u(t) \tag{3.24}$$

where

$$\tilde{A} = \sum_{i=0}^{n} a_i D^i \quad \text{and} \quad \tilde{B} = \sum_{j=0}^{m} b_j D^j \tag{3.25}$$

From eqn. (3.24) it formally follows that

$$y(t) = \tilde{A}^{-1}\tilde{B}\, u(t) \tag{3.26}$$

Introducing

$$\tilde{A}^{-1}\tilde{B}\, u(t) = \tilde{B}\, x(t) \tag{3.27}$$

we have

$$\tilde{A}^{-1}\, u(t) = x(t) \tag{3.28}$$

As can be seen, eqn. (3.24) has been replaced by two relations:

$$\tilde{A} x(t) = u(t) \tag{3.29}$$

$$\tilde{B} x(t) = y(t) \tag{3.30}$$

If we now introduce the state variables

$$\begin{aligned} x(t) &= x_1(t) \\ x^{(1)}(t) &= x_2(t) = x_1^{(1)}(t) \\ &\vdots \\ x^{(n-1)}(t) &= x_n(t) = x_{n-1}^{(1)}(t) \end{aligned} \tag{3.31}$$

it is plain that

$$x^{(n)}(t) = x_n^{(1)}(t) \tag{3.32}$$

and eqn. (3.29) may be rewritten in the form

$$x_n^{(1)}(t) = \frac{1}{a_n} u(t) - \frac{a_{n-1}}{a_n} x_n(t) - \ldots - \frac{a_1}{a_n} x_2(t) - \frac{a_0}{a_n} x_1(t) \tag{3.33}$$

Equations (3.31) and (3.33) can be written as a single equation by using constant matrix operators and vectors of state variables

$$\begin{bmatrix} x_1^{(1)}(t) \\ x_2^{(1)}(t) \\ \vdots \\ x_n^{(1)}(t) \end{bmatrix} = \begin{bmatrix} 0, & 1, & 0, & \ldots, & 0 \\ 0, & 0, & 1, & \ldots, & 0 \\ \vdots & & & & \\ -\dfrac{a_0}{a_n}, & -\dfrac{a_1}{a_n}, & \ldots, & \ldots, & -\dfrac{a_{n-1}}{a_n} \end{bmatrix} \begin{bmatrix} x_1(t) \\ x_2(t) \\ \vdots \\ x_n(t) \end{bmatrix} + \begin{bmatrix} 0 \\ 0 \\ \vdots \\ \dfrac{1}{a_n} \end{bmatrix} u(t) \tag{3.34}$$

From eqn. (3.30) we obtain, using the state variables (3.31), the equation defining the transformation of the state variables $x_i(t)$, $i = 1, 2, \ldots, n$, to the output variable $y(t)$:

$$b_0 x_1(t) + b_1 x_2(t) + \ldots + b_{n-1} x_n(t) + b_n x_n^{(1)}(t) = y(t), \quad n = m \tag{3.35}$$

Substituting the relation (3.33) for $x_n^{(1)}(t)$, eqn. (3.35) changes into

$$y(t) = \left(b_0 - \frac{a_0}{a_n} b_n \right) x_1(t) + \left(b_1 - \frac{a_1}{a_n} b_n \right) x_2(t) + \ldots$$

$$\ldots + \left(b_{n-1} - \frac{a_{n-1}}{a_n} b_n \right) x_n(t) + \frac{b_n}{a_n} u(t) \tag{3.36}$$

In eqns. (3.34) and (3.36) we already recognize the state equations (3.19) and (3.20) where

$$F = \begin{bmatrix} 0, & 1, & 0, & \ldots, & 0 \\ 0, & 0, & 1, & \ldots, & 0 \\ \multicolumn{5}{c}{\dotfill} \\ -\dfrac{a_0}{a_n}, & -\dfrac{a_1}{a_n}, & \ldots, & \ldots, & -\dfrac{a_{n-1}}{a_n} \end{bmatrix}, \quad g = \begin{bmatrix} 0 \\ 0 \\ \vdots \\ \dfrac{1}{a_n} \end{bmatrix}$$

$$h^T = \left[b_0 - \frac{a_0}{a_n} b_n, \quad b_1 - \frac{a_1}{a_n} b_n, \ldots, b_{n-1} - \frac{a_{n-1}}{a_n} b_n \right]$$

$$k = \frac{b_n}{a_n}$$

The matrix of dynamics, F, and its corresponding transpose are in the so-called Frobenius form. The simplifications obtained for $b_n = 0$ and $a_n = 1$ are apparent from both these state equations.

The state equations (3.34) and (3.36) may also be expressed by a single block diagram of state equations shown in Fig. 3.7 where the solid lines apply to $m = n - 1$ while the component for $m = n$ is shown by dashed lines.

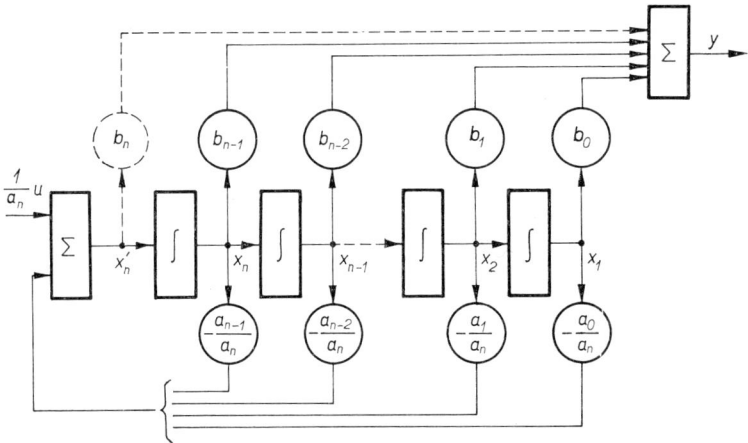

Fig. 3.7. Block diagram relating to the state equations (3.34) and (3.36).

If the constants $b_1, b_2, \ldots, b_{n-1}$ are zero and $b_0 = 1$, the state variables x_i, $i = 1, 2, \ldots, n$, are identical with the plant output variable and its derivatives, i.e. with $y^{(i-1)}(t)$, $i = 1, 2, \ldots, n$. This is a certain advantage of the definition of state variables as used in this example.

Example 3.5: Use procedure I to set up the state equations corresponding to the differential equation

$$2 y''(t) + 3 y'(t) + 2 y(t) = u(t) + 5 u'(t)$$

Solution: According to eqns. (3.34) and (3.36) we have

$$\begin{bmatrix} x_1'(t) \\ x_2'(t) \end{bmatrix} = \begin{bmatrix} 0, & 1 \\ -1, & -\frac{3}{2} \end{bmatrix} \begin{bmatrix} x_1(t) \\ x_2(t) \end{bmatrix} + \begin{bmatrix} 0 \\ \frac{1}{2} \end{bmatrix} u(t)$$

$$y(t) = [1, \ 5] \, [x_1(t), \ x_2(t)]$$

3.2.2 Procedure II — the state variables are a linear combination of the derivatives of the plant input and output variables

Let the plant dynamics be again described by the n-th order differential equation (3.21), where $m = n$. Consider the case where $m = n$ and define the state variables by the following equations:

$$x_1(t) = a_n \, y(t) - b_n \, u(t)$$

$$x_2(t) = a_{n-1} \, y(t) + a_n \, y^{(1)}(t) - b_n \, u^{(1)}(t) - b_{n-1} \, u(t)$$

$$x_3(t) = a_{n-2} \, y(t) + a_{n-1} \, y^{(1)}(t) + a_n \, y^{(2)}(t) -$$
$$\quad - b_n \, u^{(2)}(t) - b_{n-1} \, u^{(1)}(t) - b_{n-2} \, u(t)$$

$$\vdots$$

$$x_n(t) = a_1 \, y(t) + a_2 \, y^{(1)}(t) + \ldots + a_n \, y^{(n-1)}(t) -$$
$$\quad - b_n \, u^{(n-1)}(t) - b_{n-1} \, u^{(n-2)}(t) - \ldots - b_1 \, u(t) \qquad (3.37)$$

From the first equation of the set of equations (3.37) we calculate $y(t)$ and substitute it into the remaining equations; then we replace all terms including derivatives of $y(t)$ and $u(t)$ by the first derivative of the state variable defined in each case by the preceding equation. We obtain

$$x_2(t) = \frac{a_{n-1}}{a_n} x_1(t) + x_1'(t) - \left(b_{n-1} - \frac{a_{n-1}}{a_n} b_n \right) u(t)$$

$$x_3(t)^* = \frac{a_{n-2}}{a_n} x_1(t) + x_2'(t) - \left(b_{n-2} - \frac{a_{n-2}}{a_n} b_n \right) u(t)$$

$$\vdots$$

$$x_n(t) = \frac{a_1}{a_n} x_1(t) + x_{n-1}'(t) - \left(b_1 - \frac{a_1}{a_n} b_n \right) u(t) \qquad (3.38)$$

EQUATIONS OF CONTINUOUSLY WORKING SYSTEMS

Combining eqn. (3.38) with the original differential equation expressed in terms of the state variables introduced and arranging the result in the form

$$0 = \frac{a_0}{a_n} x_1(t) + x'_n(t) - \left(b_0 - \frac{a_0}{a_n} b_n \right) u(t) \qquad (3.39)$$

we may write eqns. (3.38) and (3.39) in terms of matrix operators as follows:

$$\begin{bmatrix} x'_1(t) \\ x'_2(t) \\ \vdots \\ x'_n(t) \end{bmatrix} = \begin{bmatrix} -\frac{a_{n-1}}{a_n}, & 1, & 0, & \ldots, & 0 \\ -\frac{a_{n-2}}{a_n}, & 0, & 1, & \ldots, & 0 \\ \vdots & & & & \\ -\frac{a_0}{a_n}, & 0, & 0, & \ldots, & 0 \end{bmatrix} \begin{bmatrix} x_1(t) \\ x_2(t) \\ \vdots \\ x_n(t) \end{bmatrix} + \begin{bmatrix} b_{n-1} - \frac{a_{n-1}}{a_n} b_n \\ b_{n-2} - \frac{a_{n-2}}{a_n} b_n \\ \vdots \\ b_0 - \frac{a_0}{a_n} b_n \end{bmatrix} u(t) \qquad (3.40)$$

In this case the transformation of the state variables $x_i(t)$, $i = 1, 2, \ldots, n$, into the output variable $y(t)$ is trivial. It is defined by the first of eqns. (3.37). For the purpose of writing consistency we rewrite this equation in the matrix form and obtain

$$y(t) = \begin{bmatrix} \frac{1}{a_n}, & 0, & 0, & \ldots, & 0 \end{bmatrix} x(t) + \frac{b_n}{a_n} u(t) \qquad (3.41)$$

The state equations (3.40) and (3.41) may again be briefly expressed by writing eqns. (3.19) and (3.20).

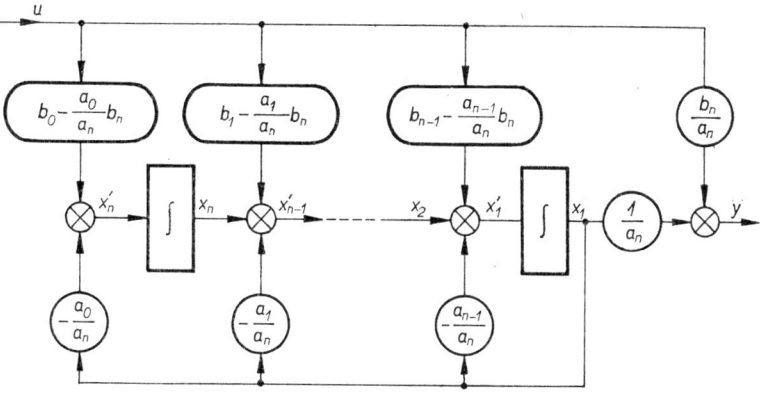

Fig. 3.8. Block diagram relating to the state equations (3.40) and (3.41).

The block diagram of the state equations (3.40) and (3.41) is shown in Fig. 3.8.

The advantage of choosing the state variables according to the procedure indicated in this paragraph is that $x_1(t) = y(t)$ for all values of the constants $b_0, b_1, \ldots, b_{n-1}$ and for $b_n = 0$ and $a_n = 1$.

In numerical calculations we then do not have to perform the transformation (3.41) while in the determination of the output variable $y(t)$ according to Par. 3.2.1 the transformation (3.36) is always necessary.

Example 3.6: Set up the state equation corresponding to the differential equation of Example 3.5 according to eqns. (3.40) and (3.41).

Solution:

$$\begin{bmatrix} x_1'(t) \\ x_2'(t) \end{bmatrix} = \begin{bmatrix} -\frac{3}{2}, & 1 \\ -1, & 0 \end{bmatrix} \begin{bmatrix} x_1(t) \\ x_2(t) \end{bmatrix} + \begin{bmatrix} 5 \\ 1 \end{bmatrix} u(t)$$

$$y(t) = \tfrac{1}{2} x_1(t)$$

Dividing the original equation by the constant $a_n = a_2 = 2$ we obtain

$$\begin{bmatrix} x_1'(t) \\ x_2'(t) \end{bmatrix} = \begin{bmatrix} -\frac{3}{2}, & 1 \\ -1, & 0 \end{bmatrix} \begin{bmatrix} x_1(t) \\ x_2(t) \end{bmatrix} + \begin{bmatrix} \frac{5}{2} \\ \frac{1}{2} \end{bmatrix} u(t)$$

$$y(t) = x_1(t)$$

3.2.3 State equation under the action of a controlling and a disturbing variables

If the controlling variable $u(t)$ is acting at the control plant input simultaneously with the disturbing variable $w(t)$, the equation of a linear time-invariant plant may be written as

$$\sum_{i=0}^{n} a_i y^{(i)}(t) = \sum_{i=0}^{m} b_i u^{(i)}(t) + \sum_{i=0}^{l} c_i w^{(i)}(t), \quad m < n, \quad l < n \quad (3.42)$$

For $m = l = n - 1$ and in accordance with procedure II (see Par. 3.2.2) we calculate

$$\begin{bmatrix} x_1'(t) \\ x_2'(t) \\ \vdots \\ x_n'(t) \end{bmatrix} = \begin{bmatrix} -\dfrac{a_{n-1}}{a_n}, & 1, 0, \ldots, 0 \\ -\dfrac{a_{n-2}}{a_n}, & 0, 1, \ldots, 0 \\ \cdots \\ -\dfrac{a_0}{a_n}, & 0, 0, \ldots, 0 \end{bmatrix} \begin{bmatrix} x_1(t) \\ x_2(t) \\ \vdots \\ x_n(t) \end{bmatrix} + \begin{bmatrix} b_{n-1}, & c_{n-1} \\ b_{n-2}, & c_{n-2} \\ \cdots \\ b_0, & c_0 \end{bmatrix} \begin{bmatrix} u(t) \\ w(t) \end{bmatrix} \quad (3.43)$$

$$y(t) = \begin{bmatrix} \dfrac{1}{a_n}, & 0, 0, \ldots, 0 \end{bmatrix} x(t) \quad (3.44)$$

or briefly

$$x'(t) = F x(t) + G u^*(t) \quad (3.45)$$

$$y(t) = h^T x(t) \quad (3.46)$$

where the meaning of the individual symbols follows from a comparison of eqns. (3.43) and (3.44). Here the state variables are defined as follows:

$$x_1(t) = a_n y(t) \tag{3.47}$$

$$x_2(t) = \frac{a_{n-1}}{a_n} x_1(t) + x_1'(t) - b_{n-1} u(t) - c_{n-1} w(t)$$

$$x_3(t) = \frac{a_{n-2}}{a_n} x_1(t) + x_2'(t) - b_{n-2} u(t) - c_{n-2} w(t)$$

$$\vdots$$

$$x_n(t) = \frac{a_1}{a_n} x_1(t) + x_{n-1}'(t) - b_1 u(t) - c_1 w(t) \tag{3.48}$$

With the state variables defined in this way the original differential equation (3.42) assumes the form

$$x_n'(t) + \frac{a_0}{a_n} x_1(t) - b_0 u(t) - c_0 w(t) = 0 \tag{3.49}$$

Equations (3.48) and (3.49) define the state equation (3.45) and eqn. (3.47) determines the transformation of the state variables into the plant output variable.

According to the procedure indicated in Par. 3.2.2 we could even derive the state equations for the case $m = l = n$ or for a plant with more than one disturbing variable acting at its input in addition to the controlling variable.

If a single disturbing variable is acting at the plant output, we include it in the output equation (3.44) as an additive term.

Equations (3.43) and (3.45) describe the case where the disturbing variable $w(t)$ is a scalar quantity and its effect on the individual state variables is determined by the parameters c_i, $i = 0, 1, \ldots, n - 1$. Besides this model a case is also possible where mutually independent disturbing variables are acting on the individual state variables $x_i(t)$, $i = 1, 2, \ldots, n$, and on the output variable $y(t)$. For plants with one input $u(t)$ and one output $y(t)$, the state equations (3.45) and (3.46) then assume the form

$$x'(t) = F x(t) + g u(t) + w(t) \tag{3.50}$$

$$y(t) = h^T x(t) + k u(t) + v(t) \tag{3.51}$$

where $w(t)$ is a vector of the dimension $(n; 1)$ and $v(t)$ a scalar disturbing variable. Model (3.50) and (3.51) is usually chosen when the disturbing variables $w(t)$ and $v(t)$ are random variables.

3.3 Continuous multi-input/multi-output controlled plant

The dynamics of a continuous multi-input/multi-output linear time-invariant plant is described by the following set of linear differential equations with constant coefficients:

$$\sum_{j=1}^{N} a_{ij}(D) y_j(t) = \sum_{j=1}^{N} b_{ij}(D) u_j(t), \quad i = 1, 2, \ldots, N \quad (3.52)$$

In this set of equations the number of controlling variables is equal to N controlled variables. The case of unequal number of controlling and controlled variables will be discussed in Sec. 11.3. The factors $a_{ij}(D)$ and $b_{ij}(D)$ are polynomials in D of the n_{ij}-th and m_{ij}-th degree, respectively, $m_{ij} \leq n_{ij}$ $(i, j = 1, 2, \ldots, N)$ and $D = d/dt$. The set of equations (3.50) may also be written in matrix form:

$$F^*(D) y(t) = G^*(D) u(t) \quad (3.53)$$

Here $F^*(D)$ and $G^*(D)$ are matrices set up of the elements $a_{ij}(D)$ and $b_{ij}(D)$, $i, j = 1, 2, \ldots, N$, respectively.

The matrices $F^*(D)$ and $G^*(D)$ in eqn. (3.53) may be written as the sum of matrices multiplied by the power of the operator D. For $s = \max(n_{ij})$ we have

$$(F_s^* D^s + F_{s-1}^* D^{s-1} + \ldots + F_1^* D + F_0^*) y(t) =$$
$$= (G_s D^s + G_{s-1}^* D^{s-1} + \ldots + G_1^* D + G_0^*) u(t) \quad (3.54)$$

Henceforward the process of determination of state equations may be essentially identical with that given in Pars. 3.2.1 and 3.2.2. The only difference is that now we shall be operating with matrices and vectors.

We use procedure II Par. 3.2.2 and by introducing the vectors of state variables

$$x_v^T = [x_{v1}, x_{v2}, \ldots, x_{vN}], \quad v = 1, 2, \ldots, s$$

we obtain the equation of dynamics

$$\begin{bmatrix} x_1'(t) \\ x_2'(t) \\ \vdots \\ x_s'(t) \end{bmatrix} = \begin{bmatrix} -F_{s-1}, & E, & 0, & \ldots, & 0 \\ -F_{s-2}, & 0, & E, & \ldots, & 0 \\ \cdots & \cdots & \cdots & \cdots & \cdots \\ -F_0, & 0, & 0, & \ldots, & 0 \end{bmatrix} \begin{bmatrix} x_1(t) \\ x_2(t) \\ \vdots \\ x_s(t) \end{bmatrix} + \begin{bmatrix} G_{s-1} \\ G_{s-2} \\ \vdots \\ G_0 \end{bmatrix} u(t) \quad (3.55)$$

where E is a unit matrix,

$$F_{s-l} = F_{s-l}^* F_s^{*-1}, \quad G_{s-l} = G_{s-l}^* - F_{s-l}^* F_s^{*-1} G_s^*, \quad l = 1, 2, \ldots, s$$

and where the vector

$$u^T(t) = [u_1(t), \ldots, u_N(t)]$$

EQUATIONS OF CONTINUOUSLY WORKING SYSTEMS 35

We transform the state variables into the vector of output variables according to the equation

$$y(t) = F_s^{*-1} x(t) + F_s^{*-1} G_s^* u(t) \qquad (3.56)$$

where the vector

$$y^T(t) = [y_1(t), y_2(t), \ldots, y_N(t)]$$

The second term on the right-hand side of eqn. (3.56) is zero if $m_{ij} < n_{ij}$ for all values of $i, j = 1, 2, \ldots, N$. The state equations (3.55) and (3.56) may formally be written in the form

$$x' = Fx + Gu \qquad (3.57)$$

$$y = Hx + Ku \qquad (3.58)$$

where the matrices have, on the given assumption, the dimensions $F(Ns; Ns)$, $G(Ns; N)$, $H(N; Ns)$ and $K(N; N)$. However, if we designate, regardless of the introduced assumptions, the order of the multivariable system according to n state variables, where $n = Ns$, the matrix dimensions in eqns. (3.57) and (3.58) will be $F(n; n)$, $G(n; r)$, $H(p; n)$, $K(p; r)$, where r is the number of plant input variables, p the number of plant controlled (output) variables; for a dynamic system, $p = r$ or even $p \geqslant r$. The dimensions of the vectors follow from the dimensions of the corresponding matrices.

The transformation of the set of differential equations (3.52) into the state equations (3.55) and (3.56) by the method indicated is evidently possible only if the matrix F_s^* is non-singular. If it is singular, the equations defining the state variables according to the single-input/single-output model (3.37) remain valid. A transcription of the differential equations (3.52) to state space is, therefore, possible but no generally explicit solution can be given. For this reason the method of transcription of a set of differential equations including a singular matrix F_s^* to state equations can only be indicated by a special example.

It is always advisable to check at the end of calculation whether the resultant dimensions of the state equation matrices can be reduced, because redundant state variables might have been introduced when transcribing the differential equations to state space by the method described.

Example 3.7: Set up the state equations of a double-input/double-output plant whose dynamics is formulated by the following equations:

$$3 y_1'(t) + y_1(t) + y_2''(t) + 5 y_2'(t) + 2 y_2(t) =$$
$$= u_1(t) + 2 u_1'(t) + 3 u_2(t) + 4 u_2'(t)$$

$$2 y_1''(t) + 2 y_1'(t) + y_1(t) + 3 y_2''(t) + 6 y_2'(t) + 2 y_2(t) =$$
$$= 2 u_1(t) + 5 u_1'(t) + 2 u_2(t)$$

Solution: First we set up the matrices F^*_{s-l} and G^*_{s-l}, $l = 0, 1, 2, s = 2$:

$$F^*_0 = \begin{bmatrix} 1, & 2 \\ 1, & 2 \end{bmatrix} \qquad G^*_0 = \begin{bmatrix} 1, & 3 \\ 2, & 2 \end{bmatrix}$$

$$F^*_1 = \begin{bmatrix} 3, & 5 \\ 2, & 6 \end{bmatrix} \qquad G^*_1 = \begin{bmatrix} 2, & 4 \\ 5, & 0 \end{bmatrix}$$

$$F^*_2 = \begin{bmatrix} 0, & 1 \\ 2, & 3 \end{bmatrix} \qquad G^*_2 = \begin{bmatrix} 0, & 0 \\ 0, & 0 \end{bmatrix}$$

The matrix F^*_2 is clearly non-singular so that formulae (3.55) and (3.56) may be used:

$$F^{*-1}_2 = -\tfrac{1}{2} \begin{bmatrix} 3, & -1 \\ -2, & 0 \end{bmatrix} = \begin{bmatrix} -\tfrac{3}{2}, & \tfrac{1}{2} \\ 1, & 0 \end{bmatrix}$$

$$F_1 = F^*_1 F^{*-1}_2 = \begin{bmatrix} 3, & 5 \\ 2, & 6 \end{bmatrix} \begin{bmatrix} -\tfrac{3}{2}, & \tfrac{1}{2} \\ 1, & 0 \end{bmatrix} = \begin{bmatrix} \tfrac{1}{2}, & \tfrac{3}{2} \\ 3, & 1 \end{bmatrix}$$

$$F_0 = F^*_0 F^{*-1}_2 = \begin{bmatrix} 0, & 1 \\ 2, & 3 \end{bmatrix} \begin{bmatrix} -\tfrac{3}{2}, & \tfrac{1}{2} \\ 1, & 0 \end{bmatrix} = \begin{bmatrix} 1, & 0 \\ 0, & 1 \end{bmatrix}$$

Since $G^*_2 = 0$, then $G_0 = G^*_0$ and $G_1 = G^*_1$.

According to eqn. (3.55)

$$\begin{bmatrix} x'_{11}(t) \\ x'_{12}(t) \\ x'_{21}(t) \\ x'_{22}(t) \end{bmatrix} = \begin{bmatrix} -\tfrac{1}{2}, & -\tfrac{3}{2}, & 1, & 0 \\ -3, & -1, & 0, & 1 \\ -1, & 0, & 0, & 0 \\ 0, & -1, & 0, & 0 \end{bmatrix} \begin{bmatrix} x_{11}(t) \\ x_{12}(t) \\ x_{21}(t) \\ x_{22}(t) \end{bmatrix} + \begin{bmatrix} 2, & 4 \\ 5, & 0 \\ 1, & 3 \\ 2, & 2 \end{bmatrix} \begin{bmatrix} u_1(t) \\ u_2(t) \end{bmatrix}$$

and by (3.56) we determine

$$\begin{bmatrix} y_1(t) \\ y_2(t) \end{bmatrix} = \begin{bmatrix} -\tfrac{3}{2}, & \tfrac{1}{2} \\ 1, & 0 \end{bmatrix} \begin{bmatrix} x_{11}(t) \\ x_{12}(t) \end{bmatrix}$$

Example 3.8: The differential equations of a double-input/double-output plant are

$$y''_1(t) + 2 y'_1(t) + 2 y_1(t) + y_2(t) = 4 u_1(t) + 5 u'_1(t) + 3 u_2(t)$$
$$7 y'_1(t) + 5 y_1(t) + 3 y'_2(t) + y_2(t) = 3 u_1(t) + u_2(t)$$

Calculate the state equations of this plant.

Solution: Since the matrix F^*_2 is singular, the transcription to state equations by means of the formulae (3.55) and (3.56) cannot be effected. First we test whether the given equations describe a physically realizable system. This test may be made, for instance, by (3.26) where $y(t)$ and $u(t)$ are now two-component vectors and $\tilde{A}, \tilde{B}$ are polynomial matrices of the dimension (2; 2):

$$y(t) = \begin{bmatrix} D^2 + 2D + 2, & 1 \\ 7D + 5, & 3D + 1 \end{bmatrix}^{-1} \begin{bmatrix} 5D + 4, & 3 \\ 3, & 1 \end{bmatrix} u(t)$$

It is apparent that $\det \tilde{A}$ is a polynomial of the third degree in D and the elements of the matrix $(\text{adj } \tilde{A})\tilde{B}$ are polynomials in D of at most the second degree. Therefore, the elements of the matrix $\tilde{A}^{-1}\tilde{B}$, all of which are rational fraction functions in D, have numerators of a lower degree than the denominators and, consequently, are physically realizable. Hence it follows that a transcription to state equations does exist.

Let us rewrite the given equations in the form

$$D^2 y_1(t) + D[2 y_1(t) - 5 u_1(t)] + 2 y_1(t) + y_2(t) - 4 u_1(t) - 3 u_2(t) = 0$$
$$D[7 y_1(t) + 3 y_2(t)] + 5 y_1(t) + y_2(t) - 3 u_1(t) - u_2(t) = 0 \qquad (3.59)$$

and introduce the following state variables:

$$\begin{aligned} x_1(t) &= y_1(t) \\ x_2(t) &= Dy_1(t) + 2 y_1(t) - 5 u_1(t) \\ x_3(t) &= 7 y_1(t) + 3 y_2(t) \end{aligned} \qquad (3.60)$$

From eqns. (3.59) it further follows that

$$\begin{aligned} Dx_2(t) &= -2 y_1(t) - y_2(t) + 4 u_1(t) + 3 u_2(t) \\ Dx_3(t) &= -5 y_1(t) - y_2(t) + 3 u_1(t) + u_2(t) \end{aligned} \qquad (3.61)$$

From the third of eqns. (3.60) we calculate

$$y_2(t) = -\tfrac{7}{3} y_1(t) + \tfrac{1}{3} x_3(t) \qquad (3.62)$$

and substitute it into both equations (3.61); for $y_1(t)$ we substitute according to the first of eqns. (3.60). The second of eqns. (3.60) and both eqns. (3.61) determine the equation of dynamics

$$\begin{bmatrix} x_1'(t) \\ x_2'(t) \\ x_3'(t) \end{bmatrix} = \begin{bmatrix} -2, & 1, & 0 \\ \tfrac{1}{3}, & 0, & -\tfrac{1}{3} \\ -\tfrac{8}{3}, & 0, & -\tfrac{1}{3} \end{bmatrix} \begin{bmatrix} x_1(t) \\ x_2(t) \\ x_3(t) \end{bmatrix} + \begin{bmatrix} 5, & 0 \\ 4, & 3 \\ 3, & 1 \end{bmatrix} \begin{bmatrix} u_1(t) \\ u_2(t) \end{bmatrix}$$

The first and third of eqns. (3.60) determine the output equation

$$\begin{bmatrix} y_1(t) \\ y_2(t) \end{bmatrix} = \begin{bmatrix} 1, & 0, & 0 \\ -\tfrac{7}{3}, & 0, & \tfrac{1}{3} \end{bmatrix} \begin{bmatrix} x_1(t) \\ x_2(t) \\ x_3(t) \end{bmatrix}$$

4

Operators

4.1 Basic properties

The state equations discussed in Chap. 3, e.g. eqns. (3.19) and (3.20), express a *transformation of the vectors* $u \in U$ and $x \in X$ into the vector $x' \in X'$ or $y \in Y$; the transformations are effected by means of the *operators* F, g, h^T and k. Since transformation by means of operators is one of the most frequent operations in state space, the basic types and properties of these operators as well as some applications in connection with state equations will be set forth in this chapter.

Let X and Y be two vector spaces. Let the element $x \in X$ be uniquely assigned to each $y \in Y$; then this assignment is briefly written as

$$x = Ay \qquad (4.1)$$

and we speak of the operator A as being defined in the space Y and mapping Y into X.

The *sum of operators*, $A_1 + A_2$, is defined by the relation

$$(A_1 + A_2)y = A_1 y + A_2 y \qquad (4.2)$$

for all values of $y \in Y$.

Since addition in vector space is commutative, it follows from eqn. (4.2) that

$$(A_1 + A_2)y = (A_2 + A_1)y \qquad (4.3)$$

$$[A_1 + (A_2 + A_3)]y = [(A_1 + A_2) + A_3]y \qquad (4.4)$$

The *product of operators*, $A_1 A_2$, represents a successive use of the operator A_2 for y and of the operator A_1 for $A_2 y$.

Here the commutative law does not apply:

$$A_1 A_2 y \neq A_2 A_1 y \qquad (4.5)$$

OPERATORS

If y is an m-dimensional vector equal to the sum of m-dimensional vectors, e.g. $y = y_1 + y_2$, the distributive law does not, in general, apply to transformation by means of the operator A:

$$A(y_1 + y_2) \neq Ay_1 + Ay_2 \tag{4.6}$$

If we multiply an m-dimensional vector y by an arbitrary constant (number) a we again obtain an m-dimensional vector:

$$f = ay \tag{4.7}$$

The operator A is called *homogeneous* if

$$Aay = aAy \tag{4.8}$$

Special types of operators are: the *vector operator*, where

$$f = Ay$$
$$A = [A_1, A_2, ..., A_n] \tag{4.9}$$

and the *matrix operator*:

$$\begin{bmatrix} f_1 \\ f_2 \\ \vdots \\ f_n \end{bmatrix} = \begin{bmatrix} A_{11}, A_{12}, ..., A_{1m} \\ A_{21}, A_{22}, ..., A_{2m} \\ \cdots\cdots\cdots\cdots\cdots \\ A_{n1}, A_{n2}, ..., A_{nm} \end{bmatrix} \begin{bmatrix} y_1 \\ y_2 \\ \vdots \\ y_m \end{bmatrix} \tag{4.10}$$

where

$$f_i = \sum_{j=1}^{m} A_{ij} y_j, \quad i = 1, 2, ..., n \tag{4.11}$$

or

$$f = Ay \tag{4.12}$$

4.2 Linearity

An operator is linear if it is homogeneous and the distributive law

$$A(ay_1 + by_2) = aAy_1 + bAy_2 \tag{4.13}$$

applies. If the equality (4.13) is not satisfied the operator is nonlinear.

If the mathematical model of a plant can be expressed by linear differential equations then the state equations with linear operators assume the previously mentioned form (3.19) and (3.20):

$$\frac{dx(t)}{dt} = F(t) x(t) + G(t) u(t) \tag{4.14}$$

$$y(t) = H(t) x(t) + K(t) u(t) \tag{4.15}$$

where $u(t)$ is an r-dimensional input vector, $x(t)$ an n-dimensional state vector, $y(t)$ a p-dimensional output vector and F, G, H, K are linear matrix operators of the type $(n; n)$, $(n; r)$, $(p; n)$ and $(p; r)$, respectively.

For discrete linear systems we have instead of eqns. (4.14) and (4.15) the equations

$$x(k + 1) = A(k) x(k) + B(k) u(k) \tag{4.16}$$

$$y(k) = C(k) x(k) + D(k) u(k) \tag{4.17}$$

In eqns. (4.14) through (4.17) all linear operators are, in general, expressed as functions of the independent time variable. The given state equations describe, therefore, generally time-variant systems.

4.3 Stationarity in time

An operator A is time-invariant if it satisfies the relation

$$x(t - t_1) = A[y(t - t_1)] \tag{4.18}$$

for all values of t and t_1 and for all values of $y \in Y$. The system is time-invariant if it can be described by time-invariant operators. In other words, a system described by differential equations with constant coefficients is time-invariant and, similarly, a system described by eqns. (4.14) and (4.15) will be time-invariant if the operators F, G, H, K are constant matrices.

Analogously we may also define the stationarity of a discrete system.

4.4 System equivalence

Two systems with the input, state and output vectors u^1, x^1, y^1 and u^2, x^2, y^2, respectively, are said to be *observably equivalent* if for all values of $u \in U$ and for all values of t the equality

$$u^1(t) = u^2(t) \tag{4.19}$$

implies the relation

$$y^1(t) = y^2(t) \tag{4.20}$$

From the foregoing it is evident that the definition of observable equivalence does not consider the state of the systems being compared but involves only the external variables, i.e. the input and the output. If the equivalence or difference of the internal properties of a system is to be considered as well, a more complete definition of equivalence involving also the system state must be introduced. However, such a definition must not depend on an arbitrarily chosen coordinate system for the state variables.

In other words, the state vectors of two systems with equivalent states may at any instant t be different, but it must be possible to transform one state vector into the other by means of a linear constant operator.

From this consideration follows the definition of the so-called *strict equivalence*: Two systems are strictly equivalent if the condition of observable equivalence is satisfied and if eqn. (4.19) implies the relation

$$x^1(t) = F x^2(t) \qquad (4.21)$$

where F is a non-singular constant matrix.

5

Solution of the Continuous-Time State Equation

5.1 Solution in the time domain

Let us consider the state equation of a time-invariant linear controlled plant

$$x'(t) = F\,x(t) + G\,u(t) + w(t) \tag{5.1}$$

where F, G are constant matrices of the type $(n; n)$ and $(n; r)$, respectively, $x(t)$ is the controlled plant state vector whose components determine the dynamic state of the plant at any instant t, $u(t)$ is the vector of controlling variables and $w(t)$ the vector of disturbing variables. It will be recalled that for a single-input/single-output linear controlled plant the number of components of the state vector $x(t)$ is determined by the plant order and the vector $u(t)$ can have only one component. Assume that the initial vector $x(t_0)$ at the instant t_0 and the input signal $u(t)$ for $t \geq t_0$ are known. The task is now to calculate the vector $x(t)$ for $t > t_0$.

The homogeneous equation corresponding to eqn. (5.1) is

$$x'(t) = F\,x(t) \tag{5.2}$$

whose solution is

$$x(t) = e^{F(t-t_0)}\,x(t_0) \tag{5.3}$$

where $x(t_0)$ is the vector of the initial conditions and e^{Ft} is defined by the power series expansion

$$e^{Ft} = \sum_{k=0}^{\infty} \frac{F^k t^k}{k!} \tag{5.4}$$

Introducing

$$e^{F(t-t_0)} = \Phi(t - t_0) \tag{5.5}$$

eqn. (5.3) may be rewritten in the form

$$x(t) = \Phi(t - t_0) x(t_0) \tag{5.6}$$

Let the solution of the non-homogeneous equation (5.1) be

$$x(t) = \Phi(t - t_0) C_1(t) \tag{5.7}$$

Differentiating eqn. (5.7) with respect to t we obtain

$$x'(t) = F x(t) + \Phi(t - t_0) C_1'(t) \tag{5.8}$$

Comparing eqns. (5.1) and (5.8) we have

$$\Phi(t - t_0) C_1'(t) = G u(t) + w(t) \tag{5.9}$$

and consequently

$$C_1(t) = \int_{t_0}^{t} \Phi^{-1}(\tau - t_0) [G u(\tau) + w(\tau)] d\tau + C_2 \tag{5.10}$$

The solution of eqn. (5.1) is, therefore,

$$x(t) = \Phi(t - t_0) C_2 + \int_{t_0}^{t} \Phi(t - \tau) [G u(\tau) + w(\tau)] d\tau \tag{5.11}$$

Since $\Phi(t - t_0)$ does not depend on the integration variable τ, we could have written in eqn. (5.11)

$$\Phi(t - t_0) \Phi^{-1}(\tau - t_0) = e^{F(t-t_0)} e^{-F(\tau-t_0)} = e^{F(t-\tau)} = \Phi(t - \tau)$$

The vector C_2 is calculated from the initial conditions. For $t = t_0$ we have

$$x(t_0) = \Phi(0) C_2 = C_2 \tag{5.12}$$

so that the resultant form of solution of eqn. (5.1) is

$$x(t) = \Phi(t - t_0) x(t_0) + \int_{t_0}^{t} \Phi(t - \tau) [G u(\tau) + w(\tau)] d\tau \tag{5.13}$$

It should be noted that

$$\Phi(t) = e^{Ft}$$

is the so-called *fundamental matrix* of the plant, for which we easily derive the following properties:

$$\begin{aligned} \Phi(-t) &= e^{-Ft} = \Phi^{-1}(t) \\ \Phi(t + t_0) &= \Phi(t) \Phi(t_0) \\ \Phi(t - t_0) &= \Phi(t) \Phi^{-1}(t_0) \end{aligned} \tag{5.14}$$

5.2 Decomposition into partial fractions — Jordan form

As already stated, the state variables of a system may be chosen in various ways. In this section we shall introduce a further possibility, which essentially follows from the principle of superposition of linear systems, in other words, from the decomposition of the linear system response into individual components. For this purpose let us consider the plant transfer function

$$S(s) = \frac{C(s)}{A(s)} = \frac{\sum_{j=0}^{m} c_j s^j}{\sum_{i=0}^{n} a_i s^i}, \quad m < n \tag{5.15}$$

and assume, for simplicity, non-multiple roots of the characteristic equation. The Laplace transform of the output variable

$$Y(s) = S(s)\, U(s) + \frac{X(0)}{A(s)} \tag{5.16}$$

where $U(s)$ is the input variable transform and $X(0)$ is the linear combination of the initial values $y^{(\nu)}(0)$, $\nu = 0, 1, \ldots, n-1$ and $u^{(\mu)}(0)$, $\mu = 0, 1, \ldots, m-1$, may be expressed as a linear combination of the components

$$Y(s) = \sum_{i=0}^{n} h_i\, X_i(s) \tag{5.17}$$

$$X_i(s) = \frac{1}{s - \lambda_i} U(s) + \frac{x_i(0)}{s - \lambda_i} \tag{5.18}$$

The first term on the right-hand side of eqn. (5.18) represents, except for the constant h_i, the component of the transfer function $S(s)$, corresponding to the root factor $s - \lambda_i$, and the second term represents the component of the transform of initial conditions $X(0)/A(s)$ corresponding to the same root factor. Each component $X_i(s)$, $i = 1, 2, \ldots, n$ is the response of the system with the transfer function

$$S_i(s) = \frac{1}{s - \lambda_i}$$

to the input signal $U(s)$, which implicates that $x_i(t)$ satisfies the differential equation

$$x_i'(t) = \lambda_i\, x_i(t) + u(t) \tag{5.19}$$

Its Laplace transform is given by eqn. (5.18) where $x_i(0)$ is the initial value of $x_i(t)$, $i = 1, 2, \ldots, n$, in the differential equation (5.19). The vector $x^T(t) = [x_1(t), x_2(t), \ldots, x_n(t)]$ whose components $x_i(t)$, $i = 1, 2, \ldots, n$, represent the solution of eqn.

(5.11) for the given $x_i(0)$, $i = 1, 2, \ldots, n$, and $u\langle 0, t)$ may, therefore, be qualified as the state vector of the plant considered, the output variable of this plant being uniquely determined by the equation

$$y(t) = \sum_{i=0}^{n} h_i x_i(t) \tag{5.20}$$

The state equations of the system are

$$x'(t) = F x(t) + g u(t)$$
$$y(t) = h^T x(t) \tag{5.21}$$

where

$$F = \begin{bmatrix} \lambda_1, & 0, & \ldots, & 0 \\ 0, & \lambda_2, & \ldots, & 0 \\ \multicolumn{4}{c}{\ldots\ldots\ldots\ldots} \\ 0, & 0, & \ldots, & \lambda_n \end{bmatrix}; \quad g = \begin{bmatrix} 1 \\ 1 \\ \vdots \\ 1 \end{bmatrix}$$

$$h^T = [h_1, h_2, \ldots, h_n] \tag{5.22}$$

It should be emphasized that the decomposition of the transfer function into components, i.e. into partial fractions, leads to a diagonal form of the matrix F if the characteristic equation has non-multiple roots.

Consider now the case where the characteristic polynomial of the transfer function also includes multiple roots. Let

$$S(s) = \frac{\sum_{j=0}^{m} c_j s^j}{a_n (s - \lambda_1)^k (s - \lambda_2) \ldots (s - \lambda_n)} ; \tag{5.23}$$

$$m < k + n - 1$$

Decomposing into partial fractions we obtain

$$S(s) = \frac{h_1}{(s - \lambda_1)^k} + \ldots + \frac{h_k}{s - \lambda_1} + \frac{h_{k+1}}{s - \lambda_2} + \ldots + \frac{h_{n+k-1}}{s - \lambda_n} \tag{5.24}$$

For zero initial conditions the response transform is

$$Y(s) = \frac{h_1}{(s - \lambda_1)^k} U(s) + \ldots + \frac{h_k}{s - \lambda_1} U(s) + \frac{h_{k+1}}{s - \lambda_2} U(s) + \ldots + \frac{h_{n+k-1}}{s - \lambda_n} U(s) \tag{5.25}$$

The last sum may also be expressed as the linear combination

$$Y(s) = \sum_{i=1}^{n+k-1} h_i X_i(s) \tag{5.26}$$

and the relevant vector $x^T(t) = [x_1(t), \ldots, x_{n+k-1}(t)]$ may be considered to be a state vector since

$$X_1(s) = \frac{1}{(s-\lambda_1)^k} U(s) = \frac{1}{s-\lambda_1} X_2(s)$$

$$X_2(s) = \frac{1}{(s-\lambda_1)^{k-1}} U(s) = \frac{1}{s-\lambda_1} X_3(s)$$

$$\vdots$$

$$X_k(s) = \frac{1}{(s-\lambda_1)} U(s)$$

$$X_{k+1}(s) = \frac{1}{(s-\lambda_2)} U(s)$$

$$\vdots$$

$$X_{n+k-1}(s) = \frac{1}{(s-\lambda_n)} U(s) \tag{5.27}$$

In the time domain the set of equations (5.27) may be written as the system of differential equations

$$\begin{aligned}
x_1'(t) &= \lambda_1 x_1(t) + x_2(t) \\
x_2'(t) &= \lambda_1 x_2(t) + x_3(t) \\
&\vdots \\
x_k'(t) &= \lambda_1 x_k(t) + u(t) \\
x_{k+1}'(t) &= \lambda_2 x_{k+1}(t) + u(t) \\
&\vdots \\
x_{n+k-1}'(t) &= \lambda_n x_{n+k-1}(t) + u(t)
\end{aligned} \tag{5.28}$$

Writing this set of equations in the vector-matrix form we again arrive at eqns. (5.21) where, however, the matrices F, g and h have the form

$$F = \begin{bmatrix} \begin{matrix} \lambda_1, & 1, & 0, & \ldots, & 0 \\ 0, & \lambda_1, & 1, & \ldots, & 0 \\ \multicolumn{5}{c}{\ldots\ldots\ldots\ldots\ldots} \\ 0, & 0, & 0, & \ldots, & \lambda_1 \end{matrix} \Big\} k & 0 \\ \hline 0 & \begin{matrix} \lambda_2, & 0, & \ldots, & 0 \\ 0, & \lambda_3, & \ldots, & 0 \\ \multicolumn{4}{c}{\ldots\ldots\ldots\ldots} \\ 0, & 0, & \ldots, & \lambda_n \end{matrix} \end{bmatrix} \; ; \; g = \begin{bmatrix} \begin{matrix} 0 \\ 0 \\ \vdots \\ 0 \end{matrix} \Big\} k-1 \\ \begin{matrix} 1 \\ \vdots \\ 1 \end{matrix} \end{bmatrix}$$

$$h^T = [h_1, h_2, \ldots, h_{n+k-1}] \tag{5.29}$$

SOLUTION OF THE CONTINUOUS-TIME STATE EQUATION

The matrix F in eqns. (5.22) and (5.29) has a very important form, the so-called *Jordan canonical form*, since the elements on the main diagonal of this matrix are the roots of the characteristic polynomial of the transfer function. Some further properties of the Jordan matrix will be given in Sec. 10.2.

5.3 Relationship to Laplace transformation

The relation between the state space expression of plant dynamics by eqns. (3.19) and (3.20) and the transfer function will become evident if we write the Laplace transforms corresponding to both these equations:

$$s X(s) - x(0) = F X(s) + g U(s) \qquad (5.30)$$

$$Y(s) = h^T X(s) + k U(s) \qquad (5.31)$$

From eqn. (5.30) it follows that

$$X(s) = (sE - F)^{-1} x(0) + (sE - F)^{-1} g U(s) \qquad (5.32)$$

Substituting this result into eqn. (5.31) we have

$$Y(s) = h^T (sE - F)^{-1} x(0) + \left[h^T (sE - F)^{-1} g + k \right] U(s) \qquad (5.33)$$

where E in eqns. (5.32) and (5.33) denotes a unit matrix.

The first term on the right hand side of eqn. (5.33) is the solution of the homogeneous equation for $u(t) = 0$ and the second term represents the particular solution corresponding to the input variable $u(t)$ of the non-homogeneous equation. For $x(0) = 0$, eqn. (5.33) yields the transfer function

$$G(s) = \frac{Y(s)}{U(s)} = h^T (sE - F)^{-1} g + k \qquad (5.34)$$

From eqn. (5.34) it is evident that the transfer function $G(s)$ can also be expressed as follows:

$$G(s) = \frac{b_0 + b_1 s + \ldots + b_{n-1} s^{n-1} + k s^n}{\det (sE - F)} \qquad (5.35)$$

Comparing the solution (5.32) with the solution in the time domain (5.13) it can be seen that

$$(sE - F)^{-1} = L[e^{Ft}] = L[\Phi(t)] \qquad (5.36)$$

This expression also directly follows from the definition of the Laplace transform of the function e^{Ft}.

6

Transcription of an Ordinary Difference Equation into State Equations

6.1 Single-input/single-output plant

6.1.1 Procedure I — the state variables are defined by the relation

$$x_i(k + 1) = x_{i+1}(k)$$

In the discrete version of deriving the state equations for control loop elements, described by a linear difference equation with constant coefficients, we must distinguish between two cases, i.e. the case of discretely working elements and the case of continuously working elements. Denoting the input variable of the control loop element by u and the output variable by y (see Fig. 3.6), the relevant difference equation may be written in the form

$$a_n y(k - n) + a_{n-1} y(k - n + 1) + \ldots + a_1 y(k - 1) + a_0 y(k) =$$
$$= b_0 u(k) + b_1 u(k - 1) + \ldots + b_n u(k - n) \qquad (6.1)$$

For discretely working control loop elements realized, for instance, by a computer, the constant b_0 may be non-zero, while for continuously working elements it must be equal to zero except for the trivial case of a zero-order plant.

In eqn. (6.1) a unit period of sampling of the variables y and u was chosen in the interest of writing brevity. In general, for instance, $y[T(k - i)]$ should be written instead of $y(k - i)$. Let us introduce the shift operator E^{-i} for i periods of sampling, i.e.

$$E^{-i}[y(k)] = y(k - i) \qquad (6.2)$$

Using the notation of eqn. (6.2) we may rewrite eqn. (6.1) in the form

$$\sum_{i=0}^{n} a_i E^{-i}[y(k)] = \sum_{i=0}^{n} b_i E^{-i}[u(k)] \qquad (6.3)$$

TRANSCRIPTION OF AN ORDINARY DIFFERENCE EQUATION

$$\hat{A}\, y(k) = \hat{B}\, u(k) \tag{6.4}$$

where evidently

$$\hat{A} = \sum_{i=0}^{n} a_i E^{-i} \quad \text{and} \quad \hat{B} = \sum_{i=0}^{n} b_i E^{-i} \tag{6.5}$$

As in Par. 3.2.1, we may express eqn. (6.4) by two relations:

$$\hat{A}\, x(k) = u(k) \tag{6.6}$$

$$\hat{B}\, x(k) = y(k) \tag{6.7}$$

Introducing now the state variables

$$x(k-n) = x_1(k)$$
$$x(k-n+1) = x_1(k+1) = x_2(k)$$
$$x(k-n+2) = x_2(k+1) = x_3(k)$$
$$\vdots$$
$$x(k-1) = x_{n-1}(k+1) = x_n(k) \tag{6.8}$$
$$x(k) = x_n(k+1) \tag{6.9}$$

we may rewrite eqn. (6.6) in the form

$$x_n(k+1) = \frac{1}{a_0}\left[u(k) - a_1 x_n(k) - a_2 x_{n-1}(k) - \ldots - a_n x_1(k)\right]; \quad a_0 \neq 0 \tag{6.10}$$

With the state variables given by eqns. (6.8) and (6.9) we now obtain eqn. (6.7) in the form

$$y(k) = b_0 x_n(k+1) + b_1 x_n(k) + \ldots + b_n x_1(k) \tag{6.11}$$

Substituting for $x_n(k+1)$ in eqn. (6.11) from eqn. (6.10) we obtain

$$y(k) = \frac{b_0}{a_0}\left[u(k) - a_1 x_n(k) - a_2 x_{n-1}(k) - \ldots - a_n x_1(k)\right] +$$
$$+ b_1 x_n(k) + \ldots + b_n x_1(k) \tag{6.12}$$

$$y(k) = \frac{b_0}{a_0} u(k) + \left(b_1 - \frac{b_0 a_1}{a_0}\right) x_n(k) + \left(b_2 - \frac{b_0 a_2}{a_0}\right) x_{n-1}(k) + \ldots$$
$$\ldots + \left(b_n - \frac{b_0 a_n}{a_0}\right) x_1(k) \tag{6.13}$$

Equations (6.8) and (6.10) determine the first state equation

$$\begin{bmatrix} x_1(k+1) \\ x_2(k+1) \\ \vdots \\ x_n(k+1) \end{bmatrix} = \begin{bmatrix} 0, & 1, & 0, & \ldots, & 0 \\ 0, & 0, & 1, & \ldots, & 0 \\ \vdots & & & & \\ -\dfrac{a_n}{a_0}, & -\dfrac{a_{n-1}}{a_0}, & \ldots, & -\dfrac{a_1}{a_0} \end{bmatrix} \begin{bmatrix} x_1(k) \\ x_2(k) \\ \vdots \\ x_n(k) \end{bmatrix} + \begin{bmatrix} 0 \\ 0 \\ \vdots \\ \dfrac{1}{a_0} \end{bmatrix} u(k) \quad (6.14)$$

and eqn. (6.13) determines the second state equation

$$y(k) = \left[\left(b_n - \frac{b_0 a_n}{a_0} \right), \ldots, \left(b_1 - \frac{b_0 a_1}{a_0} \right) \right] [x_1(k), \ldots, x_n(k)] + \frac{b_0}{a_0} u(k) \quad (6.15)$$

Both equations may briefly be written as follows:

$$x(k+1) = A\,x(k) + b\,u(k) \quad (6.16)$$
$$y(k) = c^T x(k) + d\,u(k) \quad (6.17)$$

The form of the individual matrix operators and vectors follows from a comparison with eqns. (6.14) and (6.15). In the case of a controlled plant, where $b_0 = 0$, the state equation (6.15) will be simplified. From eqns. (6.14) and (6.15) it is apparently reasonable to adjust the original difference equation (6.1) so that $a_0 = 1$.

The block diagram of the state equations is shown in Fig. 6.1.

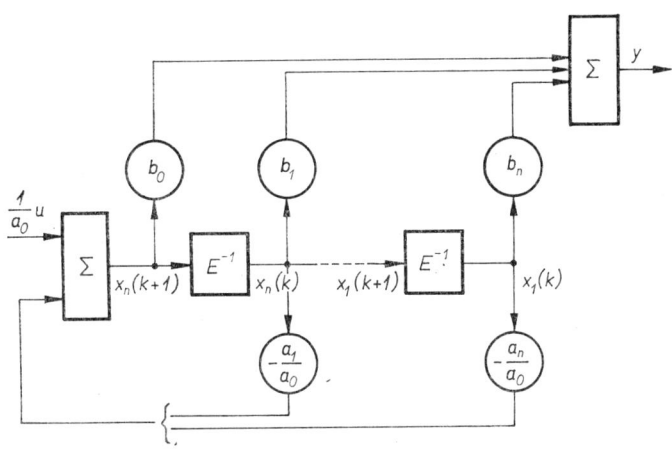

Fig. 6.1. Block diagram relating to the state equations (6.14) and (6.15).

Let us henceforth designate, as in the case of continuously working systems, the first of the set of state equations (6.16) and (6.17) as the *equation of dynamics* and the second as the *output equation*. Then A is the *matrix of dynamics*, $b(B)$ the *input column matrix (input matrix)*, $c^T(C)$ the *output row matrix (output matrix)* and

TRANSCRIPTION OF AN ORDINARY DIFFERENCE EQUATION   51

$d(D)$ the *input factor (input matrix factor)*. The symbols in brackets correspond to multivariable systems.

Example 6.1: The transfer function of the digital correcting member in Z-transformation is given by the expression

$$P(z) = \frac{E_2(z)}{E_1(z)} = \frac{2 + 3.6z^{-1} - 0.8z^{-2}}{1 + 0.7z^{-1} + 0.1z^{-2}}$$

Establish the relevant state equations.

Solution: We transform the given transfer function into the form corresponding to eqn. (6.1)

$$0.1\, e_2(k-2) + 0.7\, e_2(k-1) + e_2(k) = 2\, e_1(k) + 3.6\, e_1(k-1) - 0.8\, e_1(k-2)$$

The first state equation is established according to eqn. (6.14)

$$\begin{bmatrix} x_1(k+1) \\ x_2(k+1) \end{bmatrix} = \begin{bmatrix} 0, & 1 \\ -0.1, & -0.7 \end{bmatrix} \begin{bmatrix} x_1(k) \\ x_2(k) \end{bmatrix} + \begin{bmatrix} 0 \\ 1 \end{bmatrix} u(k)$$

and the second state equation according to eqn. (6.15):

$$y(k) = [-1,\ 2.2]\, [x_1(k),\ x_2(k)]^T + 2\, u(k)$$

6.1.2 Procedure II — the state variables are a linear combination of the values of the plant input and output variables

Let us write eqn. (6.1) in the form

$$a_n y(k) + a_{n-1} y(k+1) + \ldots + a_1 y(k+n-1) + a_0 y(k+n) =$$
$$= b_0 u(k+n) + b_1 u(k+n-1) + \ldots + b_n u(k) \tag{6.18}$$

and introduce the state variables

$$x_1(k) = a_0 y(k) - b_0 u(k) \tag{6.19}$$

$$x_2(k) = a_1 y(k) + a_0 y(k+1) - b_0 u(k+1) - b_1 u(k)$$

$$x_3(k) = a_2 y(k) + a_1 y(k+1) + a_0 y(k+2) -$$
$$\quad - b_0 u(k+2) - b_1 u(k+1) - b_2 u(k)$$

$$\vdots$$

$$x_n(k) = a_{n-1} y(k) + a_{n-2} y(k+1) + \ldots + a_0 y(k+n-1) - \ldots$$
$$\ldots - b_0 u(k+n-1) - \ldots - b_{n-2} u(k+1) - b_{n-1} u(k) \tag{6.20}$$

We calculate $y(k)$ from eqn. (6.19) and substitute it into all eqns. (6.20), and replace in these equations all terms including $y(k + i)$ and $u(k + i)$, $i = 1, 2, \ldots, k + n - 1$, by the state variables $x_v(k + 1)$, $v = 1, 2, \ldots, n - 1$. We obtain

$$y(k) = \frac{1}{a_0} x_1(k) + \frac{b_0}{a_0} u(k) \tag{6.21}$$

$$x_2(k) = \frac{a_1}{a_0} x_1(k) + x_1(k + 1) - \left(b_1 - \frac{a_1}{a_0} b_0\right) u(k)$$

$$x_3(k) = \frac{a_2}{a_0} x_1(k) + x_2(k + 1) - \left(b_2 - \frac{a_2}{a_0} b_0\right) u(k)$$

$$\vdots$$

$$x_n(k) = \frac{a_{n-1}}{a_0} x_1(k) + x_{n-1}(k + 1) - \left(b_{n-1} - \frac{a_{n-1}}{a_0} b_0\right) u(k) \tag{6.22}$$

The original difference equation (6.18) may now be expressed in terms of state variables as follows:

$$0 = \frac{a_n}{a_0} x_1(k) + x_n(k + 1) - \left(b_n - \frac{a_n}{a_0} b_0\right) u(k) \tag{6.23}$$

Equations (6.22) and (6.23) determine the first state equation

$$\begin{bmatrix} x_1(k+1) \\ x_2(k+1) \\ \vdots \\ x_n(k+1) \end{bmatrix} = \begin{bmatrix} -\frac{a_1}{a_0}, & 1, 0, & \ldots, 0 \\ -\frac{a_2}{a_0}, & 0, 1, & \ldots, 0 \\ \cdots\cdots\cdots\cdots\cdots \\ -\frac{a_n}{a_0}, & 0, 0, & \ldots, 0 \end{bmatrix} \begin{bmatrix} x_1(k) \\ x_2(k) \\ \vdots \\ x_n(k) \end{bmatrix} + \begin{bmatrix} b_1 - \frac{a_1}{a_0} b_0 \\ b_2 - \frac{a_2}{a_0} b_0 \\ \vdots \\ b_n - \frac{a_n}{a_0} b_0 \end{bmatrix} u(k) \tag{6.24}$$

The second state equation is determined by eqn. (6.21). The vector-matrix form of eqns. (6.24) and (6.21) is

$$x(k + 1) = A\, x(k) + b\, u(k)$$
$$y(k) = c^T x(k) + d\, u(k) \tag{6.25}$$

where the form of the individual matrix operators and vectors follows from a comparison with eqns. (6.24) and (6.21). Let us note for completeness that $c^T =$

TRANSCRIPTION OF AN ORDINARY DIFFERENCE EQUATION

$= [1/a_0, 0, \ldots, 0]$ and $d = b_0/a_0$. For $b = 0$, eqns. (6.24) and (6.21) will be simplified. From these equations it also follows that it is reasonable to adjust the original difference equation (6.18) so that $a_0 = 1$.

The block diagram of the state equations is shown in Fig. 6.2.

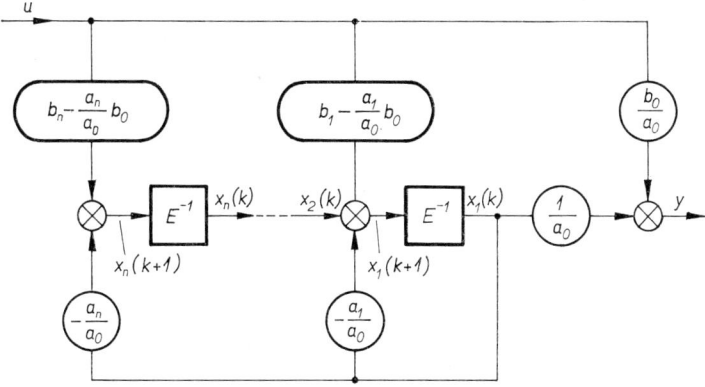

Fig. 6.2. Block diagram relating to the state equations (6.21) and (6.24).

Example 6.2: Establish the state equations relating to Example 6.1.

Solution: According to eqns. (6.24) and (6.25)

$$\begin{bmatrix} x_1(k+1) \\ x_2(k+1) \end{bmatrix} = \begin{bmatrix} -0.7, & 1 \\ -0.1, & 0 \end{bmatrix} \begin{bmatrix} x_1(k) \\ x_2(k) \end{bmatrix} + \begin{bmatrix} 2.2 \\ -1 \end{bmatrix} u(k)$$

$$y(k) = x_1(k) + 2\,u(k)$$

Example 6.3: The discrete transfer function of the continuously working part of a control loop is given as

$$G_1(z) = \frac{0.1306 z^{-1} + 0.4094 z^{-2} + 0.0792 z^{-3}}{1 - 2.2130 z^{-1} + 1.5809 z^{-2} - 0.3679 z^{-3}}$$

The control loop comprises a zero-order hold element, an integrating servomotor with a final control element and a controlled plant. Calculate the forced unit-step response of the input signal $u(k) = 1$, $k = 0, 1, 2, \ldots$.

Solution: The problem may be solved by multiplying the transfer function $G_1(z, 0)$ by the transform of the input signal $U(z) = 1/(1 - z^{-1})$ with a following division of the resultant polynomials. An alternative way of solution is to write the state equations according to eqns. (6.25) and to calculate the response on a computer:

$$\begin{bmatrix} x_1(k+1) \\ x_2(k+1) \\ x_3(k+1) \end{bmatrix} = \begin{bmatrix} 2.2130, & 1, & 0 \\ -1.5809, & 0, & 1 \\ 0.3679, & 0, & 0 \end{bmatrix} \begin{bmatrix} x_1(k) \\ x_2(k) \\ x_3(k) \end{bmatrix} + \begin{bmatrix} 0.1306 \\ 0.4094 \\ 0.0792 \end{bmatrix} u(k)$$

$$y(k) = x_1(k)$$

For $x_1(0) = x_2(0) = x_3(0) = 0$ we obtain:

k	0	1	2	3	4	5
y(k)	0	0.1306	0.8290	2.2474	4.3300	6.9537

6.1.3 Discrete-plant state equation under action of the controlling and disturbing variable

The difference equation of a plant with both the controlling and the disturbing variable w acting at its input is written, in agreement with eqn. (6.18), in the form

$$\sum_{i=0}^{n} a_{n-i} y(k+i) = \sum_{i=0}^{n} b_{n-i} u(k+i) + \sum_{i=0}^{n} c_{n-i} w(k-i) \qquad (6.26)$$

In the same way as in Par. 6.1.2 we arrive at the solution

$$\begin{bmatrix} x_1(k+1) \\ x_2(k+1) \\ \vdots \\ x_n(k+1) \end{bmatrix} = \begin{bmatrix} -\dfrac{a_1}{a_0}, & 1, & 0, & \ldots, & 0 \\ -\dfrac{a_2}{a_0}, & 0, & 1, & \ldots, & 0 \\ \vdots & & & & \\ -\dfrac{a_n}{a_0}, & 0, & 0, & \ldots, & 0 \end{bmatrix} \begin{bmatrix} x_1(k) \\ x_2(k) \\ \vdots \\ x_n(k) \end{bmatrix} +$$

$$+ \begin{bmatrix} b_1 - \dfrac{a_1}{a_0} b_0, & c_1 - \dfrac{a_1}{a_0} c_0 \\ b_2 - \dfrac{a_2}{a_0} b_0, & c_2 - \dfrac{a_2}{a_0} c_0 \\ \vdots & \vdots \\ b_n - \dfrac{a_n}{a_0} b_0, & c_n - \dfrac{a_n}{a_0} c_0 \end{bmatrix} \begin{bmatrix} u(k) \\ w(k) \end{bmatrix} \qquad (6.27)$$

$$y(k) = \frac{1}{a_0} x_1(k) + \frac{b_0}{a_0} u(k) + \frac{c_0}{a_0} w(k) \tag{6.28}$$

The state variables are defined as follows:

$$x_1(k) = a_0 y(k) - b_0 u(k) - c_0 w(k) \tag{6.29}$$

$$x_2(k) = a_1 y(k) + x_1(k+1) - b_1 u(k) - c_1 w(k)$$

$$\vdots$$

$$x_n(k) = a_{n-1} y(k) + x_{n-1}(k+1) - b_{n-1} u(k) - c_{n-1} w(k) \tag{6.30}$$

Equation (6.26) with the state variables given by eqns. (6.29) and (6.30) assumes the form

$$0 = a_n y(k) + x_n(k+1) - b_n u(k) - c_n w(k) \tag{6.31}$$

6.2 Multi-input/multi-output plant

The dynamics of a discrete linear time-invariant multi-input/multi-output controlled plant is described by the following set of linear difference equations with constant coefficients:

$$\sum_{j=1}^{N} a_{ij}(E) y_j(k) = \sum_{j=1}^{N} b_{ij}(E) u_j(k), \quad i = 1, 2, \ldots, N \tag{6.32}$$

According to eqn. (6.32) the number of controlling variables is equal to the number of controlled variables, $a_{ij}(E)$ and $b_{ij}(E)$ are polynomials in E of the n_{ij}-th degree and E denotes the shift operator in the sense of eqn. (6.2). The set of equations (6.32) may also be written in matrix form

$$A^*(E) y(k) = B^*(E) u(k) \tag{6.33}$$

where $A^*(E)$ and $B^*(E)$ are matrices with the elements $a_{ij}(E)$ and $b_{ij}(E)$, $i, j = 1, 2, \ldots, N$, respectively.

Let us decompose the matrices $A^*(E)$ and $B^*(E)$ into a matrix sum in such a way that the elements of the individual matrices be, after decomposition, multiplied by the same power of the operator E:

$$(A_n^* + A_{n-1}^* E + \ldots + A_1^* E^{n-1} + A_0^* E^n) y(k) =$$

$$= (B_0^* E^n + B_1^* E^{n-1} + \ldots + B_{n-1}^* E + B_n^*) u(k) \tag{6.34}$$

Equation (6.34) is analogous to eqn. (6.18) except that in eqn. (6.34) the matrices A_v^* and B_v^*, $v = 0, 1, 2, \ldots, n$, $n = \max(n_{ij})$, $i, j = 1, 2, \ldots, N$, appear instead of the constants a_v and b_v, and the variables y and u are replaced by the vectors y and u.

Introducing the vectors of the state variables we obtain, analogously to Par. 6.1.2, the relations

$$x_v^T = [x_{v1}, x_{v2}, \ldots, x_{vN}], \quad v = 1, 2, \ldots, n$$

$$\begin{bmatrix} x_1(k+1) \\ x_2(k+1) \\ \vdots \\ x_n(k+1) \end{bmatrix} = \begin{bmatrix} -A_1, & E, & 0, & \ldots, & 0 \\ -A_2, & 0, & E, & \ldots, & 0 \\ \ldots & \ldots & \ldots & \ldots & \ldots \\ -A_n, & 0, & 0, & \ldots, & 0 \end{bmatrix} \begin{bmatrix} x_1(k) \\ x_2(k) \\ \vdots \\ x_n(k) \end{bmatrix} + \begin{bmatrix} B_1 \\ B_2 \\ \vdots \\ B_n \end{bmatrix} u(k) \quad (6.35)$$

where

$$A_l = A_l^* A_0^{*-1}, \quad l = 1, 2, \ldots, n \quad (6.36)$$

$$B_l = B_l^* - A_l^* A_0^{*-1} B_0^*, \quad l = 1, 2, \ldots, n \quad (6.37)$$

$$y(k) = A_0^{*-1} x_1(k) + A_0^{*-1} B_0^* u(k) \quad (6.38)$$

In the state equations (6.36) through (6.38)

$$u^T(k) = [u_1(k), \ldots, u_N(k)] \quad \text{and} \quad y^T(k) = [y_1(k), \ldots, y_N(k)]$$

As can be seen from eqns. (6.36) through (6.38), the difference equations can be transcribed into state equations by the method indicated only if the matrix A_0^* is non-singular. If it is singular a state description may still exist. For the same reasons as in Sec. 3.3 we shall give only an example without explicit solution.

Since for a non-singular matrix A_0^* the solution is obtained by a simple substitution into eqns. (6.35) and (6.38), only an example of solution with a singular matrix A_0^* will be given.

Example 6.4: A double-input/double-output plant is given by the difference equations

$$0.1 \, y_1(k-2) + 0.7 \, y_1(k-1) + y_1(k) - 0.09 \, y_2(k-2) + y_2(k) =$$
$$= -0.5 \, u_1(k-1) + 0.6 \, u_1(k-2) - 2 \, u_2(k-1) + u_2(k-2)$$

$$0.3 \, y_1(k-2) - 1.1 \, y_1(k-1) + y_1(k) + 0.2 \, y_2(k-1) + y_2(k) =$$
$$= -4 \, u_1(k-1) + 2 \, u_1(k-2) - 0.2 \, u_2(k-1) - 0.5 \, u_2(k-2)$$

Establish the corresponding state equations!

Solution: According to the given difference equations we have

$$A_0^* = \begin{bmatrix} 1, & 1 \\ 1, & 1 \end{bmatrix}$$

TRANSCRIPTION OF AN ORDINARY DIFFERENCE EQUATION 57

which is a singular matrix so that eqns. (6.35) and (6.38) cannot be used. The transcription of the given set of difference equations may, however, be obtained by eliminating the variables $y_1(k)$ and $y_2(k)$ from the equations defining the state variables

$$x_1(k) = A_0^* \, y(k)$$
$$x_2(k) = A_1^* \, y(k) + x_1(k+1) - B_1^* \, u(k) \tag{6.39}$$

and from the original set of difference equations expressed by means of the defined state variables

$$0 = A_2^* \, y(k) + x_2(k+1) - B_2^* \, u(k) \tag{6.40}$$

From the first of eqns. (6.39) it follows that

$$\begin{bmatrix} x_{11}(k) \\ x_{12}(k) \end{bmatrix} = \begin{bmatrix} 1, & 1 \\ 1, & 1 \end{bmatrix} \begin{bmatrix} y_1(k) \\ y_2(k) \end{bmatrix}$$

$$x_{11}(k) = y_1(k) + y_2(k) = x_{12}(k) \tag{6.41}$$

From the second of eqns. (6.39) we obtain

$$\begin{bmatrix} x_{21}(k) \\ x_{22}(k) \end{bmatrix} = \begin{bmatrix} 0.7, & 0 \\ -1.1, & 0.2 \end{bmatrix} \begin{bmatrix} y_1(k) \\ y_2(k) \end{bmatrix} + \begin{bmatrix} x_{11}(k+1) \\ x_{11}(k+1) \end{bmatrix} + \begin{bmatrix} 0.5, & 2 \\ 4, & 0.2 \end{bmatrix} \begin{bmatrix} u_1(k) \\ u_2(k) \end{bmatrix}$$

$$x_{21}(k) = 0.7 \, y_1(k) + x_{11}(k+1) + 0.5 \, u_1(k) + 2 \, u_2(k)$$
$$x_{22}(k) = -1.1 \, y_1(k) + 0.2 \, y_2(k) + x_{11}(k+1) + 4 \, u_1(k) + 0.2 \, u_2(k) \tag{6.42}$$

Subtracting eqns. (6.42) we eliminate $x_{11}(k+1)$ and obtain

$$x_{21}(k) - x_{22}(k) = 1.8 \, y_1(k) - 0.2 \, y_2(k) - 3.5 \, u_1(k) + 1.8 \, u_2(k) \tag{6.43}$$

From eqns. (6.41) and (6.43) we now determine $y_1(k)$ and $y_2(k)$:

$$y_1(k) = 0.1 \, x_{11}(k) + 0.5 \, x_{21}(k) - 0.5 \, x_{22}(k) + 1.75 \, u_1(k) - 0.9 \, u_2(k)$$
$$y_2(k) = 0.9 \, x_{11}(k) - 0.5 \, x_{21}(k) + 0.5 \, x_{22}(k) - 1.75 \, u_1(k) + 0.9 \, u_2(k) \tag{6.44}$$

Applying the matrix description, eqn. (6.40) assumes the form

$$0 = \begin{bmatrix} 0.1, & -0.09 \\ 0.3, & 0 \end{bmatrix} \begin{bmatrix} y_1(k) \\ y_2(k) \end{bmatrix} + \begin{bmatrix} x_{21}(k+1) \\ x_{22}(k+1) \end{bmatrix} - \begin{bmatrix} 0.6, & 1 \\ 2, & -0.5 \end{bmatrix} \begin{bmatrix} u_1(k) \\ u_2(k) \end{bmatrix}$$

$$0 = 0.1 \, y_1(k) - 0.09 \, y_2(k) + x_{21}(k+1) - 0.6 \, u_1(k) - u_2(k)$$
$$0 = 0.3 \, y_1(k) \qquad\qquad + x_{22}(k+1) - 2 \, u_1(k) + 0.5 \, u_2(k) \tag{6.45}$$

If we now substitute for $y_1(k)$ and $y_2(k)$ in any of eqns. (6.42) and in eqns. (6.45) the relations (6.44), we can explicitly determine $x_{11}(k+1)$, $x_{21}(k+1)$ and $x_{22}(k+1)$

and set up the first state equation

$$\begin{bmatrix} x_{11}(k+1) \\ x_{21}(k+1) \\ x_{22}(k+1) \end{bmatrix} = \begin{bmatrix} -0.07, & 0.65, & 0.35 \\ 0.071, & -0.095, & +0.095 \\ -0.03, & -0.15, & +0.15 \end{bmatrix} \begin{bmatrix} x_{11}(k) \\ x_{21}(k) \\ x_{22}(k) \end{bmatrix} +$$
$$+ \begin{bmatrix} -1.725, & -1.37 \\ 0.2675, & 1.171 \\ 1.475, & -0.23 \end{bmatrix} \begin{bmatrix} u_1(k) \\ u_2(k) \end{bmatrix} \quad (6.46)$$

The second state equation expressing the transformation of the state variables into the variables $y_1(k)$ and $y_2(k)$ is defined by relations (6.44):

$$\begin{bmatrix} y_1(k) \\ y_2(k) \end{bmatrix} = \begin{bmatrix} 0.1, & 0.5, & -0.5 \\ 0.9, & -0.5, & 0.5 \end{bmatrix} \begin{bmatrix} x_{11}(k) \\ x_{21}(k) \\ x_{22}(k) \end{bmatrix} + \begin{bmatrix} 1.75, & -0.9 \\ -1.75, & 0.9 \end{bmatrix} \begin{bmatrix} u_1(k) \\ u_2(k) \end{bmatrix} \quad (6.47)$$

7

Solution of the Discrete-Time State Equation

7.1 Solution in the time domain

Let us consider the discrete plant state equation

$$x(k + 1) = A\,x(k) + B\,u(k) + w(k)$$

where A and B are constant $(n; n)$ and $(n; r)$ matrices, respectively. The initial vector $x(0)$ and the input signals, i.e. the controlling variables $u(k)$ and the disturbing variable $w(k)$, are known.

According to the state equation we calculate

$$x(1) = A\,x(0) + B\,u(0) + w(0)$$

$$x(2) = A\,x(1) + B\,u(1) + w(1) =$$
$$= A^2\,x(0) + AB\,u(0) + B\,u(1) + A\,w(0) + w(1)$$

$$x(3) = A\,x(2) + B\,u(2) + w(2) =$$
$$= A^3\,x(0) + A^2 B\,u(0) + AB\,u(1) + B\,u(2) + A^2\,w(0) + A\,w(1) + w(2)$$

and in general

$$x(k) = A^k\,x(0) + \sum_{j=0}^{k-1} A^j B\,u(k - j - 1) + \sum_{j=0}^{k-1} A^j\,w(k - j - 1) =$$

$$= A^k\,x(0) + \sum_{r=0}^{k-1} A^{k-r-1}[B\,u(r) + w(r)] \qquad (7.1)$$

For the output variable we have

$$y(k) = C\,x(k) + D\,u(k)$$

so that with eqn. (7.1)

$$y(k) = CA^k x(0) + \sum_{j=0}^{k-1} CA^j B\, u(k-j-1) + \sum_{j=0}^{k-1} CA^j w(k-j-1) + D\, u(k) =$$
$$= CA^k x(0) + \sum_{j=1}^{k} CA^{j-1}[B\, u(k-j) + w(k-j)] + D\, u(k) =$$
$$= CA^k x(0) + \sum_{r=0}^{k-1} CA^{k-r-1}[B\, u(r) + w(r)] + D\, u(k) \quad (7.2)$$

where $j = k - r - 1$ was substituted in the last modification of eqns. (7.1) and (7.2).

For $x(0) = 0$, $w(r) = 0$, $r \geq 0$, $u(0) = 1$ and $u(r) = 0$, $r > 0$, the formula for the impulse response of a single-input/single-output plant also follows from the last expression of eqn. (7.2) if the matrices B, C, D are replaced by b, c^T, d according to eqns. (6.16) and (6.17). Then it holds that

$$s(k) = \begin{cases} d & \text{for } k = 0 \\ c^T A^{k-1} b & \text{for } k > 0 \end{cases} \quad (7.3)$$

7.2 Weighting matrix

If we know the discrete values of the plant impulse response, $s(k)$, $k = 0, 1, 2, \ldots$, and of the controlling variable $u(k)$, $k = 0, 1, 2, \ldots$, we can calculate the values of the plant response by means of the convolution sum at the individual instants k

$$y(k) = \sum_{l=0}^{k} s(k-l)\, u(l) \quad (7.4)$$

or

$$y(k) = \sum_{l=0}^{k} s(l)\, u(k-l) = \sum_{l=1}^{k} s(l)\, u(k-l) + s(0)\, u(k) \, ; \quad k = 0, 1, 2, \ldots$$

For the impulse response we have

$$s(k) = \begin{cases} s(0) & \text{for } k = 0 \\ s(l) & \text{for } k > 0 \end{cases}$$

From a comparison with relations (7.3) it also follows that

$$s(k) = \begin{cases} s(0) = d & \text{for } k = 0 \\ s(l) = c^T A^{k-1} b & \text{for } k > 0 \end{cases} \quad (7.5)$$

Equation (7.4) applies to one-dimensional vectors y and u. For a plant whose vectors y and u have the dimensions p and r, respectively, eqn. (7.4) may be generalized in the form

$$y_i(k) = \sum_{l=0}^{k} \sum_{j=1}^{r} s_{ij}(k-l)\, u_j(l), \quad i = 1, 2, \ldots, p \quad (7.6)$$

SOLUTION OF THE DISCRETE-TIME STATE EQUATION

or in the form with the matrix $S(k - l)$ having the elements $s_{ij}(k - l)$, $i = 1, 2, \ldots, p$ and $j = 1, 2, \ldots, r$:

$$y(k) = \sum_{l=0}^{k} S(k - l) u(l)$$

$$\begin{bmatrix} y_1(k) \\ y_2(k) \\ \vdots \\ y_p(k) \end{bmatrix} = \sum_{l=0}^{k} \begin{bmatrix} s_{11}(k-l), & s_{12}(k-l), & \ldots, & s_{1r}(k-l) \\ s_{21}(k-l), & s_{22}(k-l), & \ldots, & s_{2r}(k-l) \\ \vdots & & & \\ s_{p1}(k-l), & s_{p2}(k-l), & \ldots, & s_{pr}(k-l) \end{bmatrix} \begin{bmatrix} u_1(l) \\ u_2(l) \\ \vdots \\ u_r(l) \end{bmatrix} \quad (7.7)$$

Equation (7.4) may also be written as follows:

$$\begin{bmatrix} y(0) \\ y(1) \\ y(2) \\ \vdots \\ y(k) \end{bmatrix} = \begin{bmatrix} s(0), & 0, & 0, & \ldots, & 0 \\ s(1), & s(0), & 0, & \ldots, & 0 \\ s(2), & s(1), & s(0), & \ldots, & 0 \\ \vdots & & & & \\ s(k), & s(k-1), & s(k-2), & \ldots, & s(0) \end{bmatrix} \begin{bmatrix} u(0) \\ u(1) \\ u(2) \\ \vdots \\ u(k) \end{bmatrix} \quad (7.8)$$

where the lower triangular matrix of the type $(k + 1; k + 1)$ is the *weighting matrix* of the single-input/single-output plant.

7.3 Decomposition into partial fractions — Jordan form

In the discrete case we can, analogously to Sec. 5.2, decompose the Z-transfer function of the plant into partial fractions and choose the components of this decomposition as state variables. Thus we arrive at state equations with the matrix A in the Jordan form.

Assume, for simplicity, that the Z-transforms of the plant initial conditions and of the input signal are zero and that the characteristic polynomial of the plant Z-transfer function has non-multiple roots.

On the other hand one has to keep in view that, in accordance with the conditions of physical realizability, the Z-transfer function of the controlled plant will have (except for the trivial case of a zero-order plant) its numerator polynomial at least one degree lower than the denominator polynomial, while the Z-transfer function of the correcting member usually has both polynomials of the same degree. This fact is due to the negligible sampling time, negligible analogue-to-digital and digital-to-analogue conversion times and to the negligible time needed to calculate the controlling variable as compared with the sampling period. In this chapter we shall, therefore, consider the more general case of the Z-transfer function of a digital correcting member with equal polynomial degrees in the numerator and denominator:

$$P^*(z) = \frac{\sum_{j=0}^{n} c_j^* z^j}{\sum_{i=0}^{n} a_i z^i} \qquad (7.9)$$

Since the decomposition into partial fractions requires the rational fraction function numerator to be at least one degree lower than the denominator, we first perform a partial division indicated by the fraction (7.9) and obtain

$$P^*(z) = \frac{c_n^*}{a_n} + \frac{\sum_{j=0}^{n-1} c_j z^j}{\sum_{i=0}^{n} a_i z^i} = \frac{c_n^*}{a_n} + P(z) \qquad (7.10)$$

For a given discrete transform $U(z)$ of the input variable and for zero initial conditions of both the correcting member and the input variable, the response transform is

$$Y(z) = \frac{c_n^*}{a_n} U(z) + P(z) U(z) \qquad (7.11)$$

If we now decompose $P(z)$ into partial fractions, we may rewrite eqn. (7.11) in the form

$$Y(z) = \frac{c_n^*}{a_n} U(z) + \sum_{i=1}^{n} \frac{b_i}{z - z_i} U(z) \qquad (7.12)$$

$$Y(z) = \frac{c_n^*}{a_n} U(z) + \sum_{i=1}^{n} b_i X_i(z) \qquad (7.13)$$

where the components $X_i(z)$, $i = 1, 2, \ldots, n$, are defined by the relation

$$X_i(z) = \frac{1}{z - z_i} U(z) \qquad (7.14)$$

The components expressed by the right-hand side of eqn. (7.13) represent the components of the response to the input signal $U(z)$. If $U(z)$ is a unit impulse, these components are those of the impulse response.

In the time domain the difference equation corresponding to eqn. (7.14) is

$$x_i(k+1) = z_i x_i(k) + u(k), \quad i = 1, 2, \ldots, n \qquad (7.15)$$

Equations (7.15) define for all values of $i = 1, 2, \ldots, n$, together with the inverse transform of eqn. (7.13), the state equations of the digital correcting member

$$x(k+1) = A x(k) + b u(k) \qquad (7.16)$$

$$y(k) = c^T x(k) + d u(k) \qquad (7.17)$$

SOLUTION OF THE DISCRETE-TIME STATE EQUATION

where

$$A = \begin{bmatrix} z_1, & 0, & \ldots, & 0 \\ 0, & z_2, & \ldots, & 0 \\ \ldots & \ldots & \ldots & \ldots \\ 0, & 0, & \ldots, & z_n \end{bmatrix}; \quad b = \begin{bmatrix} 1 \\ 1 \\ \vdots \\ 1 \end{bmatrix}; \quad c = \begin{bmatrix} b_1 \\ b_2 \\ \vdots \\ b_n \end{bmatrix}, \quad d = \frac{c_n^*}{a_n} \tag{7.18}$$

As can be seen, the matrix A is diagonal and its elements on the main diagonal are the roots of the characteristic polynomial of the system considered. The matrix A is a Jordan matrix and the vector $x^T(k) = [x_1(k), \ldots, x_n(k)]$, whose components are the solutions of the difference equations (7.15) for $i = 1, 2, \ldots, n$, is the vector of the state variables of this system.

If eqns. (7.16) and (7.17) should correspond to a physically realizable plant, then $d = 0$ because the degree of the numerator of the discrete transfer function of such a plant must necessarily be lower than n.

In the case of multiple roots of the characteristic transfer function polynomial of the correcting member let, for instance, be

$$P^*(z) = \frac{\sum_{j=0}^{m} c_j^* z^j}{a_n(z - z_1)^s (z - z_2) \ldots (z - z_n)} \tag{7.19}$$

Equation (7.19) has been written for real roots of the characteristic polynomial. For complex roots the further procedure would be analogous. On decomposition into partial fractions we obtain

$$P^*(z) = \frac{c_m^*}{a_n} + \frac{b_1}{(z - z_1)^s} + \ldots + \frac{b_s}{(z - z_1)} + \frac{b_{s+1}}{(z - z_2)} + \ldots$$

$$\ldots + \frac{b_{n+s-1}}{z - z_n} = \frac{c_m^*}{a_n} + P(z) \tag{7.20}$$

For zero initial conditions and a known Z-transform of the input variable the Z-transform of the response will be

$$Y(z) = \frac{c_m^*}{a_n} U(z) + P(z) U(z) \tag{7.21}$$

$$Y(z) = \frac{c_m^*}{a_n} U(z) + \sum_{i=1}^{m} b_i X_i(z) \tag{7.22}$$

where the expression $P(z) U(z)$ in eqn. (7.21) has been replaced by a linear combination of the variables $X_i(z)$ defined as follows:

$$X_1(z) = \frac{1}{(z-z_1)^s} U(z) = \frac{1}{z-z_1} X_2(z)$$

$$X_2(z) = \frac{1}{(z-z_1)^{s-1}} U(z) = \frac{1}{z-z_1} X_3(z)$$

$$\vdots$$

$$X_s(z) = \frac{1}{z-z_1} U(z)$$

$$X_{s+1}(z) = \frac{1}{z-z_2} U(z)$$

$$\vdots$$

$$X_m(z) = \frac{1}{z-z_n} U(z) \qquad (7.23)$$

In the time domain eqns. (7.23) will be expressed by the set of difference equations

$$\begin{aligned}
x_1(k+1) &= z_1 x_1(k) + x_2(k) \\
x_2(k+1) &= z_1 x_2(k) + x_3(k) \\
&\vdots \\
x_s(k+1) &= z_1 x_s(k) + u(k) \\
x_{s+1}(k+1) &= z_2 x_{s+1}(k) + u(k) \\
&\vdots \\
x_m(k+1) &= z_n x_m(k) + u(k)
\end{aligned} \qquad (7.24)$$

Equations (7.24) together with eqn. (7.22) define the state equations with the matrices A, b, c^T and d of the form

$$A = \left[\begin{array}{c|c}
\begin{matrix} z_1, & 1, & 0, & \ldots, & 0 \\ 0, & z_1, & 1, & \ldots, & 0 \\ 0, & 0, & z_1, & \ldots, & 0 \\ \multicolumn{5}{c}{\dotfill} \\ 0, & 0, & 0, & \ldots, & z_1 \end{matrix} \bigg\} s & 0 \\
\hline
0 & \begin{matrix} z_2, & 0, & \ldots, & 0 \\ 0, & z_3, & \ldots, & 0 \\ \multicolumn{4}{c}{\dotfill} \\ 0, & 0, & \ldots, & z_n \end{matrix}
\end{array}\right] ; \quad b = \left[\begin{matrix} 0 \\ 0 \\ \vdots \\ 0 \\ 1 \\ 1 \\ \vdots \\ 1 \end{matrix}\right] \begin{matrix} \\ \\ \bigg\} s-1 \\ \\ \\ \end{matrix} ; \quad c = \left[\begin{matrix} b_1 \\ b_2 \\ \vdots \\ \\ \\ \\ \\ b_m \end{matrix}\right]$$

$$d = \frac{c_m^*}{a_n} \qquad (7.25)$$

SOLUTION OF THE DISCRETE-TIME STATE EQUATION

The matrix A is again a Jordan matrix whose upper left block corresponds to the s-multiple root z_1 and the lower right block to the simple roots z_2 to z_n of the characteristic polynomial of $P(z)$. Because of the relationship between the matrix A, the characteristic equation of the system and the components following from the decomposition into partial fractions, we refer to the matrix A as being the matrix of dynamics of the system.

7.4 Connection of the solution with the Z-transformation

Let us consider the discrete state equations

$$x(k+1) = A\,x(k) + b\,u(k) + w(k)$$
$$y(k) = c^T x(k) + d\,u(k) \qquad (7.26)$$

We determine the corresponding Z-transforms of these equations.
The Z-transform of eqn. (7.26) is

$$z\,X(z) - z\,x(0) = A\,X(z) + b\,U(z) + W(z)$$
$$(zE - A)\,X(z) = z\,x(0) + b\,U(z) + W(z)$$
$$X(z) = (zE - A)^{-1} z\,x(0) + (zE - A)^{-1}[b\,U(z) + W(z)] =$$
$$= (E - z^{-1}A)^{-1} x(0) + z^{-1}(E - z^{-1}A)^{-1}[b\,U(z) + W(z)] \qquad (7.27)$$

From a comparison of eqn. (7.27) with eqn. (7.2) it follows that the Z-transform of the matrix A raised to the power of the independent time variable k is

$$Z[A^k] = (E - z^{-1}A)^{-1} \qquad (7.28)$$

Expression (7.28) may also be written as a series directly following from the definition of the Z-transform, if we apply it to the function A^k:

$$Z[A^k] = \sum_{k=0}^{\infty} A^k z^{-k} = E + A z^{-1} + A^2 z^{-2} + \ldots \qquad (7.29)$$

According to eqns. (7.26) and for $w(k) = 0$ the Z-transform of the output variable is

$$Y(z) = c^T X(z) + d\,U(z)$$

Then, on substitution for $X(z)$ we obtain, according to eqns. (7.27),

$$Y(z) = c^T(zE - A)^{-1} z\,x(0) + [c^T(zE - A)^{-1} b + d]\,U(z) \qquad (7.30)$$

Now, if the initial state $x(0) = 0$, the relation

$$G(z) = \frac{Y(z)}{U(z)} = c^T(zE - A)^{-1} b + d \tag{7.31}$$

determines the discrete transfer function of the system and the characteristic equation is

$$\det [zE - A] = 0 \tag{7.32}$$

The discrete transfer function may alternatively be expressed as the Z-transform of the discrete impulse response (7.3):

$$G(z) = d + \sum_{k=1}^{\infty} c^T A^{k-1} b z^{-k} \tag{7.33}$$

Example 7.1: Calculate the Z-transform of the difference equation

$$y(k) - 1.5 y(k-1) + 0.5 y(k-2) = u(k-1) + 3 u(k-2)$$

Solution: According to the given equation and with respect to eqns. (6.24) and (6.21) the matrices of the state equations are

$$A = \begin{bmatrix} 1.5, & 1 \\ -0.5, & 0 \end{bmatrix}; \quad b = \begin{bmatrix} 1 \\ 3 \end{bmatrix}; \quad c^T = [1, 0]$$

By eqn. (7.28) we have

$$Z[A^k] = \begin{bmatrix} 1 - 1.5z^{-1}, & -z^{-1} \\ -0.5z^{-1}, & 1 \end{bmatrix}^{-1} = \frac{1}{1 - 1.5z^{-1} + 0.5z^{-2}} \begin{bmatrix} 1, & z^{-1} \\ -0.5z^{-1}, & 1 - 1.5z^{-1} \end{bmatrix}$$

Now, according to eqns. (7.27), the sought transform is

$$X(z) = \begin{bmatrix} \dfrac{1}{\Delta(z)}, & \dfrac{z^{-1}}{\Delta(z)} \\ -\dfrac{0.5z^{-1}}{\Delta(z)}, & \dfrac{1 - 1.5z^{-1}}{\Delta(z)} \end{bmatrix} x(0) + \begin{bmatrix} \dfrac{z^{-1} + 3z^{-2}}{\Delta(z)} \\ \dfrac{3z^{-1} - 5z^{-2}}{\Delta(z)} \end{bmatrix} U(z)$$

where $\Delta(z) = 1 - 1.5z^{-1} + 0.5z^{-2}$.

For $x(0) = 0$ and with regard to the form of the matrix c^T we obtain

$$Y(z) = \frac{z^{-1} + 3z^{-2}}{1 - 1.5z^{-1} + 0.5z^{-2}} U(z)$$

where the fraction is the discrete transfer function of the plant, which is directly evident from the given difference equation.

7.5 Continuously working controlled plant – impulse sequence input signal

Let us now consider the behaviour of a continuously working plant with a sequence of impulses acting at its input. This sequence will be expressed as the Dirac impulse array modulated by the continuous input signal $u(t)$:

$$u(t, t_k) = u(t) \sum_{k=0}^{\infty} \delta(t - t_k^+) \tag{7.34}$$

where $t_{k+1} > t_k$, but otherwise t_k may have an arbitrary value.

Similarly we express the disturbing variable

$$w(t, t_k) = w(t) \sum_{k=0}^{\infty} \delta(t - t_k^+) \tag{7.35}$$

and introduce

$$t = t_k + \sigma, \quad 0 < \sigma < t_{k+1} - t_k$$

Substituting for $u(t)$ and $w(t)$ in eqn. (5.13) according to eqns. (7.34) and (7.35) we obtain

$$x(t_k + \sigma) = \Phi(\sigma) x(t_k) + \Phi(\sigma) [g\, u(t_k^+) + w(t_k^+)] \tag{7.36}$$

Equation (7.36) is the state equation of a continuously working plant with the controlling variable u and the disturbing variable w acting at its input; both variables are represented by an impulse sequence at the instants t_k^+, $k = 1, 2, \ldots$.

If the interval $t_{k+1} - t_k = T = \text{const.}$ and $\varepsilon = \sigma/T$, $0 \le \varepsilon < 1$, eqn. (7.36) changes into

$$x(kT + \varepsilon T) = \Phi(\varepsilon T) x(kT) + \Phi(\varepsilon T) [g\, u(kT^+) + w(kT^+)] \tag{7.37}$$

For $\varepsilon = 1$

$$x((k+1)T) = \Phi(T) x(kT) + \Phi(T) [g\, u(kT^+) + w(kT^+)] \tag{7.38}$$

We solve eqn. (7.37) by expressing first explicitly $x(kT)$ from eqn. (7.38) using the method given in Sec. 7.1. Substituting $x(kT)$ determined in this way into eqn. (7.37) we obtain the explicit expression of $x(kT + \varepsilon T)$.

Applying the notation of eqn. (7.38) and by eqn. (7.1) we have

$$\begin{aligned}
x(kT) &= \Phi^k(T) x(0) + \sum_{j=0}^{k-1} \Phi^j(T) \Phi(T) g\, u[(k - j - 1) T^+] + \\
&+ \sum_{j=0}^{k-1} \Phi^j(T) \Phi(T) w[(k - j - 1) T^+] = \\
&= \Phi^k(T) x(0) + \sum_{j=1}^{k} \Phi^j(T) g\, u[(k - j) T^-] + \\
&+ \sum_{j=1}^{k} \Phi^j(T) w[(k - j) T^+] = \\
&= \Phi^k(T) x(0) + \sum_{r=0}^{k-1} \Phi^{k-r}(T) g\, u(rT^+) + \sum_{r=0}^{k-1} \Phi^{k-r}(T) w(rT^+) \quad (7.39)
\end{aligned}$$

where in the last expression we substituted $j = k - r$.

Substituting eqn. (7.39) into eqn. (7.37) we obtain

$$x(kT + \varepsilon T) = \Phi(\varepsilon T)\Phi^k(T) x(0) + \Phi(\varepsilon T)\left[\sum_{r=0}^{k-1}\Phi^{k-r}(T) g\, u(rT^+) + \right.$$

$$\left. + g\, u(kT^+) + \sum_{r=0}^{k-1}\Phi^{k-r}(T) w(rT^+) + w(kT^+)\right] \quad (7.40)$$

where

$$\Phi(\varepsilon T) = e^{F\varepsilon T}$$

$$\Phi^k(T) = e^{FkT} \quad \text{etc.}$$

7.6 Continuously working controlled plant — staircase function input signal

If in eqn. (5.13) $u(\tau) = u(t_0) = \text{const.}$ and $w(\tau) = w(t_0) = \text{const.}$, $t_0 \leq \tau < t$, then

$$x(t) = \Phi(t - t_0) x(t_0) + \int_{t_0}^{t} \Phi(t - \tau) g\, u(t_0)\, d\tau + \int_{t_0}^{t} \Phi(t - \tau) w(t_0)\, d\tau \quad (7.41)$$

Denoting

$$\int_{t_0}^{t} \Phi(t - \tau) g\, d\tau = \int_{t_0}^{t} e^{F(t-\tau)} g\, d\tau =$$

$$= -F^{-1}\left[E - e^{A(t-t_0)}\right] g = F^{-1}\left[\Phi(t - t_0) - E\right] g = b(t - t_0), \quad (7.42)$$

$$\int_{t_0}^{t} \Phi(t - \tau)\, d\tau = F^{-1}\left[\Phi(t - t_0) - E\right] = L(t - t_0) \quad (7.43)$$

eqn. (7.41) may be rewritten in the form

$$x(t) = \Phi(t - t_0) x(t_0) + b(t - t_0) u(t_0) + L(t - t_0) w(t_0)$$

If the input variables are constant within one interval of time, i.e. $u(t) = u(t_k) = \text{const.}$ and $w(t) = w(t_k) = \text{const.}$, $t_k \leq t < t_{k+1}$, the last equation assumes the form

$$x(t_k + \sigma) = \Phi(\sigma) x(t_k) + b(\sigma) u(t_k) + L(\sigma) w(t_k) \quad (7.44)$$

where $0 < \sigma < t_{k+1} - t_k$.

For a constant interval, $t_{k+1} - t_k = T = \text{const.}$, and $\sigma = \varepsilon T$, $0 \leq \varepsilon < 1$ we arrive at the discrete form of the state equations. Particularly with the introduced notation

$$A(\varepsilon T) = \Phi(\varepsilon T) = e^{F\varepsilon T}$$

$$A(T) = \Phi(T) = e^{FT} \quad (7.45)$$

SOLUTION OF THE DISCRETE-TIME STATE EQUATION

the relationship to the discrete form of, say, the state equation (7.26) is directly apparent. Equation (7.44) may, therefore, be rewritten in the form

$$x(kT + \varepsilon T) = A(\varepsilon T) x(kT) + b(\varepsilon T) u(kT) + L(\varepsilon T) w(kT) \tag{7.46}$$

For $\varepsilon = 1$ we have

$$x((k + 1) T) = A(T) x(kT) + b(T) u(kT) + L(T) w(kT) \tag{7.47}$$

Equation (7.46) is solved similarly to eqn. (7.37) in Sec. 7.5:

$$x(kT) = A^k(T) x(0) + \sum_{j=0}^{k-1} A^j(T) u[(k - j - 1) T] +$$

$$+ \sum_{j=0}^{k-1} A^j(T) L(T) w[(k - j - 1) T] =$$

$$= A^k(T) x(0) + \sum_{r=0}^{k-1} A^{k-r-1}(T) b(T) u(rT) +$$

$$+ \sum_{r=0}^{k-1} A^{k-r-1}(T) L(T) w(rT) \tag{7.48}$$

where in the last expression we substituted $j = k - r - 1$.

Substituting now eqn. (7.48) into eqn. (7.46) we obtain

$$x(kT + \varepsilon T) = A(\varepsilon T) A^k(T) x(0) + A(\varepsilon T) \sum_{r=0}^{k-1} A^{k-r-1}(T) b(T) u(rT) +$$

$$+ b(\varepsilon T) u(kT) + A(\varepsilon T) \sum_{r=0}^{k-1} A^{k-r-1}(T) L(T) w(rT) + L(\varepsilon T) w(kT) \tag{7.49}$$

For $\varepsilon = 0$ the equation of the output variable y is identical for both the continuous and the discrete version of state equations because it is an algebraic equation, so that

$$h = c, \quad k = d \tag{7.50}$$

However, for $0 \leq \varepsilon < 1$

$$y(kT + \varepsilon T) = c^T x(kT + \varepsilon T) + du(kT) \tag{7.51}$$

Substituting for $x(kT + \varepsilon T)$ according to eqn. (7.49) we obtain

$$y(kT + \varepsilon T) = c^T A(\varepsilon T) A^k(T) x(0) + c^T A(\varepsilon T) \sum_{r=0}^{k-1} A^{k-r-1}(T) b(T) u(rT) +$$

$$+ c^T b(\varepsilon T) u(kT) + c^T A(\varepsilon T) \sum_{r=0}^{k-1} A^{k-r-1}(T) L(T) w(rT) +$$

$$+ c^T L(\varepsilon T) w(kT) + du(kT) \tag{7.52}$$

where

$$b(\varepsilon T) = \int_0^\varepsilon e^{FT(\varepsilon-\tau)} gT\,d\tau = -F^{-1}T^{-1}\left[E - e^{F\varepsilon T}\right]gT =$$
$$= F^{-1}(e^{F\varepsilon T} - E)g = F^{-1}[A(\varepsilon T) - E]g =$$
$$= \varepsilon T(F\varepsilon T)^{-1}(e^{F\varepsilon T} - E)g = \varepsilon T f(F\varepsilon T)g \qquad (7.53)$$

$$b(T) = \int_0^1 e^{FT(1-\tau)} gT\,d\tau = F^{-1}(e^{FT} - E)g = F^{-1}[A(T) - E]g =$$
$$= T(FT)^{-1}(e^{FT} - E)g = T f(FT)g \qquad (7.54)$$

$$L(\varepsilon T) = \int_0^\varepsilon e^{FT(\varepsilon-\tau)} T\,d\tau = F^{-1}(e^{F\varepsilon T} - E) = F^{-1}[A(\varepsilon T) - E] =$$
$$= \varepsilon T f(F\varepsilon T) \qquad (7.55)$$

$$L(T) = \int_0^1 e^{FT(1-\tau)} T\,d\tau = F^{-1}(e^{FT} - E) = F^{-1}[A(T) - E] = T f(FT) \qquad (7.56)$$

In eqns. (7.53) through (7.56) the function

$$f(\Lambda) = \Lambda^{-1}(e^\Lambda - E) = \Lambda^{-1}\left(E + \frac{\Lambda}{1!} + \frac{\Lambda^2}{2!} + \dots - E\right) =$$
$$= \left(E + \frac{\Lambda}{2!} + \frac{\Lambda^2}{3!} + \dots\right) \qquad (7.57)$$

where $\Lambda = F\varepsilon T$ or $\Lambda = FT$. From eqn. (7.57) it follows that $f(\Lambda)$ does exist even for a singular matrix F.

For various purposes of discrete control loop analysis and synthesis we need not only the explicit solution given by eqns. (7.49) and (7.52) but also the recursive forms of difference state equations. We shall derive them for a simpler case where the disturbing variable $w = 0$.

Introducing

$$A(\varepsilon T)\,x(0) = \hat{x}(0, \varepsilon T)$$
$$A(\varepsilon T)\,b(T) = e^{F\varepsilon T}Tf(FT)g = \hat{b}$$
$$c^T b(\varepsilon T) + d = c^T \varepsilon T f(F\varepsilon T)g + d = \hat{d}$$

and taking into account that $A(\varepsilon T) A^k(T) = A^k(T) A(\varepsilon T)$, then, for the disturbing variable $w = 0$, eqn. (7.52) assumes the form

$$y(kT + \varepsilon T) = c^T A^k(T)\,\hat{x}(0, \varepsilon T) +$$
$$+ c^T \sum_{r=0}^{k-1} A^{k-r-1}(T)\,\hat{b}\,u(rT) + \hat{d}\,u(kT) \qquad (7.58)$$

SOLUTION OF THE DISCRETE-TIME STATE EQUATION

Compare this equation with eqn. (7.2), attaining with the simplification $w = 0$ the form

$$y(k) = CA^k x(0) + C \sum_{r=0}^{k-1} A^{k-r-1} B u(r) + D u(k)$$

Hence it is evident that eqn. (7.58) is only a special case of eqn. (7.2) and represents the solution of the recursive system of state equations

$$\hat{x}[(k + 1 + \varepsilon) T] = A(T) \hat{x}(kT + \varepsilon T) + \hat{b} u(kT) \tag{7.59}$$

$$y(kT + \varepsilon T) = c^T \hat{x}(kT + \varepsilon T) + \hat{d} u(kT) \tag{7.60}$$

where

$$\hat{x}(kT + \varepsilon T) = A^k(T) \hat{x}(0, \varepsilon T) + \sum_{r=0}^{k-1} A^{k-r-1}(T) \hat{b} u(rT) \tag{7.61}$$

is the newly introduced vector of the state variables, different from the original vector $x(kT + \varepsilon T)$. Here, eqn. (7.61) is the solution of eqn. (7.59), as it also follows from a comparison with the solution (7.2).

Example 7.2: The differential equation of a continuously working plant (Fig. 7.1) is

$$y''(t) + 3 y'(t) + 2 y(t) = u(t)$$

Calculate the plant response at the instants $k = 0, 1, 2, \ldots$ to a general staircase signal $u(kT)$ if $y(0) = y'(0) = 0$ and $T = 1$.

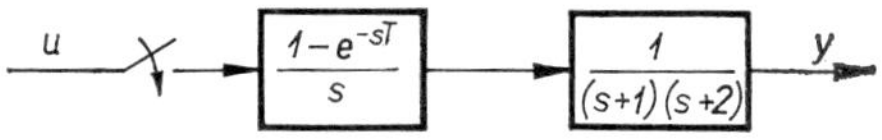

Fig. 7.1. Second-order controlled plant with a zero-order holding member.

Solution: First we write the state equations of the given plant, e.g. using eqns. (3.40) and (3.41):

$$x'(t) = \begin{bmatrix} -3, & 1 \\ -2, & 0 \end{bmatrix} x(t) + \begin{bmatrix} 0 \\ 1 \end{bmatrix} u(t)$$

$$y(t) = [1, 0] x(t)$$

According to Sec. 5.3 we then calculate

$$L[\Phi(t)] = [sE - F]^{-1} = \begin{bmatrix} s + 3, & -1 \\ 2, & s \end{bmatrix}^{-1}$$

$$L[\Phi(t)] = \frac{1}{\Delta_F(s)} \begin{bmatrix} s, & 1 \\ -2, & s + 3 \end{bmatrix}$$

where
$$\Delta_F(s) = s^2 + 3s + 2 = (s+1)(s+2)$$

is the determinant of the matrix $(sE - A)$. Determining the inverse transform of the last expression we obtain, for $t = T = 1$,

$$\Phi(T) = e^{FT} = \begin{bmatrix} -e^{-T} + 2e^{-2T}, & e^{-T} - e^{-2T} \\ -2e^{-T} + 2e^{-2T}, & 2e^{-T} - e^{-2T} \end{bmatrix} =$$

$$= \begin{bmatrix} -0.0972, & 0.2325 \\ -0.4651, & 0.6004 \end{bmatrix}$$

Applying eqn. (7.54) we calculate $b(T)$ with the matrix

$$F^{-1} = \begin{bmatrix} 0, & -\tfrac{1}{2} \\ 1, & -\tfrac{3}{2} \end{bmatrix}$$

$$b(T) = F^{-1}[A(T) - E]g = \begin{bmatrix} -e^{-T} + \tfrac{1}{2}e^{-2T} + \tfrac{1}{2} \\ -2e^{-T} + \tfrac{1}{2}e^{-2T} + \tfrac{3}{2} \end{bmatrix} = \begin{bmatrix} 0.1998 \\ 0.8319 \end{bmatrix}$$

In accordance with eqn. (7.47), the plant state equation for the staircase input signal will be

$$\begin{bmatrix} x_1[(k+1)T] \\ x_2[(k+1)T] \end{bmatrix} = \begin{bmatrix} -0.0972, & 0.2325 \\ -0.4651, & 0.6004 \end{bmatrix} \begin{bmatrix} x_1(kT) \\ x_2(kT) \end{bmatrix} + \begin{bmatrix} 0.1998 \\ 0.8319 \end{bmatrix} u(k)$$

According to the given differential equation and the definition of state variables (3.37), $x_1(t) = y(t)$ and $x_2(t) = 3y(t) + y'(t)$ so that for the given initial conditions we have $x_1(0) = x_2(0) = 0$.

Using the calculated state equation we can recursively calculate on a computer the response values at the instants $k = 1, 2, \ldots$. For instance, for $u(k) = 1$, $k = 0, 1, 2, \ldots$, the calculation result was

k	$x_1(kT)$	$x_2(kT)$
0	0	0
1	0.1998	0.8319
2	0.3738	1.2384
3	0.4514	1.4016
4	0.4818	1.4635
5	0.4932	1.4865

SOLUTION OF THE DISCRETE-TIME STATE EQUATION

From the calculated values of the variable x_1 we can easily calculate the elements of the weighting matrix (7.8):

$$s(0) = 0$$
$$s(1) = x_1(T) = 0.1998$$
$$s(2) = x_1(2T) - x_1(T) = 0.1740$$
$$s(3) = x_1(3T) - x_1(2T) = 0.0776$$
$$s(4) = x_1(4T) - x_1(3T) = 0.0304$$
$$s(5) = x_1(5T) - x_1(4T) = 0.0114, \quad \text{etc.}$$

Example 7.3: For the plant in Example 7.2 calculate the response to a general staircase signal $u(kT)$ at the instants $kT + \varepsilon T$, $k = 0, 1, 2, ...$, $\varepsilon = 0.5$, $T = 1$.

Solution: The response is calculated according to eqn. (7.46) where

$$A(\varepsilon T) = e^{F\varepsilon T} = \begin{bmatrix} -e^{-\varepsilon T} + 2e^{-2\varepsilon T}, & e^{-\varepsilon T} - e^{-2\varepsilon T} \\ -2e^{-\varepsilon T} + 2e^{-2\varepsilon T}, & 2e^{-\varepsilon T} - e^{-2\varepsilon T} \end{bmatrix} =$$

$$= \begin{bmatrix} 0.1292, & 0.2387 \\ -0.4773, & 0.8452 \end{bmatrix}$$

In accordance with eqn. (7.53)

$$b(\varepsilon T) = F^{-1}[A(\varepsilon T) - E] g = \begin{bmatrix} 0.0774 \\ 0.4709 \end{bmatrix}$$

so that with the calculated matrices $A(\varepsilon T)$ and $b(\varepsilon T)$ eqn. (7.46) becomes

$$\begin{bmatrix} x_1(kT + \varepsilon T) \\ x_2(kT + \varepsilon T) \end{bmatrix} = \begin{bmatrix} 0.1292, & 0.2387 \\ -0.4773, & 0.8452 \end{bmatrix} \begin{bmatrix} x_1(kT) \\ x_2(kT) \end{bmatrix} + \begin{bmatrix} 0.0774 \\ 0.4709 \end{bmatrix} u(k)$$

The numerical calculation is performed on a computer using the same program as in Example 7.2. Each calculated value of the vector $x(kT + \varepsilon T)$ is shifted in time with respect to the preceding one by the interval εT so that with a constant $u(k)$ and with the given initial vector we obtain successively values at the instants $\varepsilon T, 2\varepsilon T, 3\varepsilon T, \ldots$. For instance, for $u(k) = 1$, $k = 0, 1, 2, \ldots$ and $\varepsilon = 0.5$ the following values were calculated:

i	$x_1(i\varepsilon T)$	$x_2(i\varepsilon T)$
0	0	0
1	0.0774	0.4709
2	0.1998	0.8319
3	0.3018	1.1079
4	0.3738	1.2384
5	0.4216	1.3393

The resultant values for $i = 2$ and 4 of this example are necessarily identical with those of Example 7.2 for $k = 1$ and 2, respectively.

7.7 Relationships between the coefficients of continuous and discrete system models

When sampling the output variables of a continuously working system the matrix $A = \exp[FT]$ must always be calculated if the system is to be represented by discrete state equations. Then the vector b in single-input systems or the matrix B in multi-input systems can already be calculated according to eqn. (7.54). First of all it should be noted that for time-invariant systems (constant matrix F) the relation [55.1, Chap. III-4]

$$\det[e^{FT}] = e^{\text{tr}[FT]} \tag{7.62}$$

holds, where $\text{tr}[FT]$ denotes the trace of the matrix FT. For any value of $\text{tr}[FT]$ we have $\det[A] > 0$ and, therefore, the matrix A is non-singular so that A^{-1} does exist. The relation (7.62) also applies to time-variant matrices $F(t)$.

The matrix $A = \exp[FT]$ may be calculated by various methods, some of which will be discussed in the following paragraphs.

7.7.1 Calculation according to the definition of $\exp[FT]$

The numerical calculation of the matrix $A = \exp[FT]$ may be performed, for instance, according to the previously given definition (5.4)

$$A = e^{FT} = \sum_{k=0}^{\infty} \frac{F^k T^k}{k!} \tag{7.63}$$

including in the calculation as many terms of the series as needed to obtain the desired accuracy of the elements of A. The advantage of this method is that the eigenvalues of F need not be known. High values of the period of sampling, T, may cause a bad convergence of the calculation. In such a case it is advisable to calculate $(\exp[FT/N])^N$ using a sufficiently high value of N.

For time-invariant single-input/single-output systems, eqn. (7.54) may be rewritten in the reduced form

$$b = \int_0^T e^{F\tau} g \, d\tau = \int_0^T \sum_{k=0}^{\infty} \frac{F^k \tau^k g}{k!} \, d\tau = T \sum_{k=0}^{\infty} \frac{(FT)^k g}{(k+1)!} \tag{7.64}$$

In a similar manner we could calculate the matrix B for a multi-input/multi-output system.

7.7.2 Use of the relationship to Laplace transformation

According to eqns. (7.45) and (5.36)

$$A = e^{FT} = L^{-1}(sE - F)^{-1}\big|_{t=T} \tag{7.65}$$

The use of this formula was shown in Example 7.2.

The calculation of the matrix A according to eqn. (7.65) is suitable only for the simplest cases. If, however, the matrix F is of the order $n > 3$, it is advisable to use a different method. One of them, which allows the calculation of $(sE - F)^{-1}$ of any order of the matrix F, will now be given.

Let the polynomial

$$\alpha_0 + \alpha_1 s + \ldots + \alpha_p s^p$$

be the minimum plynomial of the matrix F. Then it holds that

$$f(F) = \alpha_0 E + \alpha_1 F + \ldots + \alpha_p F^p = 0$$

From the equation

$$Q = (sE - F)^{-1}$$

it follows that

$$sQ = FQ + E$$

Multiplying this equality successively from the left $(p - 1)$ times by the matrix $(sE + F)$, where p is the degree of the minimum polynomial of the matrix F, we may write the set of equations

$$Q = Q$$
$$sQ = FQ + E$$
$$s^2 Q = F^2 Q + F + sE$$
$$s^3 Q = F^3 Q + F^2 + sF + s^2 E$$
$$\vdots$$
$$s^p Q = F^p Q + F^{p-1} + sF^{p-2} + \ldots + s^{p-2} F + s^{p-1} E$$

Now we multiply the individual equations by the coefficients $\alpha_0, \alpha_1, \ldots, \alpha_p$ of the minimum polynomial and add up all equations. We obtain

$$\sum_{i=0}^{p} \alpha_i s^i Q = \sum_{i=0}^{p} \alpha_i F^i Q + \sum_{i=1}^{p} \alpha_i F^{i-1} + s \sum_{i=2}^{p} \alpha_i F^{i-2} + \ldots$$

$$\ldots + s^{p-2} \sum_{i=p-1}^{p} \alpha_i F^{i-p+1} + s^{p-1} \alpha_p E =$$

$$= \sum_{j=0}^{p-1} s^j \sum_{i=1+j}^{p} \alpha_i F^{i-j-1}$$

because

$$\sum_{i=0}^{p} \alpha_i F^i Q = f(F) Q = 0$$

From this equation it follows that

$$Q = (sE - F)^{-1} = \frac{\sum_{j=0}^{p-1} s^j \sum_{i=1+j}^{p} \alpha_i F^{i-j-1}}{\sum_{i=0}^{p} \alpha_i s^i} \tag{7.66}$$

If the minimum polynomial agrees with the characteristic polynomial of the matrix F, we may substitute into these equations $p = n$, where n is the order of the matrix F.

7.7.3 Faddeev's algorithm

The disadvantages mentioned in Par. 7.7.2 requiring an inversion of the matrix $(sE - F)$ or a determination of the minimum polynomial of F are eliminated by the modified Crammer rule [55.1] which is often referred to as the Faddeev method [66.2, p. 97]. We shall here give only the resultant instruction for the calculation relating to the single-input/single-output time-invariant plant:

$$(sE - F)^{-1} = \frac{B_0 s^{n-1} + \ldots + B_{n-2} s + B_{n-1}}{s^n + a_1 s^{n-1} + \ldots + a_{n-1} s + a_n} \tag{7.67}$$

where a_i, $i = 1, 2, \ldots, n$, are scalar values and B_i, $i = 0, 1, \ldots, n-1$, are matrices of the dimension $(n; n)$.

$$\begin{aligned}
B_0 &= E \text{ (unit matrix)}, & a_1 &= -\text{tr}\,[FB_0] \\
B_1 &= FB_0 + a_1 E, & a_2 &= -\tfrac{1}{2}\,\text{tr}\,[FB_1] \\
&\;\vdots & &\;\vdots \\
B_{n-1} &= FB_{n-2} + a_{n-1} E, & a_n &= -\frac{1}{n}\,\text{tr}\,[FB_{n-1}]
\end{aligned} \tag{7.68}$$

For checking we may use the equation

$$FB_{n-1} + a_n E = B_n \tag{7.69}$$

Theoretically, the matrix B_n should be zero, but in numerical calculations with rounded and inaccurate values $B_n \neq 0$; the accuracy of the numerical calculation may be estimated according to the non-zero elements of this matrix.

By means of eqns. (7.67) and (7.68) we can also directly calculate the coefficients in the numerator of the transfer function $G(s)$. Here we use eqn. (5.34) where we

SOLUTION OF THE DISCRETE-TIME STATE EQUATION

substitute for $(sE - F)^{-1}$ the expression (7.67) in which we express the matrices B_i, $i = 0, 1, \ldots, n - 1$ by means of eqns. (7.68). We obtain

$$G(s) = \frac{b_0 s^m + b_1 s^{m-1} + \ldots + b_{m-1} s + b_m}{s^n + a_1 s^{n-1} + \ldots + a_{n-1} s + a_n} \tag{7.70}$$

where the coefficients in the denominator are directly defined by eqns. (7.68) and the coefficients in the numerator of the transfer function (7.70) are calculated, for $m = n$, as follows

$$\begin{aligned}
b_m &= ka_n + h^T g a_{n-1} + h^T F g a_{n-2} + \ldots + h^T F^{n-2} g a_1 + h^T F^{n-1} g \\
b_{m-1} &= ka_{n-1} + h^T g a_{n-2} + h^T F g a_{n-3} + \ldots + h^T F^{n-3} g a_1 + h^T F^{n-2} g \\
&\vdots \\
b_2 &= ka_2 + h^T g a_1 + h^T F g \\
b_1 &= ka_1 + h^T g \\
b_0 &= k
\end{aligned} \tag{7.71}$$

If $m < n$, then $k = 0$, and eqns. (7.71) will be simplified correspondingly.

From eqns. (7.68), (7.69) and (7.71) it is evident that no matrix inversion need be performed when calculating the coefficients of the transfer function $G(s)$. If the minimum polynomial of the matrix F is identical with the characteristic polynomial (see Par. 10.2.2) then the polynomial in the denominator of the transfer function (7.70) is a minimum polynomial as well. This practically means that no reduction can be executed in eqn. (7.70) by any root factor.

Example 7.4: Calculate $(sE - F)^{-1}$ of Example 7.2 by the method given in Par. 7.7.3 and determine the transfer function $G(s)$ according to eqns. (7.70) and (7.71)!

Solution:

$$B_0 = \begin{bmatrix} 1, & 0 \\ 0, & 1 \end{bmatrix} \qquad a_1 = -\operatorname{tr}\begin{bmatrix} -3, & 1 \\ -2, & 0 \end{bmatrix} = 3$$

$$B_1 = \begin{bmatrix} 0, & 1 \\ -2, & 3 \end{bmatrix} \qquad a_2 = -\tfrac{1}{2}\operatorname{tr}\begin{bmatrix} -2, & 0 \\ 0, & -2 \end{bmatrix} = 2$$

$$(sE - F)^{-1} = \frac{B_0 s + B_1}{s^2 + 3s + 2} = \frac{1}{(s+1)(s+2)}\begin{bmatrix} s, & 1 \\ -2, & s+3 \end{bmatrix}$$

This result agrees with the solution in Example 7.2.

According to eqns. (7.71) we have for $n = 2$:

$$b_0 = 0 \quad \text{because} \quad k = 0$$

$$b_1 = h^T g = \begin{bmatrix} 1, & 0 \end{bmatrix} \begin{bmatrix} 0 \\ 1 \end{bmatrix} = 0$$

$$b_2 = h^T F g = \begin{bmatrix} 1, & 0 \end{bmatrix} \begin{bmatrix} -3, & 1 \\ -2, & 0 \end{bmatrix} \begin{bmatrix} 0 \\ 1 \end{bmatrix} = 1$$

$$G(s) = \frac{1}{s^2 + 3s + 2}$$

This result again agrees with the solution in Example 7.2.

Faddeev's algorithm may also be used for the calculation of A^k and for the determination of the discrete transfer function $G(z)$. All we have to do is to replace in eqns. (7.67) through (7.70) the complex variable s for z and the matrix F for A. We obtain

$$Z[A^k] = (E - z^{-1}A)^{-1} = \frac{B_0 z^n + B_1 z^{n-1} + \ldots + B_{n-1} z}{z^n + a_1 z^{n-1} + \ldots + a_{n-1} z + a_n} =$$

$$= \frac{B_0 + B_1 z^{-1} + \ldots + B_{n-1} z^{-(n-1)}}{1 + a_1 z^{-1} + \ldots + a_{n-1} z^{-(n-1)} + a_n z^{-n}} \quad (7.72)$$

The determinant of the matrix $(E - z^{-1}A)$ is

$$\det [E - z^{-1}A] = 1 + \sum_{i=1}^{n} a_i z^{-i} \quad (7.73)$$

and the adjoint matrix of the matrix $(E - z^{-1}A)$ is

$$\text{adj}\, [E - z^{-1}A] = \sum_{i=0}^{n-1} B_i z^{-i} \quad (7.74)$$

According to eqn. (6.14) or (6.24) it follows that the determinant of the matrix A is

$$\det [A] = (-1)^n a_n \quad (7.75)$$

and from eqn. (7.69) we may calculate, for $B_n = 0$, the inverse of the matrix A:

$$A^{-1} = -\frac{1}{a_n} B_{n-1} \quad (7.76)$$

From eqns. (7.75) and (7.76) follows the expression for the adjoint matrix to the matrix A:

$$\text{adj}\, [A] = (-1)^{n-1} B_{n-1} \quad (7.77)$$

SOLUTION OF THE DISCRETE-TIME STATE EQUATION

In the discrete transfer function

$$G(z) = \frac{b_1 z^{-1} + b_2 z^{-2} + \ldots + b_n z^{-n}}{1 + a_1 z^{-1} + \ldots + a_n z^{-n}} \qquad (7.78)$$

the numerator coefficients are determined by eqns. (7.71) provided that A, b, c^T, d are substituted for F, g, h^T, k, respectively.

Example 7.5: Calculate $Z[A^k]$ and $G(z)$ of a plant if

$$A = \begin{bmatrix} 1.5, & 1 \\ -0.5, & 0 \end{bmatrix}, \quad b = \begin{bmatrix} 1 \\ 3 \end{bmatrix}, \quad c^T = [1, 0], \quad d = 0$$

are known.

Solution: According to eqns. (7.68) and (7.72) we calculate

$$B_0 = \begin{bmatrix} 1, & 0 \\ 0, & 1 \end{bmatrix}, \quad a_1 = -\operatorname{tr}\begin{bmatrix} 1.5, & 1 \\ -0.5, & 0 \end{bmatrix} = -1.5$$

$$B_1 = \begin{bmatrix} 0, & 1 \\ -0.5, & -1.5 \end{bmatrix}, \quad a_2 = -\tfrac{1}{2}\operatorname{tr}\begin{bmatrix} -0.5, & 0 \\ 0, & -0.5 \end{bmatrix} = 0.5$$

$$Z[A^k] = \frac{E + B_1 z^{-1}}{1 - 1.5 z^{-1} + 0.5 z^{-2}} =$$

$$= \frac{1}{1 - 1.5 z^{-1} + 0.5 z^{-2}} \begin{bmatrix} 1, & z^{-1} \\ -0.5 z^{-1}, & 1 - 1.5 z^{-1} \end{bmatrix}$$

The coefficients in the numerator of the transfer function $G(z)$ are determined using eqns. (7.71):

$b_0 = 0$ since $d = 0$

$b_1 = c^T b = [1, 0]\begin{bmatrix} 1 \\ 3 \end{bmatrix} = 1$

$b_2 = c^T b a_1 + c^T A b = -1.5 + [1, 0]\begin{bmatrix} 1.5, & 1 \\ -0.5, & 0 \end{bmatrix}\begin{bmatrix} 1 \\ 3 \end{bmatrix} = 3$

7.7.4 Ježek's method for the calculation of discrete and continuous transfer functions

In contrast with the preceding algorithms, the Ježek method affords the calculation of not only the discrete transfer function from a given continuous transfer function but, conversely, also the calculation of the continuous transfer function if the discrete transfer function is known; this applies even if $0 \leq \varepsilon < 1$. Consider the time-invariant state equations of a continuous linear plant,

$$x'(t) = F\,x(t) + g\,u(t) \tag{7.79}$$

$$y(t) = h^T\,x(t) + k\,u(t) \tag{7.80}$$

and their discrete version,

$$x(k+1) = A\,x(k) + b\,u(k) \tag{7.81}$$

$$y(k) = c^T\,x(k) + d\,u(k) \tag{7.82}$$

where

$$A = e^{FT}$$

$$b = \int_0^T e^{F\tau} g\,d\tau = F^{-1}(e^{FT} - E)\,g =$$

$$= T\left[E + \frac{FT}{2!} + \frac{(FT)^2}{3!} + \ldots\right]g = Tf(FT)\,g \tag{7.83}$$

Eliminating from eqns. (7.81) and (7.82) the state variables we necessarily arrive at a difference equation and its coefficients. The coefficients of the difference equation are likewise coefficients of the discrete transfer function $G(z, 0)$.

Using eqns. (7.81) and (7.82) we obtain

$$y(k) = c^T\,x(k) + d\,u(k)$$
$$y(k+1) = c^T A\,x(k) + c^T b\,u(k) + d\,u(k+1)$$
$$y(k+2) = c^T A^2\,x(k) + c^T Ab\,u(k) + c^T b\,u(k+1) + d\,u(k+2)$$
$$\vdots$$
$$y(k+n) = c^T A^n\,x(k) + c^T A^{n-1} b\,u(k) + \ldots + c^T Ab\,u(k+n-2) +$$
$$+ c^T b\,u(k+n-1) + d\,u(k+n) \tag{7.84}$$

For n equations (7.84) we may write

$$\hat{y} = P\,x(k) + Q\hat{u} \tag{7.85}$$

where

$$\hat{y} = \begin{bmatrix} y(k) \\ y(k+1) \\ \vdots \\ y(k+n-1) \end{bmatrix}, \quad \hat{u} = \begin{bmatrix} u(k) \\ u(k+1) \\ \vdots \\ u(k+n-1) \end{bmatrix}, \quad P = \begin{bmatrix} c^T \\ c^T A \\ \vdots \\ c^T A^{n-1} \end{bmatrix}$$

$$Q = \begin{bmatrix} d, & 0, & \ldots, & 0 \\ c^T b, & d, & \ldots, & 0 \\ \multicolumn{4}{c}{\dotfill 0} \\ c^T A^{n-2} b, & c^T A^{n-3} b, & \ldots, & d \end{bmatrix} \tag{7.86}$$

SOLUTION OF THE DISCRETE-TIME STATE EQUATION

The matrix P is a square matrix of the dimension $(n; n)$. If the minimum polynomial of the matrix F in eqn. (7.79) is identical with the characteristic polynomial, the minimum polynomial of the matrix A in eqn. (7.81) is also identical with the characteristic polynomial except for the case where the sampling period T satisfies the equation

$$\omega T = r\pi, \quad r = 1, 2, \ldots \tag{7.87}$$

where ω is the natural frequency of some harmonic component if any of the impulse response. In such a case the eigenvalues of frequency ω of the matrix A split and, therefore, the minimum polynomial will be of a lower degree than the characteristic polynomial of the matrix A and of the matrix F. In other words, the solution of the homogeneous equation corresponding to the matrix A will have linearly dependent components while the solution corresponding to the matrix F has, on the given assumptions, only linearly independent components.[1]

If the case expressed by eqn. (7.87) does not occur, the matrix A is unsplit. In such a case it is reasonable to check whether the matrix P is non-singular. This matrix will be non-singular if all its rows are mutually independent or, identically, if its rank is equal to the state space dimension. Then the inverse of the matrix P does exist and we may, therefore, determine from eqn. (7.85) the function

$$x(k) = P^{-1}\hat{y} - P^{-1}Q\hat{u} \tag{7.88}$$

Substituting eqn. (7.88) into the last of eqns. (7.84) we obtain the sought difference equation permitting to determine also the discrete transfer function $G(z, 0)$.

Note 7.1: For T approaching zero, all poles of the transfer function $G(z, 0)$ are tending to $z_i = 1$, $i = 1, 2, \ldots, n$. The denominator of the transfer function, therefore, approaches the product of equal root factors $(z - 1)$ and the coefficients of the transfer function denominator are limiting to binomial coefficients. The coefficients of the transfer function numerator are approaching zero. This is the disadvantage

---

[1]) For example, for $F = \begin{bmatrix} j\omega & 0 \\ 0 & -j\omega \end{bmatrix}$ we have $A = e^{FT} = \begin{bmatrix} e^{j\omega T} & 0 \\ 0 & e^{-j\omega T} \end{bmatrix}$. If $\omega T = \pi + k2\pi$, $k = 1, 2, \ldots$, then $A = \begin{bmatrix} e^{j\pi} & 0 \\ 0 & e^{-j\pi} \end{bmatrix} = \begin{bmatrix} -1 & 0 \\ 0 & -1 \end{bmatrix}$. In this example, the matrix A has two simple eigenvalues corresponding to two mutually dependent components of the solution. The same result is also obtained when solving the corresponding ordinary differential equation. In the continuous version the matrix F corresponds to the second-order homogeneous equation $y'' + \omega^2 y = 0$, whose solution is $y(t) = K_1 e^{j\omega t} + K_2 e^{-j\omega t}$. For $t = Tn$, $n = 0, 1, 2, \ldots$, we obtain the discrete form of the solution $y(n) = K_1 e^{j\omega Tn} + K_2 e^{-j\omega Tn}$. If again $\omega T = \pi + k2\pi$, $k = 1, 2, \ldots$, then $y(n) = K_1 e^{j\pi n} + K_2 e^{-j\pi n} = K_1(-1)^n + K_2(-1)^n$. The result shows that the components are mutually linearly dependent so that in the discrete version only one component of the solution corresponds to the first-order difference equation $y(n) + y(n + 1) = 0$. The transfer function $G(z) = 1/(1 + z^{-1})$.

of the Z-transformation, since with decreasing sampling period the coefficients of the transfer function numerator and denominator give less information on the system, although more frequent sampling increases the information contents of the sample sequence.

On the other hand, with an increased sampling period the information on the components with small time constants and on high-frequency components is gradually lost. Thus, with increased sampling period the order of the calculated difference equation or of the discrete transfer function gradually decreases.

The same procedure may also be used for shifted discrete values of the output variable y and the state variables x. It holds that

$$\hat{x}(k+1, \varepsilon) = A\,\hat{x}(k\,\varepsilon) + \hat{b}\,u(k) \tag{7.89}$$

$$y(k\,\varepsilon) \quad = c^T\,\hat{x}(k, \varepsilon) + \hat{d}\,u(k) \tag{7.90}$$

where according to Sec. 7.6

$$A = e^{FT}$$

$$\hat{b} = e^{F\varepsilon T} Tf(FT)\,g$$

$$c^T = h^T$$

$$\hat{d} = c^T \varepsilon T f(F\varepsilon T)\,g + d$$

and the vector of the state variables is

$$x(k, \varepsilon) = A^k\,\hat{x}(0, \varepsilon) + \sum_{r=0}^{k-1} A^{k-r-1}\hat{b}\,u(rT)$$

$$\hat{x}(0, \varepsilon) = e^{F\varepsilon T} x(0)$$

Since the set of equations (7.81) and (7.82) is formally identical with the set (7.89) and (7.90), the method for the calculation of the coefficients of the discrete transfer function $G(z, \varepsilon)$ can also be used if $0 \leq \varepsilon < 1$.

If, on the other hand, it is desired to determine the continuous transfer function or differential equation from a corresponding discrete transfer function or difference equation, we can proceed essentially in a similar manner. In the calculation it is assumed that eqns. (7.81) and (7.82) are known. Using relations (7.83) we calculate

$$F = \frac{1}{T} \ln [A]$$

$$g = \frac{1}{T}(A - E)^{-1} \ln [A]\,b$$

$$h = c, \quad k = d \tag{7.91}$$

SOLUTION OF THE DISCRETE-TIME STATE EQUATION

To eliminate the vector of the state variables, $x(t)$, we use eqns. (7.79) and (7.80) and express the derivatives $y^{(i)}(t)$, $i = 0, 1, 2, \ldots, n$:

$$y(t) = h^T x(t) + h\, u(t)$$
$$y'(t) = h^T F\, x(t) + h^T g\, u(t) + k\, u'(t)$$
$$\vdots$$
$$y^{(n)}(t) = h^T F^n x(t) + h^T F^{(n-1)} g\, u(t) + \ldots + h^T Fg\, u^{(n-2)}(t) +$$
$$+ h^T g\, u^{(n-1)}(t) + k\, u^{(n)}(t) \tag{7.92}$$

The first n equations of the system (7.92) may be rewritten in the form

$$\hat{y}(t) = P\, x(t) + Q\, \hat{u}(t) \tag{7.93}$$

where

$$\hat{y}(t) = \begin{bmatrix} y(t) \\ y^{(1)}(t) \\ \vdots \\ y^{(n-1)}(t) \end{bmatrix}, \quad \hat{u}(t) = \begin{bmatrix} u(t) \\ u^{(1)}(t) \\ \vdots \\ u^{(n-1)}(t) \end{bmatrix}$$

$$P = \begin{bmatrix} h^T \\ h^T F \\ \vdots \\ h^T F^{n-1} \end{bmatrix}, \quad Q = \begin{bmatrix} k, & 0, & \ldots, & 0 \\ h^T g, & k, & \ldots, & 0 \\ \multicolumn{4}{c}{\ldots\ldots\ldots\ldots\ldots\ldots\ldots} \\ h^T F^{n-2} g, & h^T F^{n-3} g, & \ldots, & k \end{bmatrix}$$

From eqn. (7.93) we calculate

$$x(t) = P^{-1}\, \hat{y}(t) - P^{-1} Q\, \hat{u}(t) \tag{7.94}$$

Substituting eqns. (7.94) into the last equation of the set of equations (7.92) we obtain the sought differential equation.

For $0 \leq \varepsilon < 1$ we use instead of eqns. (7.91) the relations

$$F = \frac{1}{T} \ln [A]$$

$$g = \frac{1}{T} [f(\ln [A])]^{-1}\, e^{-\varepsilon \ln[A]} \hat{b} = \frac{1}{T} (A - E)^{-1} \ln [A]\, e^{-\varepsilon \ln[A]} \hat{b}$$

$$h = c$$

$$k = \hat{d} - c^T \varepsilon T f(\varepsilon \ln [A])\, g = \hat{d} - c^T (A^\varepsilon - E)(A - E)^{-1}\, e^{-\varepsilon \ln[A]} \hat{b}$$
$$\tag{7.95}$$

Otherwise the calculation procedure of the coefficients of the differential equation and, consequently, also of the continuous transfer function, is identical with the procedure described by eqns. (7.92) through (7.94).

Notes 7.2:

(a) The roots of the characteristic equation of the continuous transfer function are connected with the roots of the characteristic equation of the discrete transfer function according to the relation

$$s_i = \frac{1}{T} \ln [z_i] \qquad (7.96)$$

If z_i is a complex number we may write

$$\ln [z_i] = \ln [r_i] + j2k\varphi_i, \quad -\pi < \varphi_i < +\pi$$

where $k = 1, 2, \ldots$ and

$$z_i = r_i e^{j\varphi_i}$$

From the foregoing it follows that in the case considered $\ln [z_i]$ is not unique. Of all the possible values we prefer that one which corresponds to the oscillatory component of the lowest frequency, corresponding to the so-called main value of the logarithm.

(b) If the discrete characteristic equation has a real negative root, then this root corresponds in the time domain to a component whose successive discrete values alternate in sign. In the continuous version this will be the oscillatory component defined not by one but by two complex conjugate roots.

In this case the matrix A has one negative eigenvalue equal to the root of the discrete characteristic equation and for such a matrix no logarithm exists. Therefore, no continuous transfer function of the same order as that of the corresponding discrete transfer function exists.

From what has just been said it follows that the order of the continuous transfer function will be higher than the order of the discrete transfer function. If we knew the Jordan form corresponding to the matrix A we could add to each real negative eigenvalue in the Jordan matrix a further field with the same eigenvalue. That is to say, for a matrix established in this way, a main value of the logarithm exists enabling us to determine the sought continuous transfer function. Since, however, the structure of the Jordan matrix corresponding to the matrix A is not known, we expand the matrix A to a double form according to the model

$$\bar{A} = \begin{bmatrix} A, & 0 \\ 0, & A \end{bmatrix}$$

and the vectors b and c^T to the form

$$\bar{b} = \begin{bmatrix} b \\ 0 \end{bmatrix}, \quad \bar{c}^T = [c^T, 0]$$

Thus the matrix of the new system is established. Since the newly introduced state variables do not affect the output variable, the behaviour of the new system is the same as that of the original system. If this system is converted into the continuous version, then two negative real eigenvalues of the matrix $\bar{A}$ will correspond to two complex conjugate eigenvalues of the matrix $\bar{F}$. The reduction of the matrix $\bar{F}$ to the order corresponding to the continuous version and to the number of negative real eigenvalues of the matrix A is carried out automatically in the course of elimination of the state variables. If the matrix A appertains to a system of the n-th order, then the continuous version of this system is of the n'-th order, where

$$n \leqq n' \leqq 2n$$

(c) If the discrete characteristic equation has one or more roots $z_i = 0$, then no solution exists in the continuous version. The presence of the roots $z_i = 0$ in the characteristic equation of the discrete transfer function is recognized after the number of zero coefficients with z^k, $k \leqq n$.

(d) In eqn. (7.91) determining the matrix g for $\varepsilon = 0$, the expression $(A - E)^{-1}$. $\ln [A]$, where E is a unit matrix, must be evaluated. This expression is not defined for the roots $z_i = 1$. However, $\ln [A]$ may be expressed by the power series and then we obtain

$$(A - E)^{-1} \ln [A] = (A - E)^{-1} \left[(A - E) - \frac{(A - E)^2}{2} + \frac{(A - E)^3}{3} - \ldots \right] =$$

$$= E - \frac{A - E}{2} + \frac{(A - E)^2}{3} - \ldots \qquad (7.97)$$

The value of $\ln [A]$ can then be determined with regard to the desired accuracy by using an appropriate number of terms of the resultant power series. Since the convergence of this series is too slow, a large number of terms will be necessary for numerical calculations.

(e) With the method indicated above the calculation of $\ln [A]$ makes considerable difficulties. It is advantageous to calculate $\ln [A]$ by the Newton method according to which the matrix

$$X_{i+1} = X_i - f(X_i) [f'(X_i)]^{-1} \qquad (7.98)$$

where the function

$$f(X) = e^X - A \qquad (7.99)$$

follows from the requirement that $\ln [A] = X$. Substituting the function (7.99) into the Newton formula we obtain

$$X_{i+1} = X_i - E + A e^{-X_i} \qquad (7.100)$$

For $X_0 = 0$ we obtain the rough estimate of the logarithm

$$X_1 = A - E$$

where E is again a unit matrix. In the vicinity of the unit matrix this estimate is very good and, therefore, we always first divide the matrix A by such a number M that the norm of A be smaller than one. In this way the calculation of e^{X_i}, which must be repeated for each step of the iteration method, is somewhat simplified. Then the logarithm of the original matrix A will be

$$\ln [A] = \ln \left[\frac{A}{M}\right] + E \ln [M] \qquad (7.101)$$

Additional details of the methods described in Par. 7.7.4 may be found in Ref. [69.3].

7.8 Special types of sampling

So far we have been describing systems with only one sampling member or with several sampling members operating synchronously. In state space other types of sampling [59.1] may be simply described as well. Before showing the solution of selected special types of sampling we shall describe the general approach to the solution of this kind of problems.

Consider the solution of the difference equation (7.46) with the shift-interval εT, $0 \leq \varepsilon < 1$, and with a zero disturbing variable w:

$$x(kT + \varepsilon T) = A(\varepsilon T) x(kT) + b(\varepsilon T) u(kT) \qquad (7.102)$$

So long as we are interested only in the solution at the instants of sampling, it holds, according to eqn. (7.47), that

$$x[(k + 1) T] = A(T) x(kT) + b(T) u(kT) \qquad (7.103)$$

and the vector of the state variables, x, is quite sufficient for the description of the considered system at the instants of sampling. If, however, eqn. (7.102) is used to calculate the state vector $x(kT + \varepsilon T)$, $0 \leq \varepsilon < 1$, this state vector will not suffice for the description of the system behaviour at the instants $t \in [kT + \varepsilon T, (k + 1) T]$. Besides the vector $x(kT + \varepsilon T)$ we must know the value of the input variable $u(kT)$ which is constant during the entire sampling interval k and is acting at the system input. The complete state vector necessary for the determination of the system behaviour after the instant $kT + \varepsilon T$ is, therefore, for $T = 1$,

$$v(k) = \begin{bmatrix} x(k) \\ x_u(k) \end{bmatrix}, \quad x_u(k) = u(k) \qquad (7.104)$$

SOLUTION OF THE DISCRETE-TIME STATE EQUATION

We may now distinguish between two transitions:

(i) The step-change of the input variable u at the instants of sampling, which takes place within a very short time interval (k^-, k^+). Let us designate this change as the *transition at the instant of sampling*. It is described by the equation

$$v(k^+) = \begin{bmatrix} E, & 0 \\ 0, & 0 \end{bmatrix} v(k^-) + \begin{bmatrix} 0 \\ 1 \end{bmatrix} u(k) \tag{7.105}$$

(ii) The change within the sampling interval, which we designate as the *transition within the interval of sampling*. This transition is described by the equation

$$v(k + \varepsilon) = \begin{bmatrix} A(\varepsilon), & b(\varepsilon) \\ 0, & 1 \end{bmatrix} v(k^+) \tag{7.106}$$

Substituting eqn. (7.105) into eqn. (7.106) we obtain

$$v(k + \varepsilon) = \begin{bmatrix} A(\varepsilon), & 0 \\ 0, & 1 \end{bmatrix} v(k^+) + \begin{bmatrix} b(\varepsilon) \\ 1 \end{bmatrix} u(k) \tag{7.107}$$

In the resultant equation we may omit in the state vector the component x_u, because it is redundant. Thus we obtain again the previously given form of the state equation (7.102).

In this manner we can describe systems with several nonsynchronously or non-simultaneously sampling members. Here the vector of the state variables is extended by the output variables of all holding members. Then the interval of transition is the interval between two successive instants of sampling, then follows the transition at the instant of sampling just attained, then the transition to the next instant of sampling, etc. If the individual basic sampling periods are in a rational relation, a so-called *main interval of sampling* will exist, which can be expressed as an integer multiple of any basic period. Trivial equations relating to the sampling transition at the beginning and at the end of the main interval will then appear in the resultant difference equation expressed for the main sampling interval. The state variables corresponding to these transitions may be omitted as in eqn. (7.107).

7.8.1 Asynchronous sampling

Asynchronous sampling may be calculated directly according to the instructions given in the introductory part of Sec. 7.8.

Example 7.6: Establish the state equation of the control loop shown in Fig. 7.2. The control loop has two sampling members of which one, A, is sampling at the instants $t = kT$ and the second one, B, at the instants $t = kT + \varepsilon T$, $k = 0, 1, 2, \ldots,$ $0 \leq \varepsilon < 1$.

Solution: Figure 7.2 indicates the state variables x_1, x_2 which, if there were no requirement for asynchronous sampling, would be sufficient for the description of the given second-order system. In addition, the figure introduces the state variables x_3 and x_4 appropriate to the outputs of the holding members, H. The state variable

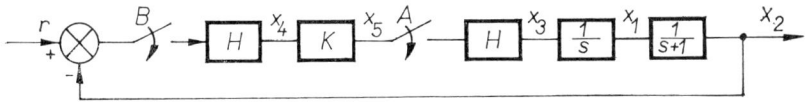

Fig. 7.2. Control loop, relating to Example 7.6, with asynchronous sampling.

x_5 need not be included in the vector of the state variables since a proportional controller with an amplification K is described by the simple relation

$$x_5(k) = K x_4(k)$$

The control loop is described by the equations

$$x_2' = -x_2 + x_1$$

$$x_1' = x_3$$

$$x_3' = x_4' = 0$$

The open loop is described by the homogeneous equation

$$x' = \begin{bmatrix} 0, & 0, & 1, & 0 \\ 1, & -1, & 0, & 0 \\ 0, & 0, & 0, & 0 \\ 0, & 0, & 0, & 0 \end{bmatrix} \quad \text{where} \quad x = \begin{bmatrix} x_1 \\ x_2 \\ x_3 \\ x_4 \end{bmatrix}$$

$$x' = Fx$$

Let us first calculate the fundamental matrix $\exp[Ft]$:

$$e^{Ft} = L^{-1}[(sE - F)^{-1}] = \begin{bmatrix} 1, & 0, & t, & 0 \\ 1 - e^{-t}, & e^{-t}, & t - 1 + e^{-t}, & 0 \\ 0, & 0, & 1, & 0 \\ 0, & 0, & 0, & 1 \end{bmatrix}$$

SOLUTION OF THE DISCRETE-TIME STATE EQUATION

At the instant $t = kT$, the sampling member A is defined by

$$x_3(k^+) = x_5(k^-) = K\, x_4(k^-)$$

$$x(k^+) = V_A\, x(k^+)$$

$$V_A = \begin{bmatrix} 1, & 0, & 0, & 0 \\ 0, & 1, & 0, & 0 \\ 0, & 0, & 0, & K \\ 0, & 0, & 0, & 1 \end{bmatrix}$$

At the instant $t = kT, + \varepsilon T$ the sampling member B is defined by

$$x_4(k + \varepsilon^+) = -x_2(k + \varepsilon^-) + r(k + \varepsilon)$$

$$x(k + \varepsilon^+) = V_B\, x(k + \varepsilon^-) + b\, r(k + \varepsilon)$$

$$V_B = \begin{bmatrix} 1, & 0, & 0, & 0 \\ 0, & 1, & 0, & 0 \\ 0, & 0, & 1, & 0 \\ 0, & -1, & 0, & 0 \end{bmatrix}$$

The resultant difference equation in state space is established using the previous relations

$$x(k + 1^+) = V_A\, x(k + 1^-) =$$
$$= V_A\, A(T - \tau)\, x(k + \varepsilon^+) =$$
$$= V_A\, A(T - \tau)\, [V_B\, x(k + \varepsilon^-) + b\, r(k + \varepsilon)] =$$
$$= V_A\, A(T - \tau)\, [V_B\, A(\varepsilon)\, x(k^+) + b\, r(k + \varepsilon)]$$

where $\tau = \varepsilon T$.

$$x(k + 1^+) = \begin{bmatrix} 1, & 0, & T, & 0 \\ 1 - e^{-T}, & e^{-T}, & T - 1 + e^{-T}, & 0 \\ -K(1 - e^{-\tau}), & -K e^{-\tau}, & -K(\tau - 1 + e^{-\tau}), & 0 \\ -(1 - e^{-\tau}), & -e^{-\tau}, & -(\tau - 1 + e^{-\tau}), & 0 \end{bmatrix} x(k^+) +$$

$$+ \begin{bmatrix} 0 \\ 0 \\ K \\ 1 \end{bmatrix} r(k + \varepsilon)$$

Since at the instants $k^+ T$ it holds that

$$x_3(k^+) = K\, x_4(k^+)$$

the state variable x_4 in the last equation may be omitted. With the state vector

$$\hat{x}(k) = \begin{bmatrix} x_1(k) \\ x_2(k) \\ x_3(k) \end{bmatrix}$$

we obtain

$$\hat{x}(k+1^+) = \begin{bmatrix} 1, & 0, & T \\ 1-e^{-T}, & e^{-T}, & T-1+e^{-T} \\ -K(1-e^{-\tau}), & -Ke^{-\tau}, & -K(\tau-1+e^{-\tau}) \end{bmatrix} \hat{x}(k^+) + \begin{bmatrix} 0 \\ 0 \\ K \end{bmatrix} r(k+\varepsilon)$$

7.8.2 Finite sampling time

In the design of discrete control loops where the controller is realized by the operation of a computer we usually assume that the sampling time is so short in comparison with the sampling interval that it may be neglected. Sometimes, however, this assumption may not be justified. This may be so for several reasons. For instance, a short sampling interval with respect to the short time constants of the controlled system and to the input signals with high-frequency components, slow analogue-to-digital and digital-to-analogue converters, a relatively long time needed for the calculation of the value of the controlling variable, etc. All these time shifts may formally be considered as a sampling time τ not negligible in comparison with the sampling interval T. In such a case the state of the control loop may be described, taking account of the finite time of sampling, according to the general instructions given in the introductory part of Sec. 7.8.

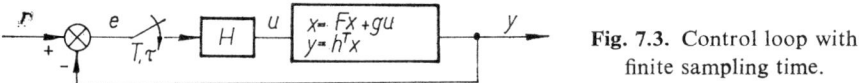

Fig. 7.3. Control loop with finite sampling time.

Let us consider the control loop shown in Fig. 7.3, whose continuously working part, i.e. the controlled plant, is described by the equations

$$x'(t) = F x(t) + g u(t)$$
$$y(t) = h^T x(t)$$

The input signal is given by the output u of the holding member H which, with the sampling member open, maintains the signal u at a constant value equal to the last value before opening. The sampling member operates with a sampling interval T and the sampling time at the beginning of this interval is τ. It holds that

$$u(t) = \begin{cases} e(t) & \text{for } kT \le t < kT + \tau \\ e(kT+\tau) & \text{for } kT + \tau \le t < (k+1)T \end{cases} \qquad (7.108)$$

SOLUTION OF THE DISCRETE-TIME STATE EQUATION

During the sampling, i.e. when the sampling member is closed, we have

$$x'(t) = F x(t) + g\, e(t) = F x(t) + g[r(t) - y(t)] =$$
$$= (F - gh^T) x(t) + g\, r(t) \qquad (7.109)$$

Let us denote $\hat{F} = F - gh^T$. Then $\hat{A}(\tau) = \exp[\hat{F}\tau]$ and

$$x(kT + \tau) = \hat{A}(\tau) x(kT) + \int_{kT}^{kT+\tau} \hat{A}(kT + \tau - v)\, g\, r(v)\, dv \qquad (7.110)$$

In the second part of the sampling interval, with the sampling member open, i.e. within the interval $kT + \tau < t < (k+1)T$, it holds that

$$x'(t) = F x(t) + g\, e(kT + \tau) \qquad (7.111)$$

where $e(kT + \tau) = r(kT + \tau) - h^T x(kT + \tau) = \text{const.}$

Introducing

$$A(v) = \exp[Fv]$$
$$b(v) = \int_0^v A(v)\, g\, dv$$

eqn. (7.111) may be rewritten in the form

$$x[(k+1)T] = A(T - \tau) x(kT + \tau) + b(T - \tau) [r(kT + \tau) - h^T x(kT + \tau)] =$$
$$= [A(T - \tau) - b(T - \tau) h^T] x(kT + \tau) + b(T - \tau) r(kT + \tau) \qquad (7.112)$$

Substituting eqn. (7.110) into the last equation we finally obtain

$$x[(k+1)T] = [A(T - \tau) - b(T - \tau) h^T] \left[\hat{A}(\tau) x(kT) + \int_{kT}^{kT+\tau} \hat{A}(kT + \tau - v)\, g\, r(v)\, dv \right] + b(T - \tau) r(kT + \tau) \qquad (7.113)$$

8

Mutual Configuration of Discrete Systems

In the preceding paragraphs the state description was mainly applied to only one dynamical system, e.g. to a controlled plant. Only in Pars. 7.8.1 and 7.8.2 the solution referred to a closed control loop, i.e. to a feedback system. However, systems may also have a much more complex structure where each member may be a dynamical member described by state equations. The individual dynamical members of a complex system structure may have either a parallel or a series or a feedback configuration. The description of these three basic configurations of dynamical members by means of state equations represents the basic algebra for the determination of the state equations corresponding to systems composed of a small or even a larger number of members. This algebra of state equations will be derived in the subsequent paragraphs.

8.1 Parallel configuration

Let us consider two systems described by the state equations

$$x_1(k + 1) = A_1 x_1(k) + b_1 u_1(k)$$
$$y_1(k) = c_1^T x_1(k) + d_1 u_1(k) \qquad (8.1)$$

and

$$x_2(k + 1) = A_2 x_2(k) + b_2 u_2(k)$$
$$y_2(k) = c_2^T x_2(k) + d_2 u_2(k) \qquad (8.2)$$

The systems are connected in parallel. From Fig. 8.1 it follows that in this case $u_1 = u_2 = u$ and the resultant output signal $y = y_1 + y_2$. Using the state vector

$$x = \begin{bmatrix} x_1 \\ x_2 \end{bmatrix} \qquad (8.3)$$

we may write

$$x(k+1) = \begin{bmatrix} A_1, & 0 \\ 0, & A_2 \end{bmatrix} x(k) + \begin{bmatrix} b_1 \\ b_2 \end{bmatrix} u(k)$$

$$y(k) = [c_1^T, c_2^T] x(k) + (d_1 + d_2) u(k) \qquad (8.4)$$

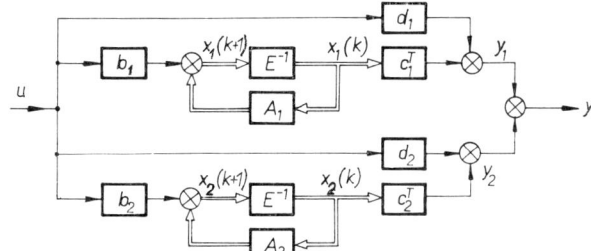

Fig. 8.1. Parallel configuration of discrete state system models.

Equations (8.4) represent the resultant state equation for two systems connected in parallel, which could easily be generalized for an arbitrary number of parallel systems.

8.2 Series configuration

Let us now establish the state equation for a series connection of the two systems described by eqns. (8.1) and (8.2). With this arrangement let the output of the first system be the input of the second system, i.e. $y_1 = u_2$. In this case

$$x_1(k+1) = A_1 x_1(k) + b_1 u_1(k)$$
$$x_2(k+1) = A_2 x_2(k) + b_2 [c_1^T x_1(k) + d_1 u_1(k)] \qquad (8.5)$$

Using the state vector (8.3) we may rewrite eqns. (8.5) in the form

$$x(k+1) = \begin{bmatrix} A_1, & 0 \\ b_2 c_1^T, & A_2 \end{bmatrix} x(k) + \begin{bmatrix} b_1 \\ b_2 d_1 \end{bmatrix} u_1(k) \qquad (8.6)$$

The output of the second system is given by the relation

$$y(k) = y_2(k) = [d_2 c_1^T, c_2^T] x(k) + d_2 d_1 u_1(k) \qquad (8.7)$$

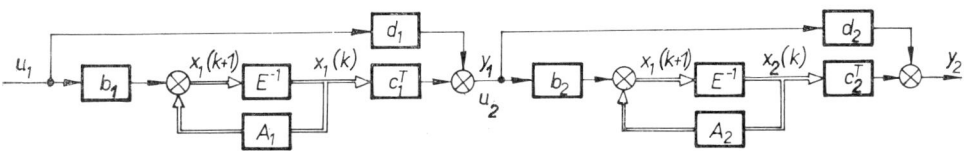

Fig. 8.2. Series configuration of discrete state system models.

The correctness of this relation well follows from the block diagram in Fig. 8.2. Here the individual components of the output variable y are the responses of the state vectors x_1 and x_2 and of the input variable u_1.

8.3 Feedback configuration

The feedback arrangement of the two systems described by eqns. (8.1) and (8.2) corresponds to the block diagram in Fig. 8.3. From this block diagram it follows that

$$u_1 = r - y_2 \tag{8.8}$$

$$u_2 = y_1$$

$$u_1 = r - c_2^T x_2 - d_2 u_2 \tag{8.9}$$

$$u_1 = r - c_2^T x_2 - d_2 c_1^T x_1 - d_2 d_1 u_1 \tag{8.10}$$

$$u_1 = Kr - K d_2 c_1^T x_1 - K c_2^T x_2 \tag{8.11}$$

where

$$K = (1 + d_2 d_1)^{-1} \tag{8.12}$$

assuming that an inversion does exist.

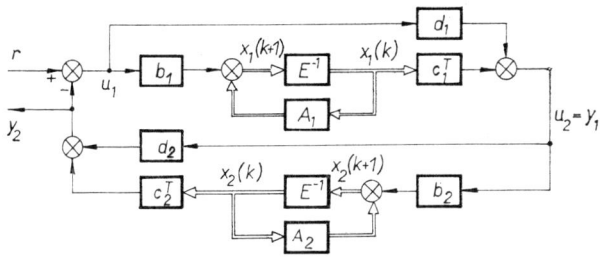

Fig. 8.3. Feedback configuration of discrete state system models.

For the sake of clarity the argument k relating to the variables and vectors of variables in eqns. (8.8) through (8.12) was omitted. Using eqn. (8.11) the first of eqns. (8.1) assumes the form

$$x_1(k+1) = (A_1 - b_1 K d_2 c_1^T) x_1(k) - b_1 K c_2^T x_2(k) + b_1 K r(k) \tag{8.13}$$

Substituting for $u_2(k)$ in the first of eqns. (8.2) we obtain

$$x_2(k+1) = (b_2 c_1^T - b_2 d_1 K d_2 c_1^T) x_1(k) + (A_2 - b_2 d_1 K c_2^T) x_2(k) + b_2 d_1 K r(k) \tag{8.14}$$

MUTUAL CONFIGURATION OF DISCRETE SYSTEMS

Using again the vector (8.3), the state equations of the feedback system may be expressed as follows:

$$x(k+1) = \begin{bmatrix} A_1 - b_1 K d_2 c_1^T, & -b_1 K c_2^T \\ b_2 c_1^T - b_2 d_1 K d_2 c_1^T, & A_2 - b_2 d_1 K c_2^T \end{bmatrix} x(k) + \begin{bmatrix} b_1 K \\ b_2 d_1 K \end{bmatrix} r(k) =$$

$$= A\, x(k) + b\, r(k) \tag{8.15}$$

The output equation is, according to the second of eqns. (8.2) and with eqn. (8.11),

$$y_2(k) = c_2^T x_2(k) + d_2 [c_1^T x_1(k) + d_1 u_1(k)] =$$
$$= c_2^T x_2(k) + d_2 c_1^T x_1(k) +$$
$$+ d_2 d_1 [K r(k) - K c_2^T x_2(k) - K d_2 c_1^T x_1(k)] =$$
$$= [d_2(1 - d_1 K d_2)\, c_1^T, (1 - d_2 d_1 K)\, c_2^T]\, x(k) +$$
$$+ d_2 d_1 K r(k) = c^T x(k) + d\, r(k) \tag{8.16}$$

This equation may alternatively be written in the form

$$y_2(k) = [K d_2 c_1^T,\, K c_2^T]\, x(k) + K d_2 d_1\, r(k) \tag{8.17}$$

which also applies to multi-input/multi-output systems, where c_1^T, c_2^T, d_1, d_2 are rectangular matrices, since

$$d_2 c_1^T - d_2 d_1 K d_2 c_1^T = K^{-1} K d_2 c_1^T - d_2 d_1 K d_2 c_1^T =$$
$$= (K^{-1} - d_2 d_1)\, K d_2 c_1^T = K d_2 c_1^T \tag{8.18}$$

$$c_2^T - d_2 d_1 K c_2^T = K^{-1} K c_2^T - d_2 d_1 K c_2^T = K c_2^T$$

$$d_2 d_1 K = K d_2 d_1 \tag{8.19}$$

The latter equality is proved as follows:

$$K^{-1} d_2 d_1 K - d_2 d_1 = (1 + d_2 d_1)\, d_2 d_1 K - d_2 d_1 =$$
$$= d_2 d_1 (1 + d_2 d_1)\, K - d_2 d_1 =$$
$$= d_2 d_1 K^{-1} K - d_2 d_1 = 0 \tag{8.20}$$

Should the feedback system of Fig. 8.3 represent a control loop with the controlling variable u and the controlled variable y_2, then the system described by eqns. (8.1) would be the controller and the system described by eqns. (8.2) the controlled plant. In such a case, to meet the requirements of physical realizability of the controlled plant, d_2 would have to be zero. On the contrary, for a digital correcting member we permit that $d_1 \ne 0$.

The state equations derived for a parallel, series and feedback configuration of two systems apply to two single-input/single-output systems. The generalization for a larger number of systems makes no particular difficulties. The resultant expressions derived for single-input/single-output systems apply for multi-input/multi-output systems as well provided that we replace b_1, b_2, c_1^T, c_2^T, d_1, d_2 by appropriate rectangular matrices. If we denote (in brackets) the dimensions of the individual matrices as $A_i(n_i; n_i)$, $B_i(n_i; r_i)$, $C_i(p_i; n_i)$, $D_i(p_i, r_i)$, $i = 1, 2$, where n_i is the order of the relevant system, p_i the number of output variables and r_i the number of input variables, then for the feedback system the equalities $p_1 = r_2$ and $r_1 = p_2$ must be satisfied. The resultant matrices have the dimensions $A(n; n)$, $B(n; p_2)$, $C(r_1; n)$, $D(r_1; r_1)$, where $n = n_1 + n_2$. The derivation of the above given state equations for the alternative arrangement has been left to the reader.

9

System Properties

9.1 Reachability, controllability and stabilizability

When solving control problems in system state theory, advantage is taken of some fundamental properties of dynamical systems, which do not occur in the classical control theory, based only on the input and output of the system considered. The properties of concern are the reachability, controllability and stabilizability of systems. All these three concepts as well as others mentioned in Secs. 9.2 and 9.3 are defined mathematically. They represent a tool facilitating a brief expression and formulation of the conditions needed when solving problems of synthesis, i.e. for the design of a controller providing optimum control in a given sense.

At this point it should be noted that the system state theory, just like any control theory, is not concerned with real objects but only with their mathematical models. Therefore, the tools used can only be mathematical and such are the solutions which this theory yields. When designing a controller providing optimum control of a given plant in the desired sense, the system state theory frequently requires the controlled plant to be controllable and observable [69.4]. The fulfilment of these requirements enables us to calculate the optimum control often by rather simple mathematical operations. This, however, does not mean that an uncontrollable and unobservable system cannot be controlled suboptimally in a practical sense. It may be said beforehand that the plant stabilizability and detectability is decisive for the realizability of its control. These basic concepts are defined in this chapter.

We shall successively define the basic system properties and show how they may be verified.

DEFINITION 9.1: *The state $x(t_1)$ of a linear system is reachable if there exists a time instant $t_0 < t_1$, where $(t_1 - t_0)$ is a finite interval, and such an input $u(t)$ by which the initial system state $x(t_0) = 0$ is transferred into the desired state $x(t_1)$* [69.4].

DEFINITION 9.2: *The state $x(t_1)$ of a linear system is controllable if there exists a time instant $t_2 > t_1$ and such an input $u(t)$ by which the system state $x(t_1)$ is transferred into the state $x(t_2) = 0$, the interval $t_2 - t_1$ being finite.*

The concept of so-called complete system reachability[1]) (or complete controllability) at the time t was additionally introduced, which means that any state $x(t) \in X$ may be reachable (or controllable). In time-invariant discrete systems the state reachability and controllability does not depend on the instant t_1. If, in addition, the system is a continuously working one, then each reachable state is controllable as well so that in the literature dealing with continuously working systems it is usually spoken just of controllability as it is not necessary to distinguish between state reachability and state controllability. The initial state $x(t_0)$ of a controllable continuously working linear system may be transferred in a finite length of time to any other state by a suitable input signal. For a discrete system this applies only if in the state equation

$$x(k+1) = A x(k) + b u(k) \tag{9.1}$$

the determinant

$$\det [A] \neq 0 \tag{9.2}$$

From eqn. (7.62) and from the definition of the matrix $A = \exp [FT]$ it follows that the matrix A of the discretized continuously working system is always such that $\det [A] \neq 0$. On the other hand, for discrete systems whose physical nature of operation is discontinuous as, for instance, for digital correcting members, discrete automata, etc., the condition (9.2) is not always satisfied.

The requirement that a system should be transferable from any arbitrary state to another arbitrary state may also be formulated by stipulating that the system transferred from a zero initial state to the non-zero state $x(t_1)$ should be transferable back to the original zero state by the action of a suitable input signal. If a system satisfies such a requirement we denote it as a *reversible system*.

For instance, the system

$$x(k+1) = \begin{bmatrix} 0, & 1 \\ 0, & 0 \end{bmatrix} x(k) + \begin{bmatrix} 0 \\ 0 \end{bmatrix} u(k)$$

representing two first-order members connected in series, is not reversible. The state $x(1)$ of the system is controllable because from the initial state $x^T(0) = [x_1(0), x_2(0)]$, $x_1(0) \neq 0$, $x_2(0) \neq 0$ we obtain $x^T(1) = [x_2(0), 0]$ and $x^T(2) = [0, 0]$. The system is not reachable, however, because it cannot be transferred from the zero initial state to an arbitrary other state by any input $u(k)$.

---

[1]) For simplicity, the term system reachability (controllability), etc. will henceforth be used instead of the term complete system reachability (controllability), etc.

SYSTEM PROPERTIES

Further relationships may be read from the conditions of reachability and from the conditions of controllability.

From eqn. (7.1) it follows that

$$x(k) = A^k x(0) + A^{k-1} b\, u(0) + A^{k-2} b\, u(1) + \ldots + Ab\, u(k-2) + b\, u(k-1) =$$
$$= A^k x(0) + [b, Ab, \ldots, A^{k-1}b]\,[u(k-1), u(k-2), \ldots, u(0)]^T \qquad (9.3)$$

When investigating the reachability of the state $x(k)$ let us put $x(0) = 0$ so that the first term on the left-hand side of eqn. (9.3) vanishes. Then any state from the space spanned by the vectors $b, Ab, \ldots, A^{n-1}b$ will be reachable in n steps under the action of the input set $u(0), u(1), \ldots, u(n-1)$. These vectors must be linearly independent[1]) in order that the system (9.1) might be transferred to the desired state $x(n)$, where n is the state space dimension. From the above consideration it follows:

THEOREM 9.1: *The state $x(n)$ of the system (9.1) is reachable if and only if the rank h of the matrix $b, Ab, \ldots, A^{n-1}b$ is equal to the state space dimension n.*

It is plain that no general state $x(k)$ can be reached when $k < n$. When $k = n$, the system will be reachable but may not in general be reachable for $k > n$ as, for instance, in the case of linear plants with bounded input.

THEOREM 9.2: *If a vector $A^i b$ linearly dependent on the vectors $b, Ab, \ldots, A^{i-1}b$ occurs in the set of vectors $b, Ab, \ldots, A^{i-1}b$, then each following vector $A^m b, m > i$, also depends on the same vectors.*

Theorem 9.2 will now be proved. According to the assumption we have

$$A^i b = \sum_{j=0}^{i-1} {}^0 c_j A^j b \qquad (9.4)$$

The next vector may then be expressed as

$$A^{i+1} b = \sum_{j=0}^{i-1} {}^0 c_j A^{j+1} b = \sum_{k=1}^{i} {}^0 c_{k-1} A^k b \qquad (9.5)$$

where the last sum was obtained by substituting $j + 1 = k$. Expressing the last term of this sum separately we obtain

$$A^{i+1} b = \sum_{k=1}^{i-1} {}^0 c_{k-1} A^k b + {}^0 c_{i-1} A^i b =$$
$$= \sum_{k=1}^{i-1} {}^0 c_{k-1} A^k b + {}^0 c_{i-1} \sum_{j=0}^{i-1} {}^0 c_j A^j b = \sum_{j=0}^{i-1} {}^1 c_j A^j b \qquad (9.6)$$

---

[1]) See Appendix B for the test of linear dependence of vectors.

Supposing that the theorem also applies to $A^{i+2}b, \ldots, A^{i+m-1}b$, let us prove that it applies to $A^{i+m}b$ as well:

$$A^{i+m}b = \sum_{j=0}^{i-1} {}^0c_j A^{j+m}b = \sum_{k=m}^{i+m-1} {}^0c_{k-m} A^k b =$$

$$= \sum_{k=m}^{i-1} {}^0c_{k-m} A^k b + {}^0c_{i-m} A^i b + {}^0c_{i-m+1} A^{i+1} b + \ldots + {}^0c_{i-1} A^{i+m-1} b =$$

$$= \sum_{k=m}^{i-1} {}^0c_{k-m} A^k b + {}^0c_{i-m} \sum_{j=0}^{i-1} {}^0c_j A^j b + \ldots + {}^0c_{i-1} \sum_{j=0}^{i-1} {}^{(m-1)}c_j A^j b =$$

$$= \sum_{j=0}^{i-1} {}^{(m)}c_j A^j b \tag{9.7}$$

The first term in the last but one row is equal to zero if $i - 1 < m$. Equation (9.7) proves Theorem 9.2.

According to Theorem 9.1 the system state (9.1) is reachable if the matrix

$$Q_D = [b, Ab, \ldots, A^{n-1}b] \tag{9.8}$$

has the rank $h = n$. For this case we can, using eqn. (9.3), calculate the set of values of the controlling variable:

$$u(k) = Q_D^{-1} [x(k) - A^k x(0)], \quad k = 1, 2, \ldots, n \tag{9.9}$$

The conditions of controllability may again be derived from eqn. (9.3). According to Theorem 9.2 it holds that $x(k) = x(n) = 0$ and $x(0) \neq 0$. We obtain

$$A^n x(0) = -[b, Ab, \ldots, A^{n-1}b] [u(n-1), \ldots, u(0)]^T \tag{9.10}$$

$$x(0) = -[A^{-n}b, A^{-n+1}b, \ldots, A^{-1}b] [u(n-1), \ldots, u(0)]^T = -Q_R u(k) \tag{9.11}$$

In order that the system be controllable, i.e. transferable according to Theorem 9.2 in n steps from the state $x(0) \neq 0$, under the action of the input sequence $u(0), \ldots, u(n-1)$, to the state $x(n) = 0$, where n is the state space dimension, then evidently the state $x(0)$ may be a state from the space spanned by the vectors $A^{-n}b$, $A^{-n+1}b, \ldots, A^{-1}b$. Here the vectors must be linearly independent since otherwise the state $x(n) = 0$ could not be reached. From the above consideration it follows:

THEOREM 9.3: *The state $x(0) \neq 0$ of the system 9.1 is controllable if and only if the rank h of the matrix $[A^{-1}b, A^{-2}b, \ldots, A^{-n}b]$ is equal to the state space dimension n.*

It is clear that the above theorem is sensible only for invertible (i.e. non-singular) matrices A. As already stated, this requirement is always met for systems generated by sampling of continuously working systems. For systems which, according to their

SYSTEM PROPERTIES

operation, are discrete systems, the matrix A may not be invertible. In this case the conditions of controllability would have to be defined according to eqn. (9.10).

If in Theorem 9.2 an inversion of the matrix A exists, then this matrix is non-singular and $\det A \neq 0$. This substantiates the condition (9.2). On the other hand it should be noted that the rank of the matrix Q_R defined by Theorem 9.3 remains unchanged if Q_R is multiplied by a non-singular matrix. If it is multiplied by A^n from the left we obtain the matrix Q_D with a reversed array of columns. Consequently, if the matrix A is non-singular, the conditions of reachability and controllability are equivalent.

Comparing the conditions of reachability (Theorem 9.1) with the conditions of controllability (Theorem 9.3) we can see that the requirement of reversibility necessarily implies that the matrix A should be invertible as already expressed by the requirement (9.2).

For the vectors of Q_R, like for the vectors of Q_D, it can be proved that if the vector $A^{-1}b$ is linearly dependent on the vectors $A^{-1}b, \ldots, A^{-(i-1)}b$, then all the following vectors $A^{-m}b, m > i$, are also linearly dependent on these vectors. It can also be proved that if the columns of the matrix Q_D are mutually linearly independent, the columns of the matrix Q_R are mutually linearly independent as well.

Since in our further considerations we shall mostly deal with discrete systems generated by sampling continuously working systems, where the condition $\det A \neq 0$ is always satisfied, it will only be necessary to investigate the satisfaction of the conditions of reachability.

With continuously working systems the requirement of state reachability becomes identical with the requirement of controllability and, therefore, only the term controllability of continuously working systems is used in technical literature. Most frequently the condition of controllability is defined as in Theorem 9.1, i.e. as the condition of state reachability in discrete systems. A continuously working system whose input is a staircase function and whose output is sampled synchronously with the changes at its input, may not be reachable even if the continuously working system itself is controllable. This case occurs if the characteristic equation of the continuously working system has at least one pair of complex roots with an equal real part, i.e. if $s_{1,2} = \alpha \pm j\omega$. In the discrete version the conditions of reachability will be disturbed if the period of sampling $T = k\pi/\omega, k = 1, 2, 3, \ldots$.

At this point it should be noted that a system which is not reachable (and controllable) in the sense of the conditions mentioned may, on certain assumptions, be still acted upon by a controller in a closed control loop in such a way as to obtain a stable resultant control process. In such a case we speak of a stabilizable system. According to Refs. [68.10, 72.5] the following definition holds:

DEFINITION 9.3: *A system (A, B) is stabilizable if there exists a real matrix K such that the matrix $A - BK$ is stable, i.e. that for all eigenvalues λ_i of the matrix $A - BK$ it holds that $|\lambda_i| < 1, i = 1, 2, \ldots, n$.*

Alternatively it holds that the system (A, B) is stabilizable if and only if the output components (modes) corresponding to unstable eigenvalues of the matrix A are reachable.

In other words, Definition 9.3 also says that all what is needed for stabilization is a controller whose output (controlling) variables correspond to a linear transformation of the state vector, i.e.

$$u(k) = -K\,x(k) \tag{9.12}$$

The conditions of stabilizability are in fact the conditions for the stability of a closed control loop containing an unstable plant, where the input of the feedback member (controller) is determined by the state vector variables of the plant to be controlled. We shall be returning to this possibility of design of a closed-loop controller in the later chapters dealing with synthesis where the aim will be the determination of the matrix K.

9.2 Observability, reconstructability and detectability

In many cases the state of the system (9.1) is not measurable and, therefore, control in the sense of the relation (9.12) cannot be directly realized. Thus the question arises whether the state vector can be determined from a measurable output or measurable outputs of a multi-input/multi-output plant. In this connection we distinguish [69.4] between the case of observable state and the case of reconstructable state.

DEFINITION 9.4: *The state $x(t_0)$ of a system is observable if it can be determined by means of the future values $y(t)$, $t > t_0$, of the output variable and if the interval $t - t_0$ is finite.*

DEFINITION 9.5: *The state $x(t_0)$ of a system is reconstructable if it can be determined by means of the past values $y(t)$, $t < t_0$, of the output variable and if the interval $t_0 - t$ is finite.*

The conditions of observability and reconstructability may be derived by means of the output equation

$$y(k) = c^T x(k) + d\,u(k) \tag{9.13}$$

and the state equation (9.1). By a successive calculation of the value of the output variable for $k, k+1, \ldots, k+n-1$ we obtain

$$\begin{aligned}
y(k) &= c^T x(k) + d\,u(k) \\
y(k+1) &= c^T A\,x(k) + c^T b\,u(k) + d\,u(k+1) \\
y(k+2) &= c^T A^2 x(k) + c^T A b\,u(k) + c^T b\,u(k+1) + d\,u(k+2) \\
&\vdots \\
y(k+n-1) &= c^T A^{n-1} x(k) + c^T A^{n-2} b\,u(k) + \ldots \\
&\quad \ldots + c^T A b\,u(k+n-3) + \\
&\quad + c^T b\,u(k+n-2) + d\,u(k+n-1)
\end{aligned} \tag{9.14}$$

SYSTEM PROPERTIES

or in the vector-matrix form

$$\begin{bmatrix} y(k) \\ y(k+1) \\ \vdots \\ y(k+n-1) \end{bmatrix} = \begin{bmatrix} c^T \\ c^T A \\ \vdots \\ c^T A^{n-1} \end{bmatrix} \begin{bmatrix} x_1(k) \\ x_2(k) \\ \vdots \\ x_n(k) \end{bmatrix} +$$

$$+ \begin{bmatrix} d, & 0, & 0, & \ldots, & 0 \\ c^T b, & d, & 0, & \ldots, & 0 \\ c^T Ab, & c^T b, & d, & \ldots, & 0 \\ \cdots\cdots\cdots\cdots\cdots\cdots\cdots \\ c^T A^{n-2} b, & c^T A^{n-3} b, & b, & \ldots, & d \end{bmatrix} \begin{bmatrix} u(k) \\ u(k+1) \\ u(k+2) \\ \vdots \\ u(k+n-1) \end{bmatrix} \quad (9.15)$$

or in the reduced form

$$y_n(k) = Q_P \, x(k) + P_P \, u_n(k) \tag{9.16}$$

In general it can be shown that for linear systems, where the law of superposition applies, the conditions of observability can be formulated only by means of the outputs $y_n(k)$.

That is to say, if we use the symbolic notation of Definition 2.1, the output representation of linear systems will be

$$\eta(t, \varphi(t, \tau, x, \omega)) = \eta(t, \varphi, (t; \tau, x, 0)) + \eta(t, \varphi(t; \tau, 0, \omega)) \tag{9.17}$$

where the first term on the right-hand side determines the output component following from the initial state and the second term the component following from the input signal. Therefore

$$\eta(t, \varphi(t; \tau, x_1, \omega)) - \eta(t, \varphi(t; \tau, x_2, \omega)) =$$
$$= \eta(t, \varphi(t; \tau, x_1, 0)) - \eta(t, \varphi(t; \tau, x_2, 0)) =$$
$$= \eta(t, \varphi(t; \tau, x_1 - x_2, 0)) \tag{9.18}$$

Equation (9.18) shows that the components corresponding to the input ω are cancelled and the resultant output representation depends only on the state difference $x_1 - x_2$. Considering the state x_1 as the initial state and x_2 as the final state and putting $x_2 = 0$, it follows from eqn. (9.18) that the initial state may be determined, in a linear case, as the linear transformation of future outputs. According to this conclusion the second term on the right-hand side of eqn. (9.16) may be omitted.

If the matrix Q_P is non-singular, the inverse Q_P^{-1} will exist. Then it also holds that $\det[Q_P] \neq 0$ or that the individual rows of the matrix Q_P are mutually linearly independent. Then it follows from eqn. (9.16) that

$$x(k) = Q_P^{-1} \, y_n(k) \tag{9.19}$$

Thus, we can now define the following conditions of observability:

THEOREM 9.4: *A linear system described by eqns. (9.1) and (9.13) is observable if and only if the rank of the matrix*

$$Q_P = \begin{bmatrix} c^T \\ c^T A \\ \vdots \\ c^T A^{n-1} \end{bmatrix}$$

is equal to the state space dimension n.

The conclusions for the mutual linear dependence of the rows of Q_P are similar to those for the columns of Q_D formulated by Theorem 9.2.

The conditions of reconstructability may also be derived by means of eqn. (9.16) considering that

$$x(k + n) = A^n x(k) \tag{9.20}$$

Using the relation (9.20), eqn. (9.16) assumes the form

$$y_n(k) = Q_P A^{-n} x(k + n) = Q_K x(k + n) \tag{9.21}$$

where

$$Q_K = \begin{bmatrix} c^T A^{-n} \\ c^T A^{-(n-1)} \\ \vdots \\ c^T A^{-1} \end{bmatrix} \tag{9.22}$$

$$x(k + n) = Q_K^{-1} y_n(k) \tag{9.23}$$

The conditions of reconstructability may be formulated similarly to the conditions of observability:

THEOREM 9.5: *Assuming that* $\det [A] \neq 0$, *the linear system described by eqns. (9.1) and (9.13) is reconstructable if and only if the rank of the matrix*

$$Q_K = \begin{bmatrix} c^T A^{-n} \\ c^T A^{-(n-1)} \\ \vdots \\ c^T A^{-1} \end{bmatrix}$$

is equal to the state space dimension n.

Again, the conclusions for the mutual linear dependence of the rows of Q_K are similar to those for the columns of Q_D formulated by Theorem 9.2.

SYSTEM PROPERTIES

The matrix A is non-singular for systems generated by sampling continuously working systems as already stated in Sec. 7.7 and, therefore, in these cases det $[A]$ is always unequal to zero. If the matrix Q_K is multiplied from the right by the matrix A^n we obtain the matrix Q_P. Consequently, if the matrix A is non-singular, the conditions of observability and reconstructability are equivalent.

If some of the system output components are unobservable but stable we call such a system detectable. In other words, a system is detectable if and only if the output components (modes) corresponding to unstable eigenvalues of the matrix A are observable. According to Refs. [68.10, 72.5] we have the following definition:

DEFINITION 9.6: *The pair (C, A) of a system is detectable if there exists a real matrix R such that $A - RC$ is stable, i.e. for all eigenvalues λ_i of the matrix $A - RC$ it holds that $|\lambda_i| < 1$, $i = 1, 2, \ldots, n$.*

Definition 9.6 is an analogy of Definition 9.3 which we become acquainted with in connection with the questions of reachability. It can be said that, in a certain sense, reconstructability is the opposite or, in other words, the dual property of controllability while observability is the dual property of reachability and detectability is the dual property of stabilizability. That is to say, if we have a system $\Sigma_1(A, b, c^T, d)$ and a system $\Sigma_2(A^T, c, b^T, d)$, where the superscript T denotes transposition, then the system Σ_2 is controllable if the system Σ_1 is reconstructable, Σ_2 is observable if Σ_1 is reachable, Σ_2 is detectable if Σ_1 is stabilizable and all these properties have a vice versa validity.

9.3 Identifiability

In the technical literature we may find many different methods for the determination of the dynamic properties of controlled plants or, simply, various methods of identification. In some cases of time-invariant plants it is attempted to calculate the mathematical model of the plant with the aid of measured values of the input and output variables of a real plant only once by applying a "one-shot" or "one-step" procedure and use the solution for the design of the controller, in other words, for the synthesis. In other cases the identification may be realized automatically in real time, and serves for controller updating, i.e. for adaptive control. Then the mathematical model of the controlled plant need not be explicitly evaluated and the measured plant input and output variables can be used directly for the determination of the controller parameters or even of the input (controlling) variable of the controlled plant. In any case, however, the controlled plant must be detectable and stabilizable. If it is desired to determine the mathematical model of the controlled plant, the plant must in addition be identifiable. If the conditions relating to the mentioned properties are not satisfied, an optimum control based on identification of the plant cannot be realized.

In Sec. 9.1 and 9.2 we discussed some properties of plants. In addition it might be useful to give some attention to the conditions of identifiability.

Let us consider a dynamic system described by the equation

$$x(k + 1) = A x(k) \tag{9.24}$$

This is a so-called *free system* which is isolated from the environment so that it is not affected by external variables.

Assume that the initial state $x(k)$ is known and that the system is observable. Let us investigate the conditions under which the system is also identifiable or, the conditions under which we can determine the matrix A. We have

$$x(k + 1) = A x(k)$$
$$x(k + 2) = A x(k + 1) = A^2 x(k)$$
$$\vdots$$
$$x(k + n) = A x(k + n - 1) = A^n x(k) \tag{9.25}$$

Since it was assumed that all state variables are observable we may, after n measurement steps, establish the matrix

$$[x(k + 1), x(k + 2), \ldots, x(k + n)] = A[x(k), x(k + 1), \ldots, x(k + n - 1)] \tag{9.26}$$

which enables us to define the conditions of identifiability.

THEOREM 9.6: *The matrix A of the system (9.24) is identifiable if and only if the matrix $[x(k), x(k + 1), \ldots, x(k + n - 1)] \equiv [x(k), A x(k), \ldots, A^{n-1} x(k)]$ is non-singular.*

For writing brevity let us denote $[x(k), A x(k), \ldots, A^{n-1} x(k)] = Q_\mathrm{I}$.

In a physical interpretation this condition means that the initial state $x(k)$ must excite all components (modes) of the system.

9.4 Canonical decomposition

Time-invariant linear systems having a finite number of independent output variables and a finite order of the individual subsystems may in a general case be divided into four partial systems [63.2, 63.4] as follows:

A: Reachable and observable partial system
B: Reachable and unobservable partial system
C: Unreachable and observable partial system
D: Unreachable and unobservable partial system.

SYSTEM PROPERTIES

This classification into partial systems is shown in Fig. 9.1 and is designated as the canonical decomposition.

Only the reachable and observable partial system can be described by methods based on the knowledge of the system input and output, e.g. the frequency response, impulse response, step response, etc. Also, only the reachable and observable part of the system can be described by a mathematical model calculated by some method

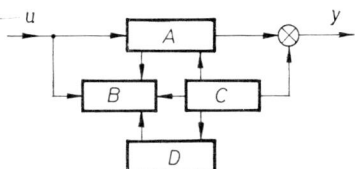

Fig. 9.1. Canonical system decomposition.

of identification based on the measurement of the system input and output variables. If the initial state of the system is non-zero, a measurement of the output variable y may enable us to detect the so far uncontrollable part C. Assume that this partial system C is stable and not affected by any disturbing variables.

Let us consider the discrete impulse response (7.3)

$$s(k) = \begin{cases} d & \text{for } k = 0 \\ c^T A^{k-1} b & \text{for } k > 0 \end{cases} \qquad (9.27)$$

of a linear discrete system having a finite dimension and order. Then each tetrad (A, b, c^T, d) which, according to eqn. (9.27), satisfies a given impulse response is referred to as the *impulse response realization*. If, at the same time, the matrix A has a minimum dimension in the set of all possible realizations, then the tetrad (A, b, c^T, d) is the so-called *minimum realization*. Thus the minimum order n and the constant $d = s(0)$ are uniquely determined. Then A, b, c^T depend on the state space base chosen. If (A_1, b_1, c_1^T, d) is the minimum realization, then the tetrad $(A_2, b_2, c_2^T, d) = (TA_1 T^{-1}, Tb_1, c_1^T T^{-1}, d)$, resulting from a transformation of the tetrad (A_1, b_1, c_1^T, d) by an arbitrary non-singular matrix T, is a minimum realization as well (see Sec. 10.1).

If a non-minimum realization were determined for the given discrete impulse response (9.27), it would comprise, besides the partial system A, at least one of the partial systems B, C, D. However, these partial systems cannot be determined with the aid of the measured values of input and output variables and, therefore, we usually take only interest in the minimum realization. In this connection let us quote [63.4] it the following important theorem:

THEOREM 9.7: *The realization of a dynamical system is minimum if and only if is reachable and observable.*

From the foregoing it follows that the description of a dynamical system by state equations is more general than that following from a measured input and output. It can comprise the partial systems B, C, D, provided they exist in the real system, while the description obtained from the input and output can comprise only the partial system A.

It should, however, be emphasized that only the partial system A can be controlled and, therefore, in control synthesis the dynamic system must be required to be reachable and observable. This, in fact, restricts the mathematical description to the partial system A. With this in view it is possible to justify the methods used in the classical control theory employing system descriptions based only on the measured values of input and output variables.

So far we have considered the cases where the partial systems B, C, D, e.g. of a controlled plant, enter into the mathematical model when the state equations are derived by means of design and physical data of the plant, i.e. when using the mathematical-physical analysis. On the other hand, in the controlled plant description following from the input and output we admitted that we know nothing about the internal structure of the plant. However, one must take into consideration even the case where the system is an open or closed control loop which may consist of several subsystems. In this case, if we use the description in state space, the order of the resultant system is equal to the sum of the subsystem orders. The same order of the linear system transfer function is obtained only if no reduction is performed. A minimum realization can only be obtained if a transfer function reduction is possible. This means, however, that the original system must have had not only the partial system A but also one of the partial systems B, C or D.

Let us show such cases by an example. Consider the partial transfer functions, corresponding to the subsystems F_1, F_2 and F_3,

$$F_1(z) = \frac{B_1(z)}{(z - \alpha) A_1(z)}$$

$$F_2(z) = \frac{(z - \alpha) B_2(z)}{A_2(z)}$$

$$F_3(z) = \frac{B_3(z)}{(z - \alpha) A_3(z)} \tag{9.28}$$

where $B_1(z)$ through $B_3(z)$ and $A_1(z)$ through $A_3(z)$ are finite polynomials, any two of which have no common root factors and none of which contains the root factor $(z - \alpha)$.

With the series arrangement $F_1 \rightarrow F_2$ a partial system with the root factor $(z - \alpha)$ (with the eigenvalue $z_1 = \alpha$) appears. This partial system is reachable but not observable. It is, therefore, a B-type partial system.

SYSTEM PROPERTIES

With the series arrangement $F_2 \to F_1$ a partial system with the same root factor appears. This partial system is observable but not reachable and represents, therefore, a C-type partial system.

Finally, with the parallel arrangement $F_1 \uparrow\uparrow F_3$ we can easily see that the partial system with the root factor $(z - \alpha)$ is neither reachable nor controllable. It, therefore, represents a D-type partial system.

In conclusion it may be noted that a system composed of reachable and observable subsystems may not necessarily be reachable and observable. If a reduction in the transfer function of a composite system is performed, the system looses one or both of the mentioned properties. On the other hand, the reachability and observability of all subsystems is an urgent prerequisite for the reachability and observability of a composite system.

10

Transformations

10.1 Linear transformation

The state equations in Sec. 3.1 were derived according to the design data and physical behaviour of the given plants. In these examples the state variables represented actual variables of the plant so that the state description was objectively related to physical reality. The state variables introduced in this way do not change their physical meaning even in the discrete version of the state description.

However, in some cases it is useful to introduce state variables defined formally as linear combinations of different physical state variables. The object of such a linear combination is to obtain certain canonical forms of state equations which permit or facilitate the recognition of certain plant and system properties or which enable us to describe them by means of fewer parameters.

Let us consider an n-dimensional vector $^1x \in {}^1X$ being a possible state vector of some system, and a non-singular matrix T having the dimension $(n; n)$. Then the vector

$$^2x = T\,^1x \tag{10.1}$$

is also a possible vector of the system considered, where $^2x \in {}^2X$.

The transformation from one state space to another will be shown by an example.

Consider the two state spaces ^1X and ^2X and the matrix operator (transformation matrix) T by which we can uniquely assign to each element $^1x \in {}^1X$ the element $^2x \in {}^2X$, according to eqn. (10.1).

Let us describe some real system by means of state equations with the state vector

$$^1x(k+1) = {}^1A\,{}^1x(k) + {}^1b\,u(k) \tag{10.2}$$

$$y(k) = {}^1c^T\,{}^1x(k) + d\,u(k) \tag{10.3}$$

and the same system by means of state equations with the state vector

$$^2x(k+1) = {}^2A\,{}^2x(k) + {}^2b\,u(k) \tag{10.4}$$

$$y(k) = {}^2c^T\,{}^2x(k) + d\,u(k) \tag{10.5}$$

TRANSFORMATIONS

From eqn. (10.1) it follows that

$$^1x(k) = T^{-1}\,{}^2x(k) \tag{10.6}$$

Substituting eqn. (10.6) into eqns. (10.2) and (10.3) we obtain

$$T^{-1}\,{}^2x(k+1) = {}^1A T^{-1}\,{}^2x(k) + {}^1b\,u(k) \tag{10.7}$$

$$y(k) = {}^1c^T T^{-1}\,{}^2x(k) + d\,u(k) \tag{10.8}$$

Multiplying now eqn. (10.8) by the matrix operator T from the left we have

$$^2x(k+1) = T\,{}^1A T^{-1}\,{}^2x(k) + T\,{}^1b\,u(k) \tag{10.9}$$

Comparing eqn. (10.4) with eqn. (10.9) and eqn. (10.5) with eqn. (10.8) we evidently obtain

$$^2A = T\,{}^1A T^{-1}$$
$$^2b = T\,{}^1b$$
$$^2c^T = {}^1c^T T^{-1}$$
$$d = d \tag{10.10}$$

DEFINITION 10.1: *The square matrices ^1A and ^2A of the n-th order are similar if there exists such a non-singular matrix T that $^2A = T\,{}^1A T^{-1}$.*

Note 10.1: The elements of the matrices ^1A, ^2A and T may be real or complex numbers.

Note 10.2: Two similar matrices ^1A and ^2A have equal characteristic polynomials and, consequently, equal eigenvalues.

These statements will now be proved. The characteristic polynomial of the matrix ^2A is

$$\det(zE - {}^2A) = \det(zE - T\,{}^1A T^{-1}) =$$
$$= \det T(zE - {}^1A) T^{-1} =$$
$$= \det T \det(zE - {}^1A) \det T^{-1} =$$
$$= \det(zE - {}^1A) \tag{10.11}$$

Q. E. D.

From the foregoing it follows that the characteristic equation of the matrix and, consequently, its eigenvalues do not depend on the basis of the state space. Two similar matrices have the same trace and the same determinant because both these values appear as coefficients in the characteristic equation.

STATE SPACE THEORY OF DISCRETE LINEAR CONTROL

THEOREM 10.1: *The necessary and sufficient condition of similarity of two matrices ^1A and ^2A is their equal Jordan form.*

The proof of Theorem 10.1 may be found in the literature [66.2].

The basic system properties, specified in Sec. 9.1 through 9.3, do not alter with a change of basis either. For instance, the rank of a matrix of reachability does not alter because

$$[^2b, {}^2A\,{}^2b, \ldots, {}^2A^{n-1}\,{}^2b] = [T\,{}^1b, T\,{}^1AT^{-1}T\,{}^1b, \ldots, T\,{}^1A^{n-1}T^{-1}T\,{}^1b] =$$
$$= T[{}^1b, {}^1A\,{}^1b, \ldots, {}^1A^{n-1}\,{}^1b] \qquad (10.12)$$

since in general

$$^2A^k = (T\,{}^1AT^{-1})^k = (T\,{}^1AT^{-1})(T\,{}^1AT^{-1})\ldots(T\,{}^1AT^{-1}) = T\,{}^1A^k T^{-1} \quad (10.13)$$

The result (10.12) may be briefly written as follows:

$$^2Q_D = T\,{}^1Q_D \qquad (10.14)$$

The rank of a matrix of observability does not alter either, because

$$\begin{bmatrix} {}^2c^T \\ {}^2c^T\,{}^2A \\ \vdots \\ {}^2c^T\,{}^2A^{n-1} \end{bmatrix} = \begin{bmatrix} {}^1c^T T^{-1} \\ {}^1c^T T^{-1} T\,{}^1AT^{-1} \\ \vdots \\ {}^1c^T T^{-1} T\,{}^1A^{n-1}T^{-1} \end{bmatrix} = \begin{bmatrix} {}^1c^T \\ {}^1c^T\,{}^1A \\ \vdots \\ {}^1c^T\,{}^1A^{n-1} \end{bmatrix} T^{-1} \qquad (10.15)$$

or briefly

$$^2Q_P T = {}^1Q_P \qquad (10.16)$$

Similarly, for the identifiability we may write

$$[^2x(k), {}^2A\,{}^2x(k), \ldots, {}^2A^{n-1}\,{}^2x(k)] =$$
$$= [T\,{}^1x(k), T\,{}^1AT^{-1}T\,{}^1x(k), \ldots, T\,{}^1A^{n-1}T^{-1}T\,{}^1x(k)] =$$
$$= T[{}^1x(k), {}^1A\,{}^1x(k), \ldots, {}^1A^{n-1}\,{}^1x(k)] \qquad (10.17)$$

or briefly

$$^2Q_I = T\,{}^1Q_I \qquad (10.18)$$

Since the matrix T is non-singular, the rank of the matrices on both sides of eqns. (10.14), (10.16) and (10.18) is the same. Similar results would also be obtained when transforming the conditions of controllability and reconstructability. The relations given are useful in such cases where we know the system state equations in two different bases and seek the relevant transformation matrix. It holds that

$$T = {}^2Q_D\,{}^1Q_D^{-1} \quad \text{or} \quad Q = T^{-1} = {}^1Q_D\,{}^2Q_D^{-1}$$
$$T = {}^2Q_P^{-1}\,{}^1Q_P \quad \text{or} \quad Q = T^{-1} = {}^1Q_P^{-1}\,{}^2Q_P$$
$$T = {}^2Q_I\,{}^1Q_I^{-1} \quad \text{or} \quad Q = T^{-1} = {}^1Q_I\,{}^2Q_I^{-1} \qquad (10.19)$$

TRANSFORMATIONS

When transforming the conditions of stabilizability, let us start from the requirement that the matrix $[^2A - {^2B}\,{^2K}]$ should be similar to the matrix $[^1A - {^1B}\,{^1K}]$. Consequently

$$[^2A - {^2B}\,{^2K}] = T(^1A - {^1B}\,{^1K})\,T^{-1} = T\,{^1A}\,T^{-1} - T\,{^1B}\,{^1K}\,T^{-1} \quad (10.20)$$

Therefore it holds that

$$^2K = {^1K}\,T^{-1}$$

A similar procedure may be followed in the case of the conditions of detectability. In the subsequent paragraphs we shall show the most important canonical forms. Since with these transformations the matrix d or D in the output equation of the state description for a single-input/single-output or a multi-input/multi-output system does not change, we put, for simplicity, $d = 0$ or $D = 0$, respectively.

10.2 Jordan matrix

10.2.1 Distinct eigenvalues

Let us consider a discrete system described by state equations and let the matrix A of this system have only distinct mutually different eigenvalues. For this case we may write the state equations according to eqns. (7.16) through (7.18):

$$x_J(k+1) = A_J\,x_J(k) + b_J\,u(k)$$
$$y(k) = c_J^T\,x_J(k) \quad (10.21)$$

where

$$A_J = \begin{bmatrix} z_1, & 0, & \ldots, & 0 \\ 0, & z_2, & \ldots, & 0 \\ 0, & 0, & \ldots, & z_n \end{bmatrix}$$

is the so-called *Jordan matrix*.

Let us mention some important properties of the state equations (10.21) with the Jordan matrix A_J.

The characteristic polynomial has the form

$$\det(zE - A_J) = \prod_{i=1}^{n}(z - z_i) \quad (10.22)$$

The conditions of reachability may be expressed by means of the matrix

$$Q_{DJ} = [b_J,\, A_J b_J,\, \ldots,\, A_J^{n-1} b_J] = \begin{bmatrix} b_{J1}, & 0, & \ldots, & 0 \\ 0, & b_{J2}, & \ldots, & 0 \\ \vdots & & & \vdots \\ 0, & 0, & \ldots, & b_{Jn} \end{bmatrix} \begin{bmatrix} 1, & z_1, & \ldots, & z_1^{n-1} \\ 1, & z_2, & \ldots, & z_2^{n-1} \\ \vdots & & & \vdots \\ 1, & z_n, & \ldots, & z_n^{n-1} \end{bmatrix} \quad (10.23)$$

The right-hand matrix in eqn. (10.23) is the Vandermond matrix. This matrix is non-singular if all the eigenvalues are mutually different [66.2]. Therefore, the system is reachable if all elements of the column matrix b_J are non-zero. Similarly we could derive that the system is observable if all elements of the row matrix c_J^T are non-zero.

Since by introduction of the Jordan form of the matrix A the system is transformed into n partial first-order systems, such a system will be reachable and observable if and only if the input u is acting on all partial systems and if the output y is given by superposition of the outputs of all partial systems.

Any system with distinct eigenvalues may be transformed to the Jordan canonical form if we introduce, according to eqn. (10.1), the transformation $x_J = Tx$ or $x = T^{-1}x_J = Qx_J$. According to eqns. (10.10) it holds that

$$AQ = QA_J \qquad (10.24)$$

Expressing the matrix Q in terms of column vectors we obtain

$$A[q_1, q_2, ..., q_n] = [q_1, q_2, ..., q_n] \begin{bmatrix} z_1, & 0, & ..., & 0 \\ 0, & z_2, & ..., & 0 \\ & \cdots \cdots \cdots & \\ 0, & 0, & ..., & z_n \end{bmatrix} =$$

$$= [q_1 z_1, q_2 z_2, ..., q_n z_n] \qquad (10.25)$$

For the individual eigenvalues we have

$$Aq_i = z_i q_i, \quad i = 1, 2, ..., n \qquad (10.26)$$

and $q_i \neq 0$ is called the *eigenvector* of the matrix A, corresponding to the eigenvalue z_i. Here the matrix A, obtained by arranging the eigenvectors one next to another, is a non-singular matrix because the eigenvectors corresponding to different eigenvalues are always linearly independent [66.2].

From eqn. (10.2) it follows that

$$[z_i E - A] q_i = 0 \qquad (10.27)$$

Since $\det [z_i E - A] = 0$, eqn. (10.26) determines only the direction of the vector q_i. Its magnitude may be suitably chosen, e.g. so that the elements of b_J or c_J are equal to one or zero.

Example 10.1: Calculate the Jordan canonical form of state equations with unit elements of the matrix b_J for a system determined by the following equations:

$$x(k+1) = \begin{bmatrix} 1, & 0, & 1 \\ -2, & 1, & 0 \\ 0, & 1, & 2 \end{bmatrix} x(k) + \begin{bmatrix} 0 \\ 0 \\ 10 \end{bmatrix} u(k)$$

$$y(k) = [1, 0, 0] x(k)$$

The system has the characteristic equation

$$\det[zE - A] = z^3 - 4z^2 + 5z = 0$$

whose roots are $z_1 = 0$, $z_2 = 2 + j$, $z_3 = 2 - j$. The eigenvector corresponding to the root z_1 is calculated according to eqn. (10.27):

$$\begin{bmatrix} 1, & 0, & 1 \\ -2, & 1, & 0 \\ 0, & 1, & 2 \end{bmatrix} \begin{bmatrix} q_{11} \\ q_{21} \\ q_{31} \end{bmatrix} = \begin{bmatrix} 0 \\ 0 \\ 0 \end{bmatrix}; \quad q_1 = \begin{bmatrix} 1 \\ 2 \\ -1 \end{bmatrix}$$

Analogously we calculate

$$q_2 = \begin{bmatrix} 1 \\ -1 + j \\ 1 + j \end{bmatrix}; \quad q_3 = \begin{bmatrix} 1 \\ -1 - j \\ 1 - j \end{bmatrix}$$

The transformation matrix Q is

$$Q = [q_1, q_2, q_3] = \begin{bmatrix} 1, & 1, & 1 \\ 2, & -1 + j, & -1 - j \\ -1, & 1 + j, & 1 - j \end{bmatrix} \begin{bmatrix} q_{11}, & 0, & 0 \\ 0, & q_{12}, & 0 \\ 0, & 0, & q_{13} \end{bmatrix}$$

where q_{11}, q_{12}, q_{13} are optional magnitudes of the vectors q_1, q_2, q_3.

$$b_J = Q^{-1} b = \begin{bmatrix} q_{11}^{-1}, & 0, & 0 \\ 0, & q_{12}^{-1}, & 0 \\ 0, & 0, & q_{13}^{-1} \end{bmatrix} \begin{bmatrix} \frac{2}{5}, & \frac{1}{5}, & -\frac{1}{5} \\ \frac{3}{10} + \frac{1}{10}j, & -\frac{1}{10} - \frac{1}{5}j, & \frac{1}{10} - \frac{3}{10}j \\ \frac{3}{10} - \frac{1}{10}j, & -\frac{1}{10} + \frac{1}{5}j, & \frac{1}{10} + \frac{3}{10}j \end{bmatrix} \begin{bmatrix} 0 \\ 0 \\ 10 \end{bmatrix} =$$

$$= \begin{bmatrix} -2 q_{11}^{-1} \\ (1 - 3j) q_{12}^{-1} \\ (1 + 3j) q_{13}^{-1} \end{bmatrix}$$

$$c_J^T = {}^1 c^T Q = [1, 0, 0] \begin{bmatrix} q_{11}, & q_{12}, & q_{13} \\ 2q_{11}, & (-1 + j) q_{12}, & (-1 - j) q_{13} \\ -q_{11}, & (1 + j) q_{12}, & (1 - j) q_{13} \end{bmatrix} = [f_{11}, f_{12}, f_{13}]$$

If we choose $q_{11} = -2$, $q_{12} = (1 - 3j)$ and $q_{13} = (1 + 3j)$, the Jordan canonical form of the state equations will be

$$x(k + 1) = \begin{bmatrix} 0, & 0, & 0 \\ 0, & 2 + j, & 0 \\ 0, & 0, & 2 - j \end{bmatrix} x(k) + \begin{bmatrix} 1 \\ 1 \\ 1 \end{bmatrix} u(k)$$

$$y(k) = [-2, (1 - 3j), (1 + 3j)] x(k)$$

If the system is reachable, the transformation matrix Q may also be calculated according to eqn. (10.19):

$$Q^* = Q_D Q_{DJ}^{-1}$$

where

$$Q_D = [b, Ab, A^2b] = \begin{bmatrix} 0, & 10, & 30 \\ 0, & 0, & -20 \\ 10, & 10, & 40 \end{bmatrix}$$

For the unit elements of the vector b we have, according to eqn. (10.23),

$$Q_{DJ}^{-1} = \begin{bmatrix} 1, & 0, & 0 \\ 1, & 2+j, & 3+4j \\ 1, & 2-j, & 3-4j \end{bmatrix}^{-1} = \begin{bmatrix} 1, & 0, & 0 \\ -\frac{4}{5}, & \frac{1}{10}(4+3j), & \frac{1}{10}(4-3j) \\ -\frac{1}{5}, & \frac{1}{10}(-1-2j), & \frac{1}{10}(-1+2j) \end{bmatrix}$$

$$Q^* = 10 \begin{bmatrix} -\frac{1}{5}, & \frac{1}{10}(1-3j), & \frac{1}{10}(1+3j) \\ -\frac{2}{5}, & \frac{2}{10}(1+2j), & \frac{2}{10}(1-2j) \\ \frac{1}{5}, & \frac{2}{10}(2-j), & \frac{2}{10}(2+j) \end{bmatrix} \begin{bmatrix} q_{11}^*, & 0, & 0 \\ 0, & q_{12}^*, & 0 \\ 0, & 0, & q_{13}^* \end{bmatrix}$$

If we choose

$$q_{11}^* = -\frac{5}{10}, \quad q_{12}^* = \frac{1}{10}(1+3j), \quad q_{13}^* = \frac{1}{10}(1-3j)$$

we obtain the same transformation matrix $Q = [q_1, q_2, q_3]$ as in the first alternative of the procedure, where we have calculated it by means of the eigenvectors of the matrix A. The further procedure for the determination of the Jordan canonical form of state equations remains unchanged.

The relation between the Jordan canonical form of state equations with distinct eigenvalues of the matrix A and the transfer function is derived according to eqn. (7.31):

$$G(z) = c_J^T (zE - A_J)^{-1} b_J = \sum_{i=1}^n \frac{c_{Ji} b_{Ji}}{z - z_i} \qquad (10.28)$$

The transfer function $G(z)$ refers only to the reachable part of the system and, therefore, none of the elements of the matrix b_J can be equal to zero. If all elements of b_J are chosen to be ones then the elements of the row matrix c_J^T are

$$c_{Ji} = \operatorname{Res} G(z)|_{z=z_i} = \lim_{z \to z_i} (z - z_i) G(z) \qquad (10.29)$$

From the above relations follows the relationship between the Jordan canonical form of state equations and the transfer function decomposition into partial fractions as already pointed out in Sec. 7.3. We shall come back to this relationship once more

TRANSFORMATIONS

in this chapter as soon as we become acquainted with the calculation of the Jordan matrix with multiple eigenvalues.

A disadvantage of the formula (10.28) and a general disadvantage of the Jordan canonical form is the need to calculate the eigenvalues of A. Basically, we prefer such calculation procedures where knowledge of the eigenvalues of A is not required. If, however, these eigenvalues for a continuously working system are known, one can easily calculate $\exp[F_J T]$ since

$$A_J = e^{F_J T} = \begin{bmatrix} e^{p_1 T}, & 0, & \ldots, & 0 \\ 0, & e^{p_2 T}, & \ldots, & 0 \\ \multicolumn{4}{c}{\dotfill} \\ 0, & 0, & \ldots, & e^{p_n T} \end{bmatrix} \qquad (10.30)$$

10.2.2 Multiple eigenvalues

Let us now be concerned with the case of multiple eigenvalues. As already stated in connection with eqn. (10.26), the matrix A with distinct eigenvalues z_i, $i = 1, 2, \ldots$ $\ldots, n$, has n linearly independent eigenvectors. If the matrix A has multiple eigenvalues and the overall order of the system is n, then, in some cases, we may also find n mutually independent eigenvectors. However, it can easily be proved that such a system is not completely reachable and observable. It therefore contains, according to the conclusions of Par. 9.1.1, two parallel-coupled members with equal denominators. The corresponding state variables of these members can be neither controlled nor observed. We shall show these properties by an example.

Example 10.2: We calculate the Jordan canonical form of the state equation of a system described by the following equations:

$$x(k+1) = \begin{bmatrix} -0.5, & 1, & -1 \\ -0.5, & 1, & -0.5 \\ 0, & 0, & 0.5 \end{bmatrix} x(k) + \begin{bmatrix} 6 \\ 5 \\ 2 \end{bmatrix} u(k)$$

$$y(k) = [2, -1, 2] x(k)$$

Using the transformation matrix

$$Q = \begin{bmatrix} 2q_{11}, & 0, & -q_{13} \\ q_{11}, & q_{12}, & 0 \\ 0, & q_{12}, & q_{13} \end{bmatrix}$$

with which

$$x(k) = Q x_J(k)$$

we calculate the Jordan canonical form of state equations

$$x_J(k+1) = \begin{bmatrix} 0, & 0, & 0 \\ 0, & 0.5, & 0 \\ 0, & 0, & 0.5 \end{bmatrix} x_J(k) + \begin{bmatrix} 3q_{11}^{-1} \\ 2q_{12}^{-1} \\ 0 \end{bmatrix} u(k)$$

$$y(k) = [3q_{11}, \ q_{12}, \ 0] \, x_J(k)$$

The result of the solution shows that the component corresponding to the eigenvalue $z_3 = 0.5$ is not reachable because $b_{J3} = 0$ and is not observable because $c_{J3} = 0$. We therefore design the control using only a second-order system described by the equations

$$x_J(k+1) = \begin{bmatrix} 0, & 0 \\ 0, & 0.5 \end{bmatrix} x_J(k) + \begin{bmatrix} 3q_{11}^{-1} \\ 2q_{12}^{-1} \end{bmatrix} u(k)$$

$$y(k) = [3q_{11}, \ q_{12}] \, x_J(k)$$

which, as may be ascertained, is reachable as well as observable.

In addition it should be noted that the individual columns of the transformation matrix used are linearly independent, but the number of transformation matrices of this type is not finite.

As in Example 10.1, we may select such values of the individual elements of the vector b_J that this vector be a unit vector. For $q_{11} = 3$ and $q_{12} = 2$ we have $b_J^T = [1, 1]$ and $c_J^T = [9, 2]$.

It may, therefore, be summarized that any system whose matrix A can be transformed into a diagonal matrix A_J, with at least two eigenvalues on the diagonal of A_J being equal, has identical partial systems which cannot be independently controlled by a simple input signal and the output variables of these equivalent partial systems cannot be mutually distinguished.

More important is the case where the matrix A of the system has multiple eigenvalues but cannot be transformed into a diagonal form because it has only m, $(m < n)$, linearly independent eigenvectors. In this case the corresponding Jordan matrix has the general form

$$J = \begin{bmatrix} J_1, & 0, & \ldots, & 0 \\ 0, & J_2, & \ldots, & 0 \\ \multicolumn{4}{c}{\ldots\ldots\ldots\ldots} \\ 0, & 0, & \ldots, & J_m \end{bmatrix}; \quad J_i = \begin{bmatrix} z_i, & 1, & \ldots, & 0 \\ & z_i, & 1, & \\ & & \ldots\ldots & \\ & & z_i, & 1 \\ 0, & \ldots & & z_i \end{bmatrix}; \quad i = 1, 2, \ldots, m \qquad (10.31)$$

The individual blocks on the diagonal of the matrix J are called *Jordan blocks*. If all Jordan blocks have the dimension equal to one, then all eigenvalues of the Jordan matrix will be non-multiple and mutually different. If the number of Jordan blocks is equal to the number of eigenvalues z_i, then such a Jordan matrix is called *cyclic* and the corresponding matrix A which was transformed into the cyclic Jordan matrix is called cyclic as well [69.4].

Note that according to this definition a cyclic matrix can have altogether only non-multiple different eigenvalues or also multiple different eigenvalues, but only one Jordan block may correspond to each eigenvalue regardless of its multiplicity.

In this section only some of the basic properties of the Jordan forms (10.31) will be mentioned. For more details see Ref. [60.3, 66.2].

First of all it is useful to know the method of calculation of the Jordan matrix (10.31). It holds that there exists at least one eigenvector q_i corresponding to each eigenvalue z_i; this eigenvector is calculated from the already known equation

$$(A - z_i E) q_i = 0 \tag{10.32}$$

If eqn. (10.32) yields for a p-multiple eigenvalue z_i less than p linearly independent eigenvectors, then there exists at least one eigenvector which starts the *chain* of the so-called *generalized eigenvectors* $q_{i+1}, q_{i+2}, \ldots, q_{i+k_i-1}$, which are calculated with the aid of the relations

$$(A - z_i E) q_{i+1} = q_i$$
$$\vdots$$
$$(A - z_i E) q_{i+k_i-1} = q_{i+k_i-2} \tag{10.33}$$

It is to be noted that for a p-multiple eigenvalue the sum of the eigenvectors and generalized eigenvectors is equal to p. These vectors form the transformation matrix Q_i. Each eigenvector gives one Jordan block of the form (10.31) whose dimension is $(k; k)$, i.e. is equal to the number of the corresponding linearly independent vectors $q_i, q_{i+1}, \ldots, q_{i+k_i-1}$.

It should be emphasized that the eigenvectors as well as the generalized eigenvectors are mutually linearly independent.

It holds that

$$AQ = QJ$$

$$A[q_1, q_2, \ldots, q_{k_i}, \ldots, q_m] =$$

$$= [q_1, q_2, \ldots, q_{k_i}, \ldots, q_m] \begin{bmatrix} \overbrace{z_1, 1, \quad \ldots, 0}^{k_1} & & \\ \quad z_1, 1, & & 0 \\ \quad \cdots\cdots\cdots\cdots & & \\ \quad\quad\quad z_1, 1 & & \\ 0, \ldots, \quad\quad z_1 & & \\ \hline & J_2, \ldots, 0 & \\ 0 & 0, \ldots, J_m & \end{bmatrix} \tag{10.34}$$

This equation gives, in accordance with eqns. (10.32) and (10.38), the relations

$$Aq_1 = z_1 q_1$$
$$Aq_2 = q_1 + z_1 q_2$$
$$\vdots$$
$$Aq_{k_1} = q_{k_1-1} + z_1 q_{k_1} \qquad (10.35)$$

The structure of the individual Jordan blocks evidently depends on the length of the generalized eigenvector chains corresponding to one eigenvector. They are calculated by attempting to find for each eigenvector appropriate to a multiple eigenvalue as many generalized eigenvectors as possible. In doing this we should keep in view that the individual vectors must be mutually linearly independent and that the sum of the eigenvectors and generalized eigenvectors must be equal to the multiplicity of the relevant eigenvalue. The method of calculation will be shown by an example.

Example 10.3: Calculate the Jordan form of the state equations

$$x(k+1) = \begin{bmatrix} 0, & 1, & 0, & 0, & 0 \\ -1, & -1, & 1, & 0, & 0 \\ 1, & 0, & -2, & 0, & 0 \\ -1, & 0, & 1, & -1, & 0 \\ 1, & 0, & -1, & 4, & 3 \end{bmatrix} x(k) + \begin{bmatrix} 1 \\ 1 \\ 2 \\ 2 \\ 1 \end{bmatrix} u(k) =$$

$$= A x(k) + b u(k)$$

$$y(k) = [3, \ 1, \ 1, \ 1, \ 2] \, x(k) = c^T x(k)$$

According to eqn. (7.32) we calculate the characteristic polynomial

$$\det [zE - A] = (z+1)^4 (z-3)$$

For the eigenvalue $z_1 = -1$ we calculate, by (10.32),

$$(A - z_1 E) q_1 = \begin{bmatrix} 1, & 1, & 0, & 0, & 0 \\ -1, & 0, & 1, & 0, & 0 \\ 1, & 0, & -1, & 0, & 0 \\ -1, & 0, & 1, & 0, & 0 \\ 1, & 0, & -1, & 4, & 4 \end{bmatrix} \begin{bmatrix} q_{11} \\ q_{12} \\ q_{13} \\ q_{14} \\ q_{15} \end{bmatrix} = 0$$

It is clear that this relation is satisfied by the following two mutually linearly independent eigenvectors

$$u = u_0 \begin{bmatrix} 1 \\ -1 \\ 1 \\ -1 \\ 1 \end{bmatrix}; \quad v = v_0 \begin{bmatrix} 0 \\ 0 \\ 0 \\ 1 \\ -1 \end{bmatrix}$$

TRANSFORMATIONS

The Jordan matrix of a given system, A_J, will therefore have two independent blocks with the eigenvalue $z_1 = -1$. Now we attempt to find for the two eigenvectors $[1, -1, 1, -1, 1]^T$ and $[0, 0, 0, 1, -1]^T$ as many linearly independent generalized eigenvectors as many of them correspond to the multiplicity of z_1 reduced by the number of linearly independent eigenvectors corresponding to the same multiple eigenvalue. In this example the number of generalized eigenvectors will be $p - 2 = 2$ since the multiplicity of z_1 is $p = 4$. From the eigenvectors and generalized eigenvectors we then determine the transformation matrix Q and finally the sought Jordan matrix

$$J = Q^{-1}AQ$$

According to the second of eqns. (10.35) we have

$$(A - z_1 E) q_2 = \begin{bmatrix} 1, & 1, & 0, & 0, & 0 \\ -1, & 0, & 1, & 0, & 0 \\ 1, & 0, & -1, & 0, & 0 \\ -1, & 0, & 1, & 0, & 0 \\ 1, & 0, & -1, & 4, & 4 \end{bmatrix} \begin{bmatrix} q_{21} \\ q_{22} \\ q_{23} \\ q_{24} \\ q_{25} \end{bmatrix} = u_1 \begin{bmatrix} 1 \\ -1 \\ 1 \\ -1 \\ 1 \end{bmatrix} + v_1 \begin{bmatrix} 0 \\ 0 \\ 0 \\ 1 \\ -1 \end{bmatrix}$$

It will be seen that the second and fourth equation of this system can only be satisfied if $v_1 = 0$. Therefore, the first eigenvector is

$$q_1 = u_1 \begin{bmatrix} 1 \\ -1 \\ 1 \\ -1 \\ 1 \end{bmatrix}$$

To the eigenvector q_1 corresponds a chain of generalized eigenvectors of which the first is

$$q_2 = u_2 \begin{bmatrix} 1 \\ -1 \\ 1 \\ -1 \\ 1 \end{bmatrix} + v_2 \begin{bmatrix} 0 \\ 0 \\ 0 \\ 1 \\ -1 \end{bmatrix} + u_1 \begin{bmatrix} 0 \\ 1 \\ -1 \\ 0 \\ 0 \end{bmatrix}$$

The next generalized eigenvector is determined from the equation

$$(A - z_1 E) q_3 = \begin{bmatrix} 1, & 1, & 0, & 0, & 0 \\ -1, & 0, & 1, & 0, & 0 \\ 1, & 0, & -1, & 0, & 0 \\ -1, & 0, & 1, & 0, & 0 \\ 1, & 0, & -1, & 4, & 4 \end{bmatrix} \begin{bmatrix} q_{31} \\ q_{32} \\ q_{33} \\ q_{34} \\ q_{35} \end{bmatrix} = u_2 \begin{bmatrix} 1 \\ -1 \\ 1 \\ -1 \\ 1 \end{bmatrix} + v_2 \begin{bmatrix} 0 \\ 0 \\ 0 \\ 1 \\ -1 \end{bmatrix} + u_1 \begin{bmatrix} 0 \\ 1 \\ -1 \\ 0 \\ 0 \end{bmatrix}$$

This equation is satisfied by the generalized eigenvector

$$q_3 = u_3 \begin{bmatrix} 1 \\ -1 \\ 1 \\ -1 \\ 1 \end{bmatrix} + v_3 \begin{bmatrix} 0 \\ 0 \\ 0 \\ 1 \\ -1 \end{bmatrix} + u_2 \begin{bmatrix} 0 \\ 1 \\ -1 \\ 0 \\ 0 \end{bmatrix} + u_1 \begin{bmatrix} 0 \\ 0 \\ 1 \\ 0 \\ 0 \end{bmatrix}$$

if $v_2 = u_1$.

The next equation

$$(A - z_1 E) q_4 = q_3$$

gives no further linearly independent generalized eigenvector. To the quadruple eigenvalue $z_1 = -1$ therefore corresponds the vector chain q_1, q_2, q_3 and a further linearly independent eigenvector q_4:

$$q_4 = u_4 \begin{bmatrix} 1 \\ -1 \\ 1 \\ -1 \\ 1 \end{bmatrix} + v_4 \begin{bmatrix} 0 \\ 0 \\ 0 \\ 1 \\ -1 \end{bmatrix}$$

The eigenvector corresponding to the eigenvalue $z_2 = 3$ is again calculated according to eqn. (10.32):

$$(A - z_2 E) q_5 = \begin{bmatrix} -3, & 1, & 0, & 0, & 0 \\ -1, & -4, & 1, & 0, & 0 \\ 1, & 0, & -5, & 0, & 0 \\ -1, & 0, & 1, & -4, & 0 \\ 1, & 0, & -1, & 4, & 4 \end{bmatrix} \begin{bmatrix} q_{51} \\ q_{52} \\ q_{53} \\ q_{54} \\ q_{55} \end{bmatrix} = 0$$

This equation is satisfied by the eigenvector

$$q_5 = u_5 \begin{bmatrix} 0 \\ 0 \\ 0 \\ 0 \\ 1 \end{bmatrix}$$

The transformation matrix Q has, therefore, the following general form

$$Q = \begin{bmatrix} u_1 & u_2 & u_3 & u_4 & 0 \\ -u_1 & -u_2 + u_1 & -u_3 + u_2 & -u_4 & 0 \\ u_1 & u_2 - u_1 & u_3 - u_2 + u_1 & u_4 & 0 \\ -u_1 & -u_2 + v_2 & -u_3 + v_3 & -u_4 + v_4 & 0 \\ u_1 & u_2 - v_2 & u_3 - v_3 & u_4 - v_4 & u_5 \end{bmatrix}$$

TRANSFORMATIONS

The coefficients in the matrix Q may be chosen arbitrarily as far as the transformation matrix remains non-singular. Let us choose $u_1 = v_2 = v_4 = u_5 = 1$ and $u_2 = u_3 = v_3 = u_4 = 0$. With these values we obtain

$$Q = \begin{bmatrix} 1, & 0, & 0, & 0, & 0 \\ -1, & 1, & 0, & 0, & 0 \\ 1, & -1, & 1, & 0, & 0 \\ -1, & 1, & 0, & 1, & 0 \\ 1, & -1 & 0 & -1 & 1 \end{bmatrix} \text{ and } Q^{-1} = \begin{bmatrix} 1, & 0, & 0, & 0, & 0 \\ 1, & 1, & 0, & 0, & 0 \\ 0, & 1, & 1, & 0, & 0 \\ 0, & -1, & 0, & 1, & 0 \\ 0 & 0 & 0 & 1 & 1 \end{bmatrix}$$

According to eqns. (10.19) the matrix $Q = T^{-1}$. Using this substitution and denoting the original triad by (A, b, c^T) and the Jordan triad by (J, b_J, c_J^T) we calculate, with the aid of formulae (10.10),

$$J = Q^{-1}AQ = \begin{bmatrix} z_1, & 1, & 0, & 0, & 0 \\ 0, & z_1, & 1, & 0, & 0 \\ 0, & 0, & z_1, & 0, & 0 \\ 0, & 0, & 0, & z_1, & 0 \\ 0, & 0, & 0, & 0, & z_2 \end{bmatrix} = \begin{bmatrix} -1, & 1, & 0, & 0, & 0 \\ 0, & -1, & 1, & 0, & 0 \\ 0, & 0, & -1, & 0, & 0 \\ 0, & 0, & 0, & -1, & 0 \\ 0, & 0, & 0, & 0, & 3 \end{bmatrix}$$

$$b_J^T = (Q^{-1}b)^T = [1, 2, 3, 1, 3]$$

$$c_J^T = c^T Q = [4, -1, 1, -1, 2]$$

The Jordan form of state equations is, therefore,

$$x_J(k+1) = \begin{bmatrix} -1, & 1, & 0, & 0, & 0 \\ 0, & -1, & 1, & 0, & 0 \\ 0, & 0, & -1, & 0, & 0 \\ 0, & 0, & 0, & -1, & 0 \\ 0, & 0, & 0, & 0, & 3 \end{bmatrix} x_J(k) + \begin{bmatrix} 1 \\ 2 \\ 3 \\ 1 \\ 3 \end{bmatrix} u(k)$$

$$y(k) = [4, -1, 1, -1, 2] x_J(k)$$

When choosing the coefficients of the transformation matrix Q we did not take account of the resultant values of the elements of b_J and c_J^T. It is, of course, useful to choose the coefficients of Q so as to ensure that at least b_J or c_J^T assumes the simplest form. Before proceeding to the specific solution of the given example we acquaint ourselves with some general results.

1. It can be shown, as in the case of the diagonal form of the Jordan matrix, that a system is reachable and observable if all Jordan blocks have different eigen-

values, in other words, if the system is cyclic. With several Jordan blocks having the same eigenvalue it can be ensured that none of the blocks, except for that of the highest order, is affected by the input without at the same time affecting the output. In Example 10.3 this means that we may call for a zero value of the elements b_{J4} and c_{J4}^T.

2. If in the characteristic polynomial of the matrix we have each eigenvalue with only the multiplicity corresponding to the appropriate Jordan block of the highest order we obtain the *minimum polynomial of the matrix*. Systems occurring in practice are almost exclusively cyclic so that the characteristic polynomial is identical with the minimum polynomial of the system.

In our Example 10.3 the characteristic polynomial is $(z + 1)^4 (z - 3)$ and the minimum polynomial is $(z + 1)^3 (z - 3)$.

3. A partial system appropriate to one Jordan block with the eigenvalue z_i is reachable if and only if the last element b_{Jk_i} of the column matrix b_{Ji} is non-zero. This statement follows from the condition of reachability which, for the i-th Jordan block, is written in the form

$$Q_{DJi} = [b_{Ji}, J_i b_{Ji}, \ldots, J_i^{k_i-1} b_{Ji}] =$$

$$= \begin{bmatrix} b_{J1}, & b_{J2}, & \ldots & b_{Jk_i} \\ b_{J2}, & b_{J3}, & \ldots & 0 \\ \cdots & \cdots & \cdots & \cdots \\ b_{Jk_i}, & 0, & \ldots & 0 \end{bmatrix} \begin{bmatrix} 1, & z_i, & z_i^2, & \cdots & z_i^{k_i-1} \\ 0, & 1, & 2z_i, & \cdots & \binom{k_i-1}{1} z_i^{k_i-2} \\ \cdots & \cdots & \cdots & \cdots & \cdots \\ 0, & 0, & 0, & \ldots 1, & \binom{k_i-1}{k_i-2} z_i \\ 0, & 0, & 0, & \cdots & 1 \end{bmatrix} \quad (10.36)$$

The matrix of reachability Q_{DJi} is non-singular if and only if $b_{Jk_i} \neq 0$. For a reachable Jordan block we may, therefore, require that $b_{J1} = b_{J2} = \ldots = b_{J(k_i-1)} = 0$ and $b_{Jk_i} = 1$. In our Example 10.3 the first and third block is reachable and, therefore, the vector b_J could have the form

$$b_J = \begin{bmatrix} 0 \\ 0 \\ 1 \\ 0 \\ 1 \end{bmatrix}$$

4. The observability of a partial system corresponding to a single Jordan block with the eigenvalue z_i is satisfied if and only if the first element c_{J1} of the vector c_{Ji}^T is non-zero. This statement may be derived with the aid of the condition of observ-

TRANSFORMATIONS

ability which, for the i-th Jordan block, is written in the form

$$Q_{PJi} = \begin{bmatrix} c_{Ji}^T \\ c_{Ji}^T J_i \\ \vdots \\ c_{Ji}^T J_i^{k_i-1} \end{bmatrix} =$$

$$= \begin{bmatrix} 1, & 0, & \dots & & & 0 \\ z, & 1, & \dots & & & 0 \\ \hdotsfor{6} \\ z^{k_i-2}, & \binom{k_i-2}{1} z^{k_i-3}, & \dots, & \binom{k_i-2}{k_i-3} z, & 1, & 0 \\ z^{k_i-1}, & \binom{k_i-1}{1} z^{k_i-2}, & \dots, & \binom{k_i-1}{k_i-2} z, & 1 \end{bmatrix} \begin{bmatrix} c_{J1}^T, & c_{J2}^T, & \dots & c_{Jk_i}^T \\ 0, & c_{J1}^T, & \dots & c_{J(k_i-1)}^T \\ \hdotsfor{4} \\ \\ 0, & 0, & \dots & c_{J1}^T \end{bmatrix}$$

(10.37)

In Example 10.3 the first and third block is observable so that we may require the vector c_J^T to be in the form

$$c_J^T = [1, 0, 0, 0, 1]$$

Example 10.4: For the system in Example 10.3 determine such coefficients of the transformation matrix Q that the vectors c_J^T and b_J assume the simplest possible form.

In solving this problem we cannot choose the coefficients in Q arbitrarily but with regard to the given task. To facilitate the solution we rewrite the matrix Q in the form of a product of two matrices

$$Q = \begin{bmatrix} 1, & 0, 0, & 0, 0 \\ -1, & 1, 0, & 0, 0 \\ 1, & -1, 1, & 0, 0 \\ -1, & 0, 0, & 1, 0 \\ 1, & 0, 0, & -1, 1 \end{bmatrix} \begin{bmatrix} u_1, u_2, u_3, u_4, 0 \\ 0, u_1, u_2, 0, 0 \\ 0, 0, u_1, 0, 0 \\ 0, v_2, v_3, v_4, 0 \\ 0, 0, 0, 0, u_5 \end{bmatrix} = {}^1Q\, {}^2Q$$

calculate the product

$$c^{T\,1}Q = [4, 0, 1, -1, 2]$$

and set up the equations

$$[4, 0, 1, -1, 2] \begin{bmatrix} u_1, u_2, u_3, u_4, 0 \\ 0, u_1, u_2, 0, 0 \\ 0, 0, u_1, 0, 0 \\ 0, v_2, v_3, v_4, 0 \\ 0, 0, 0, 0, u_5 \end{bmatrix} = [1, 0, 0, 0, 1]$$

$$\begin{bmatrix} 1 \\ 1 \\ 2 \\ 2 \\ 1 \end{bmatrix} = \begin{bmatrix} 0.25 & u_2 & u_3 & u_4 & 0 \\ -0.25 & -u_2+0.25 & -u_3+u_2 & -u_4 & 0 \\ 0.25 & u_2-0.25 & u_3-u_2+0.25 & u_4 & 0 \\ -0.25 & -u_2+v_2 & -u_3+v_3 & -u_4+v_4 & 0 \\ 0.25 & u_2-v_2 & u_3-v_3 & u_4-v_4 & 0.5 \end{bmatrix} \begin{bmatrix} b_{J1} \\ b_{J2} \\ b_{J3} \\ b_{J4} \\ b_{J5} \end{bmatrix}$$

where the values $u_1 = 0.25$ and $u_5 = 0.5$ could have been substituted in the last equation as they follow directly from the preceding equation.

The total number of unknowns is 13 and the number of equations is 11 because we also have to satisfy the condition $u_1 = v_2$ corresponding to the generalized eigenvector q_3. Thus the number of unknowns is reduced to 12. In addition we may require that $b_{J4} = 0$ and then the number of unknowns will be identical with the number of equations. By a successive solution of these equations we obtain

$$b_{J1} = 4, \quad u_1 = 0.25, \qquad\qquad v_2 = 0.25$$
$$b_{J2} = 5, \quad u_2 = 1/16 = 0.0625, \qquad v_3 = 7/48 = 0.14583$$
$$b_{J3} = 12, \quad u_3 = -5/192 = -0.0260416, \quad v_4 = 1$$
$$b_{J4} = 0, \quad u_4 = 0.25$$
$$b_{J5} = 6, \quad u_5 = 0.5$$

The transformation matrix has now the form

$$Q = \begin{bmatrix} 0.25, & 0.0625, & -0.0260416, & 0.25, & 0 \\ -0.25, & 0.1875, & 0.0885416, & -0.25, & 0 \\ 0.25, & -0.1875, & 0.1614583, & 0.25, & 0 \\ -0.25, & 0.1875, & 0.1718749, & 0.75, & 0 \\ 0.25, & -0.1875, & -0.1718749, & 1.25, & 0.5 \end{bmatrix}$$

and the matrix

$$^2Q = \begin{bmatrix} 0.25, & 0.0625, & -0.0260425, & 0.25, & 0 \\ 0, & 0.25, & 0.0625, & 0, & 0 \\ 0, & 0, & 0.25, & 0, & 0 \\ 0, & 0.25, & 0.14583, & 1, & 0 \\ 0, & 0, & 0, & 0, & 0.5 \end{bmatrix}$$

The Jordan form of state equations is now

$$x_J(k+1) = \begin{bmatrix} -1, & 1, & 0, & 0, & 0 \\ 0, & -1, & 1, & 0, & 0 \\ 0, & 0, & -1, & 0, & 0 \\ 0, & 0, & 0, & -1, & 0 \\ 0, & 0, & 0, & 0, & 3 \end{bmatrix} x_J(k) + \begin{bmatrix} 4 \\ 5 \\ 12 \\ 0 \\ 6 \end{bmatrix} u(k)$$

$$y(k) = [1, 0, 0, 0, 1] x_J(k)$$

TRANSFORMATIONS

From this result it is clear that the given system consists of a reachable and observable part with the characteristic polynomial $(z + 1)^3 (z - 3)$ and of an unreachable and unobservable part with the characteristic polynomial $(z + 1)$. For the purpose of control, of course, only the first part is of any use, which corresponds to the Jordan form of state equations

$$x_J(k + 1) = \begin{bmatrix} -1, & 1, & 0, & 0 \\ 0, & -1, & 1, & 0 \\ 0, & 0, & -1, & 0 \\ 0, & 0, & 0, & 3 \end{bmatrix} x_J(k) + \begin{bmatrix} 4 \\ 5 \\ 12 \\ 6 \end{bmatrix} u(k)$$

$$y(k) = [1, 0, 0, 1] x_J(k)$$

The block diagram of the complete system must be drawn in such a manner that the unreachable and unobservable part be in no way connected with the input and output of the reachable and observable part of the system. The relevant block diagram is shown in Fig. 10.1. The case of the unreachable and unobservable part may be physically interpreted as two identical blocks being coupled in the system in parallel.

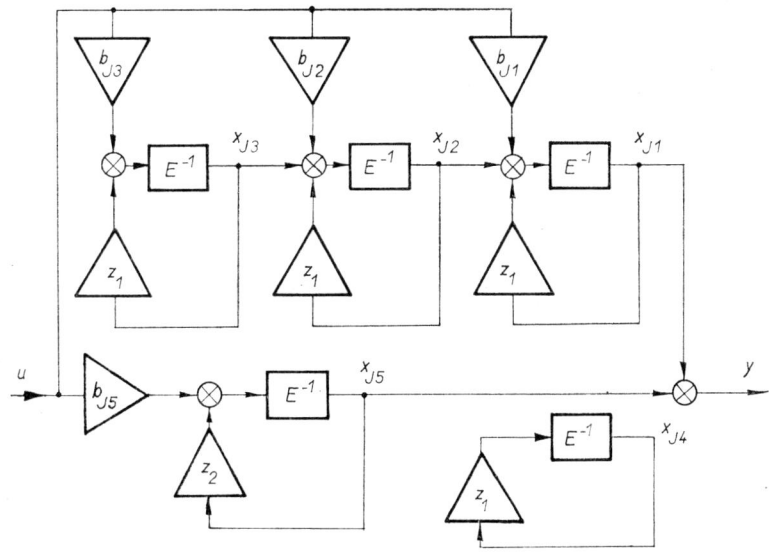

Fig. 10.1. Block diagram of the system in Example 10.4.

In addition it should be noted that the transformation of state equations to the Jordan form corresponds to the decomposition of the transfer function into partial fractions. Let us have the transfer function

$$G(z) = \frac{b_1 z^{n-1} + \ldots + b_n}{z^n + a_1 z^{n-1} + \ldots + a_n}$$

The individual coefficients a_i, b_i, $i = 1, 2, \ldots, n$, may be determined, for instance, by means of the Faddejev algorithm (see Par. 7.7.3). For the reachable and observable parts of Examples 10.3 and 10.4 we calculate

$$G(z) = \frac{4}{z+1} + \frac{5}{(z+1)^2} + \frac{12}{(z+1)^3} + \frac{6}{z-3}$$

The coefficients in the numerators of these partial fractions are equal to the elements of the column matrix b_J of the Jordan form of state equations corresponding to the reachable and observable part of the system. Conversely, the method of decomposition into partial fractions as indicated above may be used for the calculation of the Jordan form of state equations provided that the transfer function $G(z)$ is known. It should be noted that this method is more advantageous than that involving the determination of eigenvectors and generalized eigenvectors, described in this chapter, because the expressions for the coefficients a_i, b_i, $i = 1, 2, \ldots, n$, are very suitable for a computer-aided solution. However, this method can only be used for cyclic systems.

If, in a general case, the transfer function $G(z)$ has m roots with a multiplicity $n_1, n_2, \ldots, n_m$, where

$$\sum_{i=1}^{m} n_i = n$$

then

$$G(z) = \frac{b_n + b_{n-1}z + \ldots + b_1 z^{n-1}}{(z-z_1)^{n_1}(z-z_2)^{n_2} \ldots (z-z_m)^{n_m}} =$$

$$= \frac{\beta_{11}}{(z-z_1)} + \frac{\beta_{12}}{(z-z_1)^2} + \ldots + \frac{\beta_{1n_1}}{(z-z_1)^{n_1}} + \frac{\beta_{21}}{(z-z_2)} + \ldots + \frac{\beta_{mn_m}}{(z-z_m)^{n_m}}$$

and the corresponding Jordan form of state equations is

$$x(k+1) = \begin{bmatrix} z_1, 1, 0, & \ldots & 0 & 0, & \ldots & 0 & 0, & \ldots & 0 \\ 0, z_1, 1, & \ldots & 0 & & & & & & \\ \ldots & & & & & & & & & \\ 0, 0, & \ldots & z_1 & 0, & \ldots & 0 & 0, & \ldots & 0 \\ \hline 0, & \ldots & 0 & z_2, 1, & \ldots & 0 & 0, & \ldots & 0 \\ \ldots & & & & & & & & & \\ 0, & \ldots & & 0, & \ldots & z_2 & 0, & \ldots & 0 \\ \hline 0, & \ldots & 0 & 0, & \ldots & 0 & z_m, 1, & \ldots & 0 \\ \ldots & & & & & & & & & \\ 0, & \ldots & 0 & 0, & \ldots & 0 & 0, & \ldots & z_m \end{bmatrix} x(k) + \begin{bmatrix} \beta_{11} \\ \beta_{12} \\ \vdots \\ \beta_{1n_1} \\ \beta_{21} \\ \vdots \\ \beta_{mn_m} \end{bmatrix} u(k)$$

$$y(k) = [1, 0, \ldots 0 \mid 1, \ldots 0 \mid 1, \ldots 0] x(k) \qquad (10.38)$$

TRANSFORMATIONS

The vectors b and c^T in eqns. (10.38) can alternatively have the form

$$b^T = [0, \quad 0, \quad \ldots \quad 1 \quad | \quad 0, \quad \ldots \quad \| \quad \ldots \quad 1 \quad]$$
$$c^T = [\beta_{1n_1}, \quad \ldots \quad \beta_{11} \quad | \quad \beta_{2n_2}, \quad \ldots \quad \| \quad \ldots \quad \beta_{m1}] \quad (10.39)$$

10.2.3 Invariant factors

Let us now consider some additional Jordan matrix properties which will aid us in the solution of certain problems.

Let $A(z)$ be an n-th order polynomial matrix having the rank h. If we determine the *greatest common divisor* $D_r(z)$ of all r-th degree subdeterminants of the matrix $A(z)$ for $r = 1, 2, \ldots, h$, $D_0(z) = 1$, we can construct the so-called *invariant factors* (polynomials)

$$E_r(z) = \frac{D_r(z)}{D_{r-1}(z)}, \quad r = 1, 2, \ldots, n \quad (10.40)$$

being the elements of the so-called *canonical diagonal matrix* (normal diagonal matrix) of the form

$$\begin{bmatrix} E_1(z), & 0, & \ldots & 0, & 0, & \ldots, & 0 \\ 0, & E_2(z), & \ldots & 0, & 0, & \ldots, & 0 \\ \multicolumn{7}{c}{\ldots\ldots\ldots\ldots\ldots\ldots\ldots\ldots\ldots\ldots} \\ 0, & 0, & \ldots & E_h(z), & 0, & \ldots, & 0 \\ 0, & 0, & \ldots & 0, & 0, & \ldots, & 0 \\ \multicolumn{7}{c}{\ldots\ldots\ldots\ldots\ldots\ldots\ldots\ldots\ldots\ldots} \\ 0, & 0, & \ldots & 0, & 0, & \ldots, & 0 \end{bmatrix} \quad (10.41)$$

As far as the invariant factors $E_r(z)$ are non-zero, their coefficients at the highest power of z are equal to 1 and the polynomial $E_{r+1}(z_i)$ is divisible by the polynomial $E_r(z_i)$. The product

$$E_1(z) E_2(z) \ldots E_h(z) \neq 0, \quad E_{h+1}(z) = E_{h+2}(z) = \ldots = E_n(z) = 0$$

The factors $E_r(z)$ are called invariant for the following reason. If $A(z)$ and $B(z)$ are two equivalent matrices we can transform one to the other by means of elementary operations. In the course of these operations the greatest comon divisiors $D_r(z)$ and, consequently, the polynomials $E_r(z)$ do not change. We therefore speak of the factors (polynomials) $E_r(z)$, $r = 1, 2, \ldots, h$, as being invariant on transition from one matrix to another equivalent matrix. It may be noted that the matrix (10.41) is equivalent to the matrix $A(z)$ of the order n and rank h.

From eqn. (10.40) it follows that

$$D_r(z) = c\, E_1(z)\, E_2(z)\, \ldots\, E_r(z) \qquad (10.42)$$

where c is some non-zero real or complex number.

If we now decompose each (non-zero) invariant factor $E_r(z)$ into the product of the powers of different root factors $(z - z_i)^m$, then each such factor is referred to as the *elementary divisor* of the matrix $A(z)$.

Let us now consider an $(n; n)$ matrix A having the characteristic matrix

$$\mathcal{A}(z) = zE - A \qquad (10.43)$$

where $\mathcal{A}(z)$ has the rank $h = n$ and its invariant factors are the invariant factors of the matrix A.

The invariant factors of the matrix A enable us to decide whether A is cyclic, i.e. whether only one block corresponds to each eigenvalue of its Jordan form or not. It holds that:

THEOREM 10.2: *A matrix A of the order n and rank h is cyclic if its invariant factors $E_1(z) = \ldots = E_{n-1}(z) = 1$ and $E_n(z)$ is a polynomial in z.*

It can be proved that if two matrices, say A and J, are to be similar ($A = QJQ^{-1}$, where Q is a non-singular matrix), it is necessary and sufficient that they have identical invariant factors or, which is the same, identical elementary divisors, identical dimension and rank.

This property will be used to find the Jordan matrix corresponding to the matrix A. Let us take some polynomial

$$g(z) = z^m + a_{m-1}z^{m-1} + \ldots + a_1 z + a_0$$

Then we can establish the m-th order matrix

$$L = \begin{bmatrix} -a_{m-1}, & 1, & 0, & \ldots, & 0 \\ -a_{m-2}, & 0, & 1, & \ldots, & 0 \\ \multicolumn{5}{c}{\dotfill} \\ -a_1, & 0, & 0, & \ldots, & 1 \\ -a_0, & 0, & 0, & \ldots, & 0 \end{bmatrix}$$

and verify that the characteristic polynomial of L is

$$\det[zE - L] = \det \begin{bmatrix} a_{m-1} + z, & -1, & 0, & \ldots, & 0 \\ a_{m-2}, & z, & -1, & \ldots, & 0 \\ \multicolumn{5}{c}{\dotfill} \\ a_1, & 0, & 0, & \ldots, & -1 \\ a_0, & 0, & 0, & \ldots, & z \end{bmatrix} = g(z)$$

TRANSFORMATIONS

It can be seen that the minor corresponding to the characteristic matrix element a_0 is ± 1 and, therefore, the greatest common divisor

$$D_{m-1}(z) = 1$$

$$E_m(z) = D_m(z)/D_{m-1}(z) = D_m(z) = g(z)$$

$$E_1(z) = \ldots = E_{m-1}(z) = 1$$

Resolving the invariant factor $E_m(z)$ into the product of elementary divisors $\prod_{i=1}^{s}(z - z_i)^{n_{im}}$, where s is the number of different eigenvalues and n_{im} their multiplicity in the invariant factor $E_m(z)$, we can find the Jordan form of L consisting of type (10.21) or (10.31) blocks, depending on the multiplicity of the eigenvalues z_i. If we decompose all invariant factors $E_r(z)$, $r = 1, 2, \ldots, n$, of the matrix A in the way indicated, then the Jordan form of A is set up of all the Jordan matrices J_r, $r = 1, 2, \ldots, \varrho$, where ϱ is the number of elementary divisors of the matrix A. If the elementary divisors repeat in some invariant factors they must necessarily also repeat in the Jordan matrix A in order that the conditions of similarity be satisfied (see Ref. [60.3]).

Example 10.5: Consider a matrix A of the order $n = 10$ and rank $h = 10$ whose invariant factors are $E_1(z) = \ldots = E_7(z) = 1$, $E_8(z) = (z - 0.1)$, $E_9(z) = (z - 0.1)^2 (z + 0.3)^2$ and $E_{10}(z) = (z - 0.1)^2 (z + 0.3)^3$. Hence, the set of elementary divisors is $(z - 0.1)$, $(z - 0.1)^2$, $(z - 0.1)^2$, $(z + 0.3)^2$ and $(z + 0.3)^3$. The relevant Jordan matrix has the form

$$\begin{bmatrix}
0.1, & & & & & & & & & 0 \\
0, & 0.1, & 1 & & & & & & & 0 \\
0, & 0, & 0.1 & & & & & & & 0 \\
0, & \ldots & & 0.1, & 1 & & & & & 0 \\
0, & \ldots & & 0, & 0.1 & & & & & 0 \\
0, & \ldots & & & & -0.3, & 1 & & & 0 \\
0, & \ldots & & & & 0, & -0.3 & & & 0 \\
0, & \ldots & & & & & & -0,3, & 1, & 0 \\
0, & \ldots & & & & & & 0, & -0.3, & 1 \\
0, & \ldots & & & & & & 0, & 0, & -0.3
\end{bmatrix}$$

The individual blocks are demarcated by dashed lines. This Jordan matrix is not cyclic since it has blocks with equal eigenvalues, i.e. three blocks with the eigenvalue $z_1 = 0.1$ and two blocks with the eigenvalue $z_2 = -0.3$. That this matrix is non-cyclic can also be recognized from the canonical diagonal matrix which in our example has the form

$$\begin{bmatrix} 1, & 0, & \ldots & & & & & \ldots & 0 \\ 0, & 1, & \ldots & E_1(z) \text{ to } E_7(z) & & & & \ldots & 0 \\ \ldots & \ldots & & & & & & \ldots & \cdot \\ 0, & \ldots & 1 & & & & & \ldots & 0 \\ 0, & \ldots & & (z-0.1) & & & & \ldots & 0 \\ 0, & \ldots & & & (z-0.1)^2 (z+0.3)^2 & & & \ldots & 0 \\ 0, & \ldots & & & & & (z-0.1)^2 (z+0.3)^3 & & \end{bmatrix}$$

It can be seen that the last three invariant factors differ from 1.

Example 10.6: Transform the following matrix A to the Jordan form using the greatest common divisors!

$$A = \begin{bmatrix} 0.5, & 0, & 0, & 0, & \ldots & & & & \ldots & 0 \\ 0, & -0.5, & 1, & 0, & \ldots & & & & \ldots & 0 \\ 0, & -1, & 1.5, & 0, & \ldots & & & & \ldots & 0 \\ 0, & 0, & 0, & 0, & 0.5, & -0.5, & 0, & 0, & 0 \\ 0, & 0, & 0, & 0, & -0.5, & 1, & 0, & 0, & 0 \\ 0, & 0, & 0, & 0, & -1, & 1.5, & 0, & 0 & 0 \\ 0, & \ldots & & & \ldots & 0, & -0.5, & 1, & -1 \\ 0, & \ldots & & & \ldots & 0, & -0.5, & 1, & -0.5 \\ 0, & \ldots & & & \ldots & 0, & 0, & 0, & 0.5 \end{bmatrix}$$

The given matrix is quasidiagonal with three blocks which we denote by A_1, A_2, A_3 so that

$$A = \begin{bmatrix} A_1, & 0, & 0 \\ 0, & A_2, & 0 \\ 0, & 0, & A_3 \end{bmatrix}$$

In this case we can determine the greatest common divisors and the elementary divisors of each block separately. The characteristic matrix for the block A_1 is

$$A_1(z) = zE - A_1 = \begin{bmatrix} z - 0.5, & 0, & 0 \\ 0, & z + 0.5, & -1 \\ 0, & 1, & z - 1.5 \end{bmatrix}$$

It can easily be verified that the greatest common divisors and invariant factors are

$$D_1(z) = 1 \qquad E_1(z) = 1$$
$$D_2(z) = z - 0.5 \qquad E_2(z) = (z - 0.5)$$
$$D_3(z) = (z - 0.5)^3 \qquad E_3(z) = (z - 0.5)^2$$

The matrix A_1 is evidently not cyclic since $E_2 \neq 1$.

The characteristic matrix for the block A_2 is

$$A_2(z) = zE - A_2 = \begin{bmatrix} z, & -0.5, & 0.5 \\ 0, & z + 0.5, & -1 \\ 0, & 1, & z - 1.5 \end{bmatrix}$$

The greatest common divisors and invariant factors are

$$D_1(z) = 1 \qquad E_1(z) = 1$$
$$D_2(z) = 1 \qquad E_2(z) = 1$$
$$D_3(z) = z(z - 0.5)^2 \qquad E_3(z) = z(z - 0.5)^2$$

From Theorem 10.1 and from the calculated invariant factors it follows that the block A_2 represents a cyclic matrix. Finally, the characteristic matrix of the block A_3 is

$$A_3(z) = zE - A_3 = \begin{bmatrix} z + 0.5, & -1, & 1 \\ 0.5, & z - 1, & 0.5 \\ 0, & 0, & z - 0.5 \end{bmatrix}$$

The greatest common divisors and invariant factors of the characteristic matrix $A_3(z)$ are

$$D_1(z) = 1 \qquad E_1(z) = 1$$
$$D_2(z) = z - 0.5 \qquad E_2(z) = z - 0.5$$
$$D_3(z) = z(z - 0.5)^2 \qquad E_3(z) = z(z - 0.5)$$

The matrix A_3 is again not cyclic for the same reason as the matrix A_1.

The Jordan matrix of the given matrix A is determined by the blocks appropriate to the matrices A_1, A_2, A_3. We obtain

$$J = \begin{bmatrix} 0.5 & 0 & 0 & \cdots & & & & & \cdots & 0 \\ 0 & 0.5 & 1 & 0 & \cdots & & & & \cdots & 0 \\ 0 & 0 & 0.5 & 0 & \cdots & & & & \cdots & 0 \\ 0 & & & 0 & 0 & 0 & 0 & \cdots & & 0 \\ 0 & & & 0 & 0.5 & 1 & 0 & & \cdots & 0 \\ 0 & & & 0 & 0 & 0.5 & 0 & & \cdots & 0 \\ 0 & & & & & & 0 & 0 & & 0 \\ 0 & & & & & & 0 & 0.5 & & 0 \\ 0 & & & & & & 0 & 0 & & 0.5 \end{bmatrix}$$

The matrix A is not cyclic. It would be cyclic, however, if all the diagonal blocks A_1, A_2, A_3 were cyclic matrices.

10.2.4 Using the Jordan matrix for numerical calculations

The Jordan matrix form is suitable not only for the solution of theoretical problems in the analysis of system properties, but can be used to advantage even in certain numerical calculations. For instance, in the numerical calculation of $\exp[Ft]$ use is being made of the series expansion of this matrix function in the form

$$e^{Ft} = E + Ft + \frac{F^2 t^2}{2!} + \frac{F^3 t^3}{3!} + \cdots \qquad (10.44)$$

taking as many members of the series as necessary to obtain the desired result accuracy. The calculation of the powers of F is considerably simplified if the matrix F is transformed to the Jordan matrix. If the Jordan matrix is of the form (10.21) its k-th power is

$$J^k = \begin{bmatrix} z_1^k & 0 & \cdots & 0 \\ 0 & z_2^k & \cdots & 0 \\ \vdots & & & \vdots \\ 0 & 0 & \cdots & z_s^k \end{bmatrix} \qquad (10.45)$$

If the Jordan matrix is of the form (10.31), it can be divided into the sum of two matrices

$$J = I_z + I_1 \qquad (10.46)$$

TRANSFORMATIONS

where the matrix I_z has in the main diagonal the eigenvalues of the matrix F and the matrix I_1 has in its superdiagonal[1]) ones and zeros according to the matrix J. The k-th power is then calculated according to the binomial theorem

$$J^k = \sum_{i=0}^{k} \binom{k}{i} I_z^{k-i} I_1^i \qquad (10.47)$$

If the superdiagonal of I_1 contains only ones (no zeros) the diagonal of ones moves with each power to the next diagonal in the direction of the upper right-hand element. This means that the k-th power of such a matrix I_1 of the n-th order is equal to zero for $k \geq n$. If the superdiagonal of I_1 includes both ones and zeros, the k-th power of such a matrix I_1 may be zero even for $k < n$. This shows that the application of the Jordan matrix may be very useful.

The Jordan matrix may also be used to advantage in the analytical solution.

For a Jordan matrix having s fields of the first degree we evidently have

$$e^{J_0 t} = \begin{bmatrix} e^{z_1 t}, & 0, & \ldots, & 0 \\ 0, & e^{z_2 t}, & \ldots, & 0 \\ 0, & 0, & \ldots, & e^{z_s t} \end{bmatrix} \qquad (10.48)$$

and for the Jordan matrix (10.31) of the k-th degree

$$J_i = z_i E + I_1 ; \quad e^{J_i t} = e^{z_i t} e^{I_1 t} \qquad (10.49)$$

$$e^{J_i t} = e^{z_i t} \begin{bmatrix} 1, & t, & \dfrac{t^2}{2}, & \ldots, & \dfrac{t^{k-1}}{(k-1)!} \\ & 0, & 1, & t, & \ldots, & \dfrac{t^{k-2}}{(k-2)!} \\ & & \ldots\ldots\ldots\ldots\ldots \\ & 0, & 0, & 0, & \ldots, & 1 \end{bmatrix} \qquad (10.50)$$

Matrix (10.50) is easily derived by a series expansion of $\exp[I_1 t]$. The sum of k members of this series multiplied by $\exp[z_i t]$ is the matrix (10.50).

To calculate $\exp[Ft]$ we use the fact that the expression

$$\underbrace{(QMQ^{-1})^k = (QMQ^{-1})(QMQ^{-1})\ldots(QMQ^{-1})}_{k \text{ times}} = QM^k Q^{-1} \qquad (10.51)$$

applies to any matrix M.

---

[1]) A superdiagonal is the diagonal above the main diagonal.

Therefore

$$e^{tF} = e^{tQJQ^{-1}} = \sum_{k=0}^{\infty} Q^k \frac{t^k J^k}{k!} Q^{-k} = Q \left(\sum_{k=0}^{\infty} \frac{t^k J^k}{k!} \right) Q^{-1}$$

and hence

$$e^{Ft} = Q e^{Jt} Q^{-1} \qquad (10.52)$$

Example 10.7: Consider the matrix triad

$$F = \begin{bmatrix} -3, & 1 \\ -2, & 0 \end{bmatrix}; \quad g = \begin{bmatrix} 0 \\ 1 \end{bmatrix}; \quad h^T = [1, \ 0]$$

and use the Jordan matrix to establish for the matrix F a general expression for the state vector $x(kT)$ if $x(0)$ and $u(kT)$ are known.

Solution: According to eqn. (10.40) we calculate the characteristic matrix

$$F(s) = sE - F = \begin{bmatrix} s+3, & -1 \\ 2, & s \end{bmatrix}$$

The characteristic equation is

$$s^2 + 3s + 2 = 0$$

and its roots are the eigenvalues of the matrix F, i.e. $s_1 = -1$ and $s_2 = -2$.

The Jordan matrix

$$J = \begin{bmatrix} -1, & 0 \\ 0, & -2 \end{bmatrix}$$

According to eqn. (10.32) we calculate, for instance, the following operators:

$$Q = \begin{bmatrix} 1, & 1 \\ 2, & 1 \end{bmatrix}; \quad Q^{-1} = \begin{bmatrix} -1, & 1 \\ 2, & -1 \end{bmatrix}$$

The state vector $x(kT)$ is determined by eqns. (7.48) and (7.54):

$$x(kT) = A^k(T) x(0) + T \sum_{j=0}^{k-1} A^j(T) (FT)^{-1} (A(T) - E) g \, u[(k-j-1) T]$$

where

$$A(T) = e^{FT} = Q e^{JT} Q^{-1} = \begin{bmatrix} -e^{-T} + 2e^{-2T} & e^{-T} - e^{-2T} \\ \hline -2e^{-T} + 2e^{-2T} & 2e^{-T} - e^{-2T} \end{bmatrix}$$

10.2.5 Calculations involving matrices with complex elements

The disadvantage of the Jordan matrix is due not only to the fact that for its determination the eigenvalues of the original matrix must be calculated but, moreover, that the eigenvalues may be complex conjugate. With such eigenvalues all further matrix calculations become complicated. The calculations may, however, be simplified by dividing the Jordan matrix or its part including complex numbers into two complex conjugate blocks. In this way we can attain, using admissible elementary operations, a new arrangement including only real numbers.

Consider the homogeneous equation

$$\begin{bmatrix} x(k+1) \\ x^*(k+1) \end{bmatrix} = \begin{bmatrix} J, & 0 \\ 0, & J^* \end{bmatrix} \begin{bmatrix} x(k) \\ x^*(k) \end{bmatrix} \qquad (10.53)$$

where the complex conjugate Jordan matrix elements and the complex conjugate variables are marked with an asterisk.

Let us denote

$$x(k) = \alpha(k) + j\omega(k), \quad J = P + jQ$$

$$x^*(k) = \alpha(k) - j\omega(k), \quad J^* = P - jQ \qquad (10.54)$$

Equation (10.53) may now be rewritten in the form

$$\begin{bmatrix} \alpha(k+1) + j\omega(k+1) \\ \alpha(k+1) - j\omega(k+1) \end{bmatrix} = \begin{bmatrix} P + jQ, & 0 \\ 0, & P - jQ \end{bmatrix} \begin{bmatrix} \alpha(k) + j\omega(k) \\ \alpha(k) - j\omega(k) \end{bmatrix} \qquad (10.55)$$

Carrying out the multiplication on the right-hand side of eqn. (10.55), and adding up and subtracting the resulting rows, we obtain

$$\begin{bmatrix} \alpha(k+1) \\ \omega(k+1) \end{bmatrix} = \begin{bmatrix} P, & -Q \\ Q, & P \end{bmatrix} \begin{bmatrix} \alpha(k) \\ \omega(k) \end{bmatrix} \qquad (10.56)$$

As can be seen, the resultant state equation includes only real numbers.

In some cases it may be useful to perform the calculations only with the equation

$$x(k+1) = J x(k) \qquad (10.57)$$

which completely describes the relevant system; the complex conjugate part is then attached at the end of the calculation. If needed, we transform the resultant solution to a real form.

If the Jordan matrix includes only individual pairs of complex conjugate eigenvalues, for instance,

$$J = \begin{bmatrix} z_1, & \ldots & & & & 0 \\ \vdots & & & & & \\ 0, & \ldots & \alpha + j\omega, & 0, & \ldots & 0 \\ 0, & \ldots & 0, & \alpha - j\omega, & \ldots & 0 \\ \vdots & & & & & \\ 0, & \ldots & & & & z_n \end{bmatrix} \begin{matrix} \\ \\ i \\ k \\ \\ \end{matrix}$$

$$\begin{matrix} & i & k & & \end{matrix}$$

this matrix can be transformed to the form

$$\hat{J} = \begin{bmatrix} z_1, & \ldots & & & & 0 \\ \vdots & & & & & \\ 0, & \ldots & \alpha, & -\omega, & \ldots & 0 \\ 0, & \ldots & \omega, & \alpha, & \ldots & 0 \\ \vdots & & & & & \\ 0, & \ldots & & & & z_n \end{bmatrix} \begin{matrix} \\ \\ i \\ k \\ \\ \end{matrix}$$

if $x = Q\hat{x}$, where

$$Q = \begin{bmatrix} 1, & \ldots & & & 0 \\ \vdots & & & & \\ 0, & \ldots & 1, & j, & \ldots & 0 \\ 0, & \ldots & 1, & -j, & \ldots & 0 \\ \vdots & & & & \\ 0, & \ldots & & & 1 \end{bmatrix} \begin{matrix} \\ \\ i \\ k \\ \\ \end{matrix} \quad ; \quad Q^{-1} = \tfrac{1}{2}\begin{bmatrix} 2, & \ldots & & & 0 \\ \vdots & & & & \\ 0, & \ldots & 1, & 1, & \ldots & 0 \\ 0, & \ldots & -j, & j, & \ldots & 0 \\ \vdots & & & & \\ 0, & \ldots & & & 2 \end{bmatrix} \begin{matrix} \\ \\ i \\ k \\ \\ \end{matrix}$$

is substituted into the state equations.

Then

$$\hat{x}(k+1) = Q^{-1}JQ\,\hat{x}(k) + Q^{-1}B_J\,u(k)$$

$$y(k) = CQ\,\hat{x}(k) + D\,u(k)$$

$$\hat{J} = Q^{-1}JQ, \quad \hat{B}_J = Q^{-1}B_J, \quad \hat{C} = CQ$$

10.3 Canonical form of reachability

Besides the Jordan canonical form of the matrix A we may draw attention to some further forms of this matrix which can in some cases be used to advantage. These are the canonical forms noted for their simplicity and direct relationship of the matrix elements of the system state model to the coefficients of the difference equation. These canonical forms of A are closely associated with the canonical forms of

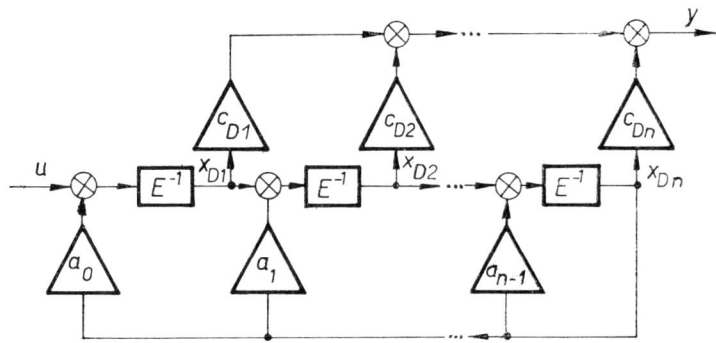

Fig. 10.2. Block diagram of state equations in the canonical form of reachability.

state equations described in Sec. 10.3 through 10.6, which define a single-input/single-output system of the order n by a total of $2n$ mutually independent coefficients. On the other hand, in the general case where the matrices in the state equations have only non-zero elements, the total number of these elements is $n(n + 2)$.

The canonical form of reachability may be derived from the block diagram shown in Fig. 10.2. It holds that

$$\begin{aligned} u(k) - a_0 x_n(k) &= x_1(k + 1) \\ x_1(k) - a_1 x_n(k) &= x_2(k + 1) \\ &\vdots \\ x_{n-1}(k) - a_{n-1} x_n(k) &= x_n(k + 1) \end{aligned} \tag{10.58}$$

$$y(k) = [c_{D1}, c_{D2}, \ldots, c_{Dn}] [x_1(k), x_2(k), \ldots, x_n(k)]^T \tag{10.59}$$

The system of equations (10.58) defines at the same time the vector of the state variables. It is directly apparent that this system of equations may also be written in the vector-matrix form

$$x_D(k + 1) = \begin{bmatrix} 0, & \ldots & 0, & -a_0 \\ 1, & \ldots & 0, & -a_1 \\ & \ldots & & \\ 0, & \ldots & 1, & -a_{n-1} \end{bmatrix} x_D(k) + \begin{bmatrix} 1 \\ 0 \\ \vdots \\ 0 \end{bmatrix} u(k) \tag{10.60}$$

$$\begin{aligned} x_D(k + 1) &= A_D x_D(k) + b_D u(k) \\ y(k) &= c_D^T x_D(k) \end{aligned} \tag{10.61}$$

Some of the important properties of this canonical form of reachability are the following:

(1) The appropriate discrete transfer function of the system described by the state equations (10.58) through (10.61) has the form

$$G(z) = \frac{b_0 + b_1 z + \ldots + b_{n-1} z^{n-1}}{a_0 + a_1 z + \ldots + a_{n-1} z^{n-1} + z_n} \tag{10.62}$$

where the coefficients a_i, $i = 0, 1, \ldots, n - 1$, of the denominator of $G(z)$ are directly the elements of the matrix A_D expressed in eqn. (10.60). The coefficients b_i, $i = 0, 1, \ldots, k, \ldots, n - 1$, of the numerator of $G(z)$ are generally defined by the relation

$$b_k = \sum_{i=1}^{n-k} c_{Di} a_{i+k} \tag{10.63}$$

The form of the matrix A_D and its transpose matrix is referred to as the Frobenius form [63.7].

(2) The matrix of reachability of a system with the Frobenius form of the matrix of dynamics is the unit matrix

$$Q_{DF} = [b_D, A_D b_D, \ldots, A_D^{n-1} b_D] = E \tag{10.64}$$

Any reachable system can be transformed to the canonical form of reachability using the transformation matrix

$$x = T^{-1} x_D = Q x_D \tag{10.65}$$

where $Q = T^{-1}$. If the system is described by the triad (A, b, c^T), then we have, according to eqns. (9.8) and (10.19),

$$Q = Q_D Q_{DF}^{-1} = Q_D = [b, Ab, \ldots, A^{n-1} b] \tag{10.66}$$

so that, as can be seen, an inversion of the matrix Q_{DF} need not be performed with this transformation because $Q_{DF} = E$. It is not even necessary to perform the inversion of the matrix $Q = Q_D$ for the calculation of $A_D = Q^{-1} A Q$ since the elements a_i, $i = 0, 1, \ldots, n - 1$, of the matrix A_D may be determined by calculating the determinant $|zE - A|$ or by the Faddeev algorithm (see Par. 7.7.2 or Sec. 10.4). The vector b_D is known.

(3) For the output matrix c_D^T we have the expression

$$c_D^T = c^T Q = [c^T b, c^T A b, \ldots, c^T A^{n-1} b] \tag{10.67}$$

If we compare the elements of the matrix c_D^T with the expression (7.5) we find that they are ordinates of the discrete impulse response.

$$c_D^T = [s(1), s(2), \ldots, s(n)] \tag{10.68}$$

(4) In contrast with the Jordan form the canonical form of reachability has only real matrix elements and is, therefore, well suited for simulation and other calculation purposes. The advantages of this form are conspicuous especially in its generalization for multivariable systems.

10.4 Canonical form of controllability

Consider again the transfer function $G(z)$ expressed by eqn. (10.62). We may also write

$$G(z) = c_R^T(zE - A_R)^{-1} b_R \qquad (10.69)$$

where the triad (A_R, b_R, c_R^T) refers to the canonical form of controllability

$$x_R(k+1) = \begin{bmatrix} 0, & 1, & \cdots & 0 \\ \vdots & & & \vdots \\ 0, & 0, & \cdots & 1 \\ -a_0, & -a_1, & \cdots & -a_{n-1} \end{bmatrix} x_R(k) + \begin{bmatrix} 0 \\ \vdots \\ 0 \\ 1 \end{bmatrix} u(k) =$$

$$= A_R x_R(k) + b_R u(k)$$

$$y(k) = [b_0, b_1, \ldots, b_{n-1}] x_R(k) = c_R^T x_R(k) \qquad (10.70)$$

The canonical form of controllability is identical with the solutions (6.14) and (6.15). The subscripts of the coefficients must, however, be changed. If we distinguish by the subscript R the coefficients in eqns. (10.70) from those in eqns. (6.14) and (6.15), then

$$a_{Rk} = a_{n-k}, \quad a_{Rn} = a_0 = 1$$
$$b_{Rk} = b_{n-k}, \quad b_{Rn} = b_0 = 0 \qquad (10.71)$$

The main properties of this form are the following:

(1) The matrix of reachability is non-singular and has the form of the lower triangular matrix

$$\begin{bmatrix} 0, & \cdots & & 0, & 1 \\ 0, & \cdots & & 1, & * \\ \vdots & & & & \\ 1, & * & \cdots & *, & * \end{bmatrix}$$

(2) The transformation matrix $Q = T^{-1}$, which enables us to transform the system defined by the triad (A, b, c^T) to the canonical form of controllability, can be calculated recursively by columns [66.4]:

$$x = Q x_R = [q_1, q_2, \ldots, q_n] x_R$$

$$AQ = Q A_R$$

$$A[q_1, q_2, \ldots, q_n] = [q_1, q_2, \ldots, q_n] \begin{bmatrix} 0, & 1, & \cdots & 0 \\ \cdots & \cdots & \cdots & \cdots \\ 0, & 0, & \cdots & 1 \\ -a_0, & -a_1, & \cdots & -a_{n-1} \end{bmatrix}$$

$$[Aq_1, Aq_2, \ldots, Aq_n] = [-q_n a_0, q_1 - q_n a_1, \ldots, q_{n-1} - q_n a_{n-1}] \quad (10.72)$$

Since b_R is known we may first calculate

$$b = Q b_R = q_n$$

and then proceed by individual columns of (10.72) from the right to the left:

$$\begin{aligned} q_n &= b \\ q_{n-1} &= A q_n + a_{n-1} q_n \\ &\vdots \\ q_1 &= A q_2 + a_1 q_n \\ 0 &= A q_1 + a_0 q_n \end{aligned} \quad (10.73)$$

The last row serves for checking. To avoid inversion of the transformation matrices we start again from the fact that we know the form of b_R, determine the coefficients a_i, $i = 0, 1, \ldots, n-1$, of the matrix A_R by means of the determinant $|zE - A|$ or by the Faddeev algorithm and finally calculate the elements of c_R^T according to the relation

$$c_R^T = c^T Q$$

where the individual columns of the matrix Q are defined by the system of equations (10.73). The elements of the matrix c_R^T can, of course, be also calculated using the formulae (7.71). However, when using either the Faddeev algorithm (Par. 7.7.3) or the formulae (7.71), the change in subscripts as defined in (10.71) must be kept in view.

(3) From the Faddeev algorithm and from eqns. (10.73) follow further relations for the columns of the transformation matrix Q:

$$\begin{aligned}
B_0 &= E, & a_{n-1} &= -\operatorname{tr} A B_0, & q_n &= B_0 b \\
B_1 &= A B_0 + a_{n-1} E, & a_{n-2} &= -\frac{1}{2} \operatorname{tr} A B_1, & q_{n-1} &= B_1 b \\
&\vdots & &\vdots & &\vdots \\
B_{n-1} &= A B_{n-2} + a_1 E, & a_1 &= -\frac{1}{n-1} \operatorname{tr} A B_{n-2}, & q_1 &= B_{n-1} b \\
0 &= A B_{n-1} + a_0 E, & a_0 &= -\frac{1}{n} \operatorname{tr} A B_{n-1}, & & & (10.74)
\end{aligned}$$

TRANSFORMATIONS

(4) The block diagram of the canonical form of controllability corresponds to Fig. 10.3. If $b_n \neq 0$, the block diagram must be extended by the dashed part of Fig. 10.3 and the output equation (10.70) changes into

$$y(k) = c_R^T x_R(k) + b_n u(k)$$

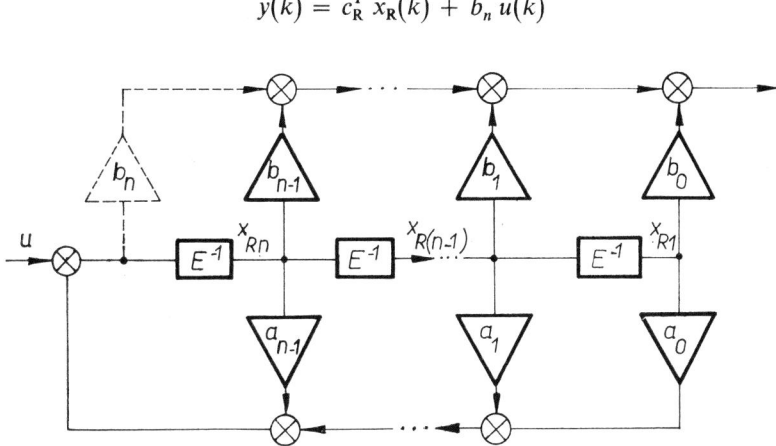

Fig. 10.3. Block diagram of state equations in the canonical form of controllability.

(5) The canonical forms of controllability and reachability are equivalent in the sense that both satisfy Theorem 9.1. The canonical form of state description satisfying in a general case Theorem 9.3 is not known.

10.5 Canonical form of observability

If it is useful to have the output matrix c^T in the simplest form we can make use of the so-called canonical form of observability where the element c_{P1} of the matrix c_P^T is equal to 1 and the other elements are zero

$$x_P(k+1) = \begin{bmatrix} 0, & 1, & \ldots & 0 \\ \hdotsfor{4} \\ 0, & 0, & \ldots & 1 \\ -a_0, & -a_1, & \ldots & -a_{n-1} \end{bmatrix} x_P(k) + \begin{bmatrix} b_{P1} \\ \vdots \\ b_{Pn} \end{bmatrix} u(k) =$$

$$= A_P x_P(k) + b_P u(k)$$

$$y(k) = [1, 0, \ldots, 0] x_P(k) = c_P^T x_P(k) \tag{10.75}$$

This is a dual form of the canonical form of reachability. This means that $A_P = A_D^T$, $b_P = c_D$, $c_P = b_D$.

The canonical form of observability has the following main properties:

(1) The coefficients a_i, $i = 0, 1, \ldots, n - 1$, in the denominator of the transfer function $G(z)$ in (10.62) are directly the elements of the matrix A_P according to (10.75). The coefficients b_i, $i = 0, 1, \ldots, n - 1$, in the numerator of the transfer function $G(z)$ are defined by the general expression

$$b_k = \sum_{i=k}^{n-1} a_{n+k-1} b_{P(n-1)} \tag{10.76}$$

(2) The matrix of observability is the unit matrix

$$Q_{PF} = \begin{bmatrix} c_P^T \\ c_P^T A_P \\ \vdots \\ c_P^T A_P^{n-1} \end{bmatrix} = E \tag{10.77}$$

(3) If the system is determined by the triad (A, b, c^T), the relevant canonical form is calculated by means of the transformation

$$x_P = Tx$$

where the transformation matrix according to eqns. (10.19) and (10.77) is

$$T = Q_P = \begin{bmatrix} c^T \\ c^T A \\ \vdots \\ c^T A^{n-1} \end{bmatrix} \tag{10.78}$$

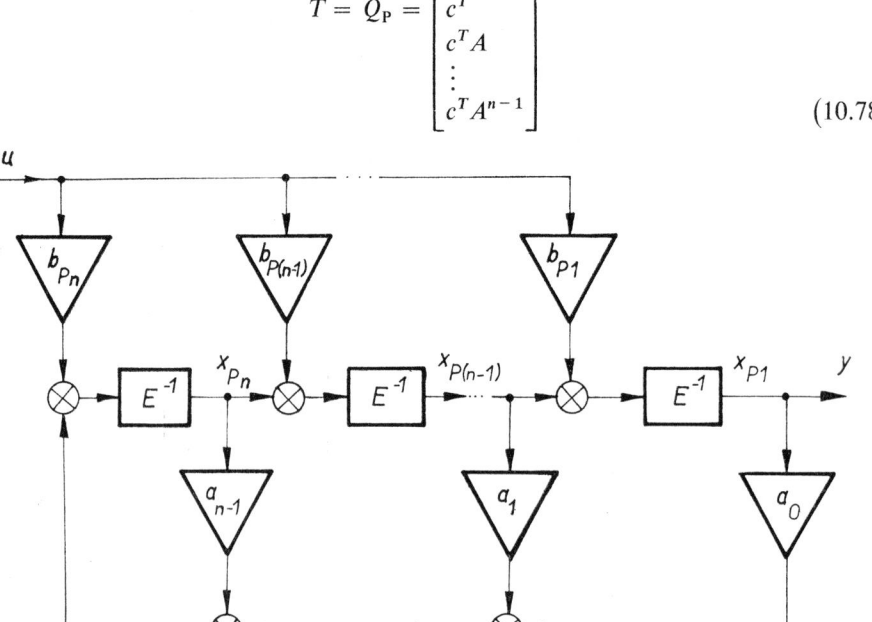

Fig. 10.4. Block diagram of state equations in the canonical form of observability.

TRANSFORMATIONS

The inversion of the matrix Q_P in the calculation of $A_P = Q_P^{-1} A Q_P$ may be avoided in the same manner as in the calculation of the matrix Q_D (see Sec. 10.3).

(4) The vector b_P is again calculated with the aid of the transformation matrix $T = Q_P$. It holds that

$$b_P = Q_P b = \begin{bmatrix} c^T b \\ c^T A b \\ \ldots \\ c^T A^{n-1} b \end{bmatrix} \tag{10.79}$$

By comparison of the matrix (10.79) with the expression (7.5) we find that the elements of the matrix b_P are ordinates of the discrete impulse response

$$b_P = [s(1), s(2), \ldots, s(n)]^T \tag{10.80}$$

The block diagram of the canonical form of observability is shown in Fig. 10.4.

10.6 Canonical form of reconstructability

The dual form of the canonical form of controllability is the canonical form of reconstructability: If we denote the matrices and vectors of this form by the subscripts K then $A_K = A_R^T$, $b_K = c_R$, $c_K = b_R$. The block diagram of this form is shown in Fig. 10.5 from which follow the relevant state equations.

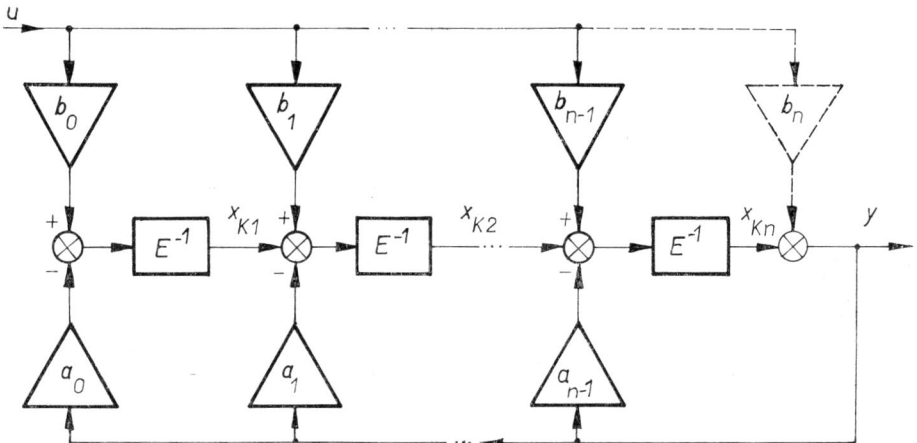

Fig. 10.5. Block diagram of state equations in the canonical form of reconstructability.

$$x_K(k+1) = \begin{bmatrix} 0, & \ldots & 0, & -a_0 \\ 1, & \ldots & 0, & -a_1 \\ \multicolumn{4}{c}{\dotfill} \\ 0, & \ldots & 1, & -a_{n-1} \end{bmatrix} x_K(k) + \begin{bmatrix} b_0 \\ b_1 \\ \vdots \\ b_{n-1} \end{bmatrix} u(k) =$$

$$= A_K\, x_K(k) + b_K\, u(k)$$
$$y(k) = [0, \ldots 0, 1] x_K(k) = c_K^T x_K(k) \tag{10.81}$$

The main properties of this form are the following:

(1) The matrix of observability is non-singular if the system is observable. The element arrangement of this matrix is analogous as in the matrix of reachability of canonical form of controllability.

(2) If the system is given by the triad (A, b, c^T), it can be transformed to the canonical form of reconstructability by substituting

$$Q x_K = x, \quad Q = T^{-1} \tag{10.82}$$

where the transformation matrix T may be calculated recursively by rows:

$$T = [t_1, t_2, \ldots, t_n]^T \tag{10.83}$$

Since we know c_K^T we first determine

$$t_n^T = c_K^T Q = c^T \tag{10.84}$$

The further rows of the transformation matrix are defined by the relation

$$AQ = QA_K$$
$$TA = A_K T \tag{10.85}$$

whence it follows that

$$\begin{bmatrix} t_1^T \\ t_2^T \\ \vdots \\ t_n^T \end{bmatrix} A = \begin{bmatrix} 0, & \ldots & 0, & -a_0 \\ 1, & \ldots & 0, & -a_1 \\ \multicolumn{4}{c}{\dotfill} \\ 0, & \ldots & 1, & -a_{n-1} \end{bmatrix} \begin{bmatrix} t_1^T \\ t_2^T \\ \vdots \\ t_n^T \end{bmatrix}$$

$$[t_1^T A, t_2^T A, \ldots, t_n^T A] = [-a_0 t_n^T,\; t_1^T - a_1 t_n^T,\; \ldots,\; t_{n-1}^T - a_{n-1} t_n^T] \tag{10.86}$$

Comparing successively from the right to the left the elements of the matrices in the last equation we obtain

$$t_{n-1}^T = t_n^T A + a_{n-1} t_n^T$$
$$t_{n-2}^T = t_{n-1}^T A + a_{n-2} t_n^T$$
$$\vdots$$
$$t_1^T = t_2^T A + a_1 t_n^T$$
$$0 = t_1^T A + a_0 t_n^T \tag{10.87}$$

TRANSFORMATIONS

The last row serves for checking. The elements a_i, $i = 0, 1, \ldots, n - 1$, of the matrix A_K may be determined by calculating the characteristic polynomial, i.e. $\det(zE - A)$, or by means of the Faddeev algorithm (10.74). The elements of the matrix b_K are calculated according to the relation

$$b_K = Tb \qquad (10.88)$$

(3) The relevant transfer function $G(z)$ is of the form (10.62) with the coefficients a_i, b_i, $i = 0, 1, \ldots, n - 1$, calculated according to the preceding relations (10.82) through (10.88). If the numerator of the transfer function (10.62) also has a coefficient $b_n \neq 0$, the block diagram of Fig. 10.5 should be extended by the dashed portion and the output equation (10.81) changes into

$$y(k) = c_K^T u(k) + b_n u(k)$$

(4) The canonical forms of reconstructability and observability are equivalent insomuch that both satisfy Theorem 9.4. The canonical form of state representation satisfying in a general case Theorem 9.5, is not known.

10.7 Relationship of the characteristic equation to impulse response

The canonical form of reachability and the canonical form of observability have in general $2n$ mutually independent coefficients, namely the coefficients a_i, $i = 0, 1, \ldots, n - 1$, of the characteristic equation and the ordinates of the discrete impulse response $s(i)$, $i = 1, 2, \ldots, n$. In addition, there exists a further relationship to the coefficients in the numerator of $G(z)$, other than already mentioned in Secs. 10.2 through 10.6.

Dividing the numerator by the denominator of the transfer function $G(z)$ of the form given by eqn. (10.62), we obtain

$$G(z) = \frac{B(z)}{A(z)} = \sum_{i=0}^{\infty} s(i) z^{-i}$$

Comparing the coefficients at the same powers of z of the expression

$$B(z) = A(z) \sum_{i=0}^{\infty} s(i) z^{-i}$$

we obtain for $n \geq i$

$$b_{n-i} = s(i) + a_{n-1} s(i-1) + a_{n-2} s(i-2) + \ldots + a_0 s(i-n) \qquad (10.89)$$

where $s(i) = 0$ and $b_i = 0$ for $i < 0$.

By means of eqn. (10.89) we may, therefore, calculate the coefficients in the numerator of the transfer function $G(z)$ if we know the coefficients of the characteristic polynomial and the ordinates of the discrete impulse response. Equation (10.89) may, of course, be likewise used for a recursive calculation of the discrete impulse response ordinates if we know a_i and b_i, $i = 0, 1, \ldots, n - 1$.

For $i > n$ we have

$$s(i) + a_{n-1} s(i-1) + \ldots + a_0 s(i-n) = 0 \tag{10.90}$$

Writing this equation for $i = n + 1, n + 2, \ldots, 2n$ we obtain a set of n linear algebraic equations the solution of which enables us to calculate the coefficients of the characteristic polynomial if we know the ordinates of the discrete impulse response.

Equation (10.90) may also be derived on the basis of the Cayley-Hamilton theorem, according to which any square matrix satisfies its own characteristic equation [66.2].

Let us consider the characteristic equation

$$P(z) = \det(zE - A) = a_0 + a_1 z + \ldots + a_{n-1} z^{n-1} + z^n = 0 \tag{10.91}$$

According to the Cayley-Hamilton theorem

$$P(A) = a_0 E + a_1 A + \ldots + a_{n-1} A^{n-1} + A^n = 0 \tag{10.92}$$

Multiplying eqn. (10.92) from the left by the expression $c^T A^{i-n-1}$ and from the right by the vector b we obtain

$$a_0 c^T A^{i-n-1} b + a_1 c^T A^{i-n} b + \ldots + a_{n-1} c^T A^{i-2} b + c^T A^{i-1} b = 0 \tag{10.93}$$

Since by eqn. (7.5)

$$s(i) = c^T A^{i-1} b \quad \text{for} \quad i > 0$$

we can use this relation for rewriting eqn. (10.93) in the form (10.90).

11

Multi-input/multi-output Systems

The multi-input/multi-output (multivariable) system is a system having $r > 1$ command variables or $p > 1$ mutually independent controlled variables. In general both $r > 1$ and $p > 1$ correspond to the usual situation.

In this chapter we shall generalize some of the previous results referred to time-invariant single-input/single-output systems for multi-input/multi-output systems. The state equations of these systems differ from the single-input/single-output case mainly in that the column and row matrices b and c^T must be replaced by the generally rectangular matrices $B(n; r)$ and $C(p; n)$, respectively, whose dimension is given in brackets. Here n denotes the dimension of the matrix $A(n; n)$, r is the number of controlling variables and p is the number of mutually independent controlled variables. Unless otherwise stated it will be assumed that the number of command variables is equal to the number of controlled variables. The input and output variables of a multi-input/multi-output system are expressed as the vectors

$$u^T = [u_1, u_2, \ldots, u_r]$$
$$y^T = [y_1, y_2, \ldots, y_p] \tag{11.1}$$

The state equations of a time-invariant continuously working multi-input/multi-output system have the form

$$\dot{x} = Fx + Gu$$
$$y = Hx + Ku \tag{11.2}$$

where we assign to the individual matrices the dimensions indicated in parentheses: $F(n; n)$, $G(n; r)$, $H(p; n)$ and $K(p; r)$.

If all components of the vector u are sampled synchronously at the instants $t = kT$, where k is an integer, and if the vector $u(t)$ within one sampling interval, i.e. $kT \leq t < (k + 1)T$, is

$$u(t) = u(kT) = \text{const.} \tag{11.3}$$

which means that a zero-order holding member has been used, then, in accordance with the results of Sec. 6.2, the discrete version of the state equations will be

$$x(k + 1) = A\,x(k) + B\,u(k)$$
$$y(k) = C\,x(k) + D\,u(k) \tag{11.4}$$

where $x \in X^n$, $u \in U^r$ and $y \in Y^p$, so that the dimensions of the matrices are, therefore, $A(n; n)$, $B(n; r)$, $C(p; n)$, $D(p; r)$. It further holds that

$$A = e^{FT} \quad \text{and} \quad B = \int_0^T e^{F\tau} G\, d\tau \tag{11.5}$$

The solution of the system of equations (11.4) has, according to eqn. (7.2) for $w = 0$, the form

$$y(k) = CA^k x(0) + C\sum_{j=1}^{k} A^{j-1} B\,u(k - j) + D\,u(k) \tag{11.6}$$

and the Z-transform of the output vector $y(k)$ is, by (7.30),

$$Y(z) = C(zE - A)^{-1} z\, x(0) + [C(zE - A)^{-1} B + D]\,U(z) \tag{11.7}$$

For the initial conditions $x(0) = 0$ we obtain, by (7.31), the matrix of transfer functions

$$G(z) = C(zE - A)^{-1} B + D \tag{11.8}$$

The expression (7.5) generalized for a multivariable case gives the matrix of discrete impulse responses

$$S(k) = \begin{cases} D & \text{for } k = 0 \\ CA^{k-1}B & \text{for } k > 0 \end{cases} \tag{11.9}$$

11.1 Asynchronous sampling

With multi-input/multi-output control loops a case may be encountered in practice where the input and output variables are not sampled synchronously but cyclically in such a manner that the sampling interval T is divided into as many subintervals ΔT as many there are variables. The individual variables are successively sampled with a time-gap ΔT beginning with the first variable, proceeding to the last one and then returning again to the first variable. The sampling period, however, is the same for all variables and equal to T. The case where the individual variables are sampled at different sampling periods will not be discussed here. In such a case one should proceed according to the rules stated in Sec. 7.8.

MULTI-INPUT/MULTI-OUTPUT SYSTEMS

Let us assume, for simplicity, that the number of command, controlling and controlled variables is the same, i.e. $p = r$.

Let us consider the sampling of r input variables of a vector u of a multivariable system, where each input variable, after sampling, is constant in the given time interval:

$$u_1(t) = u_1(kT) \qquad \text{for} \quad kT \leq t < (k+1)T$$
$$u_2(t) = u_2(kT + T/r) \qquad \text{for} \quad kT + T/r \leq t < (k+1)T + T/r$$
$$\vdots$$
$$u_r(t) = u_r(kT + T(r-1)/r) \qquad \text{for} \quad kT + T(r-1)/r \leq t < (k+1)T + T(r-1)/r \tag{11.10}$$

Since, according to Sec. 7.8, the values of the input variables $u_i(t)$, $u = 1, 2, ..., r$, stored in the holding members must be regarded as state variables, we introduce the extended vector of the state variables

$$v = \begin{bmatrix} x \\ u \end{bmatrix} \tag{11.11}$$

We calculate alternately with two transitions:

(1) The transition at the instant of sampling of a new value of the input variable u_q, $q = 0, 1, ..., (r-1)$, i.e. within the interval

$$(kT + qT/r)^- < t < (kT + qT/r)^+$$

(2) The transition within the sampling interval, i.e. in this case between two sampling instants of two subsequently sampled variables, i.e. within the interval

$$(kT + qT/r)^+ < t < (kT + (q+1)T/r)^-$$

Utilizing all these transitions we obtain the difference equation for the interval kT^- to $(k+1)T^-$. The input variable u may be eliminated from the final difference equation because it is not a state variable. We shall demonstrate this general approach by an example of a double-input/double-output control loop.

Example 11.1: Let us set up the difference state equation for the double-input/double-output unit feedback system shown in Fig. 11.1, whose continuously working plant S is described by the state equations

$$\dot{x} = Fx + Gu$$
$$y = Cx = \begin{bmatrix} c_1^T \\ c_2^T \end{bmatrix} x \tag{11.12}$$

The controlled plant has two input variables which are the outputs of the zero-order holding members H, two output variables y_1, y_2 and two command variables r_1, r_2. The input variables u_1 and u_2 are the staircase functions

$$u_1(t) = e_1(kT) \quad \text{for} \quad kT \leq t < (k+1)T$$
$$u_2(t) = e_2(kT + T/2) \quad \text{for} \quad kT + T/2 \leq t < (k+1)T + T/2 \quad (11.13)$$

where e_i, $i = 1, 2$, are the errors $e_i = r_i - y_i$.

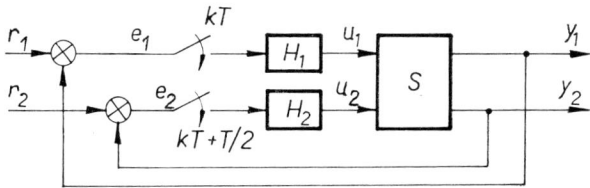

Fig. 11.1. Double-input/double-output system with unit feedback.

The general solution of eqn. (11.12) is

$$x(t) = e^{F(t-t_0)} x(t_0) + \int_0^{t-t_0} e^{F\tau} Gu \, d\tau$$

where $u = \text{const.}$ within the intervals $\langle kT, kT + T/2 \rangle$ and $\langle kT + T/2, (k+1)T \rangle$.

Since both these intervals are of equal length we can manage in the discrete version of state equations with the matrices

$$A = e^{FT/2} \quad \text{and} \quad B = \int_0^{T/2} e^{F\tau} G \, d\tau = [b_1, b_2] \quad (11.14)$$

The extended state vector is

$$v = [x^T, u_1, u_2]^T \quad (11.15)$$

In the system considered occur altogether the following four transitions:

(1) The transition within the interval $kT^- < t < kT^+$ when the error $e_1(kT) = u_1(kT)$ is being stored in the memory of the holding member H_1; according to eqn. (7.105) it holds that

$$v(kT^+) = \begin{bmatrix} E, & 0, & 0 \\ 0, & 0, & 0 \\ 0, & 0, & 1 \end{bmatrix} v(kT^-) + \begin{bmatrix} 0, & 0 \\ 1, & 0 \\ 0, & 0 \end{bmatrix} e(kT) \quad (11.16)$$

MULTI-INPUT/MULTI-OUTPUT SYSTEMS

where the vector $e(kT)$ of the dimension $(2; 1)$ is

$$e(kT) = \begin{bmatrix} e_1(kT) \\ e_2(kT) \end{bmatrix}$$

The number 2 denotes the dimension of the square matrix F.

(2) The transition within the interval $kT^+ < t < (kT + T/2)^-$ when the dynamic process is taking place; according to eqn. (7.106) it holds that

$$v(kT + T/2)^- = \begin{bmatrix} A, & B \\ 0, & E \end{bmatrix} v(kT^+) \tag{11.17}$$

(3) The transition within the interval $(kT + T/2)^- < t < (kT + T/2)^+$ when the error $e_2(kT + T/2) = u_2(kT + T/2)$ is being stored in the memory of the holding member H_2; according to eqn. (7.105) it again holds that

$$v(kT + T/2)^+ = \begin{bmatrix} E, & 0, & 0 \\ 0, & 1, & 0 \\ 0, & 0, & 0 \end{bmatrix} v(kT + T/2)^- + \begin{bmatrix} 0, & 0 \\ 0, & 0 \\ 0, & 1 \end{bmatrix} e(kT + T/2) \tag{11.18}$$

(4) The transition within the interval $(kT + T/2)^+ < t < [(k + 1)T]^-$ when the dynamic process is taking place; according to eqn. (7.106) it again holds that

$$v[(k + 1)T^-] = \begin{bmatrix} A, & B \\ 0, & E \end{bmatrix} v(kT + T/2)^+ \tag{11.19}$$

By successive substitution of the third, second and first transition equation into the fourth one we obtain

$$v[(k + 1)T^-] = \begin{bmatrix} A^2, & 0, & Ab_2 \\ 0, & 0, & 0 \\ 0, & 0, & 0 \end{bmatrix} v(kT^-) + \begin{bmatrix} Ab_1 + b_1, & 0 \\ 1, & 0 \\ 0, & 0 \end{bmatrix} e(kT) +$$

$$+ \begin{bmatrix} 0, & b_2 \\ 0, & 0 \\ 0, & 1 \end{bmatrix} e(kT + T/2) \tag{11.20}$$

In this equation, the input variable u_1 may be eliminated since at the instant $t = kT$ it is equal to the error $e_1(kT)$. Introducing a new vector of the dimension $(2 + 1; 1)$

$$\hat{x} = \begin{bmatrix} x \\ u_2 \end{bmatrix} \tag{11.21}$$

the difference equation will be

$$\hat{x}[(k+1)T^-] = \begin{bmatrix} A^2, & Ab_2 \\ 0, & 0 \end{bmatrix} \hat{x}(kT^-) + \begin{bmatrix} Ab_1 + b_1, & 0 \\ 0, & 0 \end{bmatrix} e(kT) +$$

$$+ \begin{bmatrix} 0, & b_2 \\ 0, & 1 \end{bmatrix} e(kT + T/2) \tag{11.22}$$

The control loop is closed using the equation defining the control error

$$e(kT) = r(kT) - y(kT) \tag{11.23}$$

$$e(kT) = r(kT) - [C, 0]\,\hat{x}(kT)^- \tag{11.24}$$

$$e(kT + T/2) = r(kT + T/2) - [C, 0]\,\hat{x}(kT + T/2)^- \tag{11.25}$$

The vector $x(kT + T/2)^-$ is obtained by substituting the first transition equation into the second one and by eliminating the variable $u_1(kT)$,

$$\hat{x}(kT + T/2)^- = \begin{bmatrix} A, & b_2 \\ 0, & 1 \end{bmatrix} \hat{x}(kT)^- + \begin{bmatrix} b_1, & 0 \\ 0, & 0 \end{bmatrix} e(kT)$$

so that

$$e(kT + T/2) = r(kT + T/2) - [C, 0]\left\{\begin{bmatrix} A, & b_2 \\ 0, & 0 \end{bmatrix} \hat{x}(kT)^- + \begin{bmatrix} b_1, & 0 \\ 0, & 0 \end{bmatrix} e(kT)\right\} =$$

$$= r(kT + T/2) - [CA,\ Cb_2]\,\hat{x}(kT)^- - [Cb_1,\ 0]\{r(kT) -$$

$$- [C, 0]\,\hat{x}(kT)^-\} =$$

$$= r(kT + T/2) - [Cb_1, 0]\,r(kT) - [C(A - b_1 c_1^T),\ Cb_2]\,\hat{x}(kT)^- \tag{11.26}$$

Equation (11.22) may now be written in its final form

$$\hat{x}[(k+1)T^-] = \begin{bmatrix} A^2 - (Ab_1 + b_1)c_1^T - b_2 c_2^T(A - b_1 c_1^T), & Ab_2 - b_2 c_2^T b_2 \\ c_2^T(A - b_1 c_1^T), & -c_2^T b_2 \end{bmatrix} \hat{x}(kT) +$$

$$+ \begin{bmatrix} (A + E - b_2 c_2^T)b_1, & b_2 \\ -c_2^T b_1, & 1 \end{bmatrix} \begin{bmatrix} r_1(kT) \\ r_2(kT + T/2) \end{bmatrix} \tag{11.27}$$

11.2 Index of reachability and observability of multi-input/multi-output systems

In this section we shall generalize the knowledge about reachability and observability for multi-input/multi-output systems. Let us consider the system (11.4) of the

MULTI-INPUT/MULTI-OUTPUT SYSTEMS

n-th order described by the matrix triad (A, B, C), having r input variables and p output variables; the rank of the matrix B is

$$h(B) = r \leq n \tag{11.28}$$

and that of the matrix C

$$h(C) = p \leq n \tag{11.29}$$

where n is the order of the system.

The equality (11.28) means that the input variables u_i, $i = 1, 2, \ldots, r$ are, as regards their action, mutually distinguishable or, in other words, mutually linearly independent in the sense that none of the variables of the vector u can be replaced by a linear combination of the remaining variables.

The equality (10.29) implies that the output variables y_i, $i = 1, 2, \ldots, p$, are mutually linearly independent or, in other words, none of them can be replaced by a linear combination of the remaining components of the vector y.

According to eqn. (7.1) and in analogy with eqn. (9.3), the solution of the equation of dynamics of a multi-input/multi-output system

$$x(k + N) = A^N x(k) + [B, AB, \ldots, A^{N-1}B] u_N(k) \tag{11.30}$$

where

$$u_N^T(k) = [u(k + N - 1), u(k + N - 2), \ldots, u(k)]$$

THEOREM 11.1: *The state $x(k + N) \in X^n$ of the system (11.4) is reachable in N steps from the initial state $x(k) \in X^n$ if and only if the rank of the matrix $Q_D = [B, AB, \ldots, A^{N-1}B]$ is equal to the state space dimension n.*

If the condition expressed by Theorem 11.1 is satisfied then the vectors $u(k)$, $u(k + 1), \ldots, u(k + N - 1)$, $u \in U^r$, satisfying eqn. (11.30), exist for all values of $x(k) \in X^n$ and $x(k + N) \in X^n$.

For each column b_i of the matrix B we may write the matrix

$$Q_{Di} = [b_i, Ab_i, A^2 b_i, \ldots, A^{N-1} b_i] \tag{11.31}$$

and to each such matrix Theorem 9.2 applies. It should be noted that a multi-input/multi-output system may be reachable even if $N < n$, where n is the state space dimension, because the matrix Q_D defined in Theorem 11.1 has altogether $N \cdot r$ columns so that the theorem can be satisfied even if $N < n$. Therefore, in contrast with single-input/single-output systems, the state $x(k + N)$ of multi-input/multi-output system may be reachable from the initial state $x(k)$ even after N steps, where $N < n$.

With regard to these facts the measure of reachability of multi-input/multi-output systems must be expressed in a somewhat more detailed manner than in the case of single-input/single-output systems. For this purpose we use the so-called index of reachability [72.1].

DEFINITION 11.1: *The index of reachability v of a system described by eqns. (11.4) is the smallest number of steps required to transfer the system from the initial state $x(k)$ to the desired state $x(k + N)$ or, in other words, the lowest value of N within the interval $\langle 1, n \rangle$ which satisfies Theorem 11.1.*

Since the matrix Q_D defined in Theorem 11.1 has $Nr \geq n$ columns, the smallest possible N is equal to $v \geq n/r$. The highest possible value of the index of reachability is obtained by adding to the r mutually independent columns of the matrix B only one further column from each subsequent matrix $AB, A^2B, \ldots, A^{N-1}B$. These further columns must necessarily be linearly independent with regard to the preceding ones. Hence $r + (N - 1) \leq n$ or $v \leq n - r + 1$. From both inequalities it follows that the index of reachability is within the range

$$n/r \leq v \leq n - r + 1 \tag{11.32}$$

Notice that for a single-input/single-output system, where $r = 1$, eqn. (11.32) gives $n \leq v \leq n$. Consequently, the result $v = n$ is in agreement with Theorem 9.1.

Let the relations

$$h(Q_{Di}) = h_i, \quad h_i < n$$
$$h(Q_{Dj}) = h_j, \quad h_j < n, \quad i \neq j \tag{11.33}$$

apply to the system (11.4) with $x \in X^n$ and $u \in U^r$, where $h(\cdot)$ denotes the rank and Q_{Di} and Q_{Dj} are matrices of the form (11.31) corresponding to the i-th and j-th input variable, respectively. The following theorems may now be formulated:

THEOREM 11.2: *If $h(Q_{Di}, Q_{Dj}) = h_{ij}$, $h_{ij} < h_i + h_j$, then a part of the system is influenced by both the input variable u_i and the input variable u_j, $i \neq j$.*

THEOREM 11.3: *The difference $\pi_{ij} = h_i + h_j - h_{ij}$, $i \neq j$, following from Theorem 11.2, determines the number of modes influenced by both input variables u_i and u_j.*

THEOREM 11.4: *If $h(Q_{Di}, Q_{Dj}) = h_{ij}$, $h_{ij} = h_i + h_j$, $i \neq j$, then the input variable u_i influences that part of the system which is not affected by the input variable u_j and vice versa.*

THEOREM 11.5: *If $h(Q_{Di}, Q_{Dj}) = h_{ij}$, $i \neq j$, and the difference $n - h_{ij} > 0$, then there exists a part of the system influenced neither by the input variable u_i nor by the variable u_j. If, however, $n - h_{ij} = 0$ then the input variables u_i and u_j just affect all modes of the system.*

THEOREM 11.6: *If r_n is the number of input variables which can just influence all modes of a given system, but $r_n < r$, then some modes of the system are affected by further input variables. The number of these input variables is $\varrho = r - r_n$.*

Note 11.1: If the case defined in Theorem 11.6 occurs, the remaining ϱ input variables may be omitted because the system is reachable with r_n selected input variables, or the remaining ϱ variables may be used for improving the quality of control. In such a case the system is said to have auxiliary input variables.

It is evident that the Theorems 11.2 through 11.5 were formulated for the two input variables u_i and u_j, $i \neq j$. A generalization of these theorems for a larger number of input variables is apparently possible but yields no new information.

Theorem 11.5 is associated with the following definition.

DEFINITION 11.2: *A system is minimally reachable with r_m variables if $r_m \leq r$ is the smallest number of input variables which guarantee the reachability of all modes of the system.*

Similarly, we may also generalize the conditions of observability and analyse the structure of multivariable systems with regard to these conditions.

THEOREM 11.7: *The system (11.4) is observable if and only if the rank of the matrix*

$$Q_P = \begin{bmatrix} C \\ CA \\ \vdots \\ CA^{N-1} \end{bmatrix}$$

is equal to n, where n is the state space dimension.

DEFINITION 11.3: *The index of observability μ of a system described by eqns. (11.4) is the lowest value of N satisfying Theorem 11.7.*

With the rank $h(C)$ according to (11.29) the matrix C has altogether p linearly independent rows and the matrix Q_P has altogether $pN \geq n$ rows. Therefore, the lowest possible value of N is $\mu \geq n/p$. The highest possible value of the index of observability is obtained by adding to the p mutually independent rows of the matrix C only one further row from each subsequent matrices $CA, CA^2, ..., CA^{N-1}$. These further rows must necessarily be linearly independent with regard to the preceding ones. Hence, $p + (N - 1) \leq n$ or $\mu \leq n - p + 1$. From both inequalities it follows that the index of observability is within the range

$$n/p \leq \mu \leq n - p + 1 \tag{11.34}$$

Consider the relations

$$h(Q_{Pi}) = h_i, \quad h_i < n$$

$$h(Q_{Pj}) = h_j, \quad h_j < n, \quad i \neq j$$

$$h(Q_{Pij}) = h_{ij}, \tag{11.35}$$

applying to the system (11.4) with $x \in X^n$ and $y \in Y^p$, where $h(\cdot)$ denotes again the rank and the matrices Q_{Pi}, Q_{Pj} and Q_{Pij} are of the form

$$Q_{Pi} = \begin{bmatrix} c_i^T \\ c_i^T A \\ \vdots \\ c_i^T A^{n-1} \end{bmatrix} \quad Q_{Pj} = \begin{bmatrix} c_j^T \\ c_j^T A \\ \vdots \\ c_j^T A^{n-1} \end{bmatrix} \quad Q_{Pij} = \begin{bmatrix} Q_{Pi} \\ Q_{Pj} \end{bmatrix} \tag{11.36}$$

then, as regards the system observability, the following theorems similar to Theorems 11.2 through 11.6 can be formulated:

THEOREM 11.8: *If $h(Q_{Pij}) = h_{ij}$, $h_{ij} < h_i + h_j$, then certain modes of the system are observable by both the output variable y_i and the variable y_j, $i \neq j$.*

THEOREM 11.9: *The difference $\pi_{i,j} = h_i + h_j - h_{ij}$ following from Theorem 11.8 determines the number of modes observable by both the output variable y_i and the variable y_j.*

THEOREM 11.10: *If $h(Q_{Pij}) = h_{ij}$, $h_{ij} = h_i + h_j$, $i \neq j$, then that part of the system which is observable by the output variable y_i is not observable by the output variable y_j and vice versa.*

THEOREM 11.11: *If $h(Q_{Pij}) = h_{ij}$, $i \neq j$, and the difference $n - h_{i,j} > 0$, then there exists a part of the system observable neither by the output variable y_i nor by the variable y_j. If $n - h_{ij} = 0$, then all modes of the system are just observable by the output variables y_i and y_j.*

THEOREM 11.12: *If p_n is the number of output variables by which all modes of a given system are just observable, but $p_n < p$, then some modes of the system are observable by further output variables. The number of these output variables is $\varrho = p - p_n$.*

Note 11.2: If the case defined in Theorem 11.12 occurs, the remaining ϱ output variables may be omitted because the system is observable by p_n selected variables, or the remaining variables may be used for improving the quality of control. In such a case the system is said to have auxiliary variables measured on the plant.

MULTI-INPUT/MULTI-OUTPUT SYSTEMS

Theorem 11.11 is associated with the following definition:

DEFINITION 11.4: *A system is minimally observable with p_m variables if $p_m \leq p$ is the smallest number of output variables which guarantee the observability of all modes of the system.*

Example 11.1: Consider the Jordan matrix of the equation of dynamics of a system whose matrix of dynamics has altogether different eigenvalues, and analyse the system structure as regards the reachability.

$$x(k+1) = \begin{bmatrix} z_1, & 0, & \cdots\cdots, & 0 \\ 0, & z_2, & \cdots\cdots, & 0 \\ \vdots & & z_3, & \vdots \\ \vdots & & & z_4, & \vdots \\ \vdots & & & & z_5, & 0 \\ 0, & \cdots\cdots & & & 0, & z_6 \end{bmatrix} x(k) + \begin{bmatrix} b_{11}, & 0, & 0, & b_{14} \\ b_{21}, & 0, & 0, & 0 \\ b_{31}, & b_{32}, & 0, & 0 \\ b_{41}, & b_{42}, & 0, & 0 \\ 0, & b_{52}, & 0, & 0 \\ 0, & 0, & b_{63}, & 0 \end{bmatrix} u(k) \quad (11.37)$$

The reachability analysis is carried out according to Theorems 11.2 through 11.6. Equation (11.37) may formally be rewritten in the form

$$x(k+1) = J x(k) + [b_1, b_2, b_3, b_4] [u_1, u_2, u_3, u_4]^T \quad (11.38)$$

Calculating now the vector array according to eqn. (11.31) for the first two vectors b_1 and b_2 we obtain

$$b_1, Jb_1, J^2b_1, J^3b_1, J^4b_1, \ldots, \quad (11.39)$$

where the first four vectors of the array (11.39) are mutually linearly independent since the vector b_1 has four non-zero elements so that the input variable $u_1(k)$ is acting upon four modes of the given system while the fifth and all subsequent vectors of the array (10.39) depend on the preceding ones. It is always possible to determine the coefficients k_1 through k_4 so as to obtain

$$k_1 b_1 + k_2 J b_1 + k_3 J^2 b_1 + k_4 J^3 b_1 = J^m b_1, \quad m \geq 4$$

Let us introduce the notation $Q_{D1} = [b_1, Jb_1, J^2b_1, J^3b_1]$ with the matrix rank $h(Q_{D1}) = h_1 = 4$.

Similarly, for the vector b_2 of the matrix B we have, according to (11.31),

$$b_2, Jb_2, J^2b_2, J^3b_2, \ldots, \quad (11.40)$$

where, for the same reason as in (11.39), the first three vectors of the array (11.40) are mutually independent so that the input variable u_2 is acting on three modes of the given system while the fourth and all subsequent vectors of the array (11.40) depend on the preceding ones. Since the matrix of dynamics of the given system is in the Jordan form, it is directly apparent which partial system (i.e. with regard to eigenvalues) is controlled by the variables u_1 and u_2.

Now we calculate

$$h(Q_{D1}, Q_{D2}) = h[b_1, Jb_1, J^2b_1, J^3b_1, b_2, Jb_2, J^2b_2] = 5 \qquad (11.41)$$

since we can always find such coefficients k_1 through k_8 that the matrix $[Q_{D1}, Q_{D2}]$ can be written in the form with five mutually independent columns

$$\begin{bmatrix} k_1b_{11}, & k_3z_1b_{11}, & k_5z_1^2b_{11}, & z_1^3b_{11}, & 0, \\ k_1b_{21}, & k_3z_2b_{21}, & k_5z_2^2b_{21}, & z_2^3b_{21}, & 0, \\ 0, & 0, & 0, & z_3^3b_{31}, & 0, \\ 0, & 0, & 0, & z_4^3b_{41}, & 0, \\ -k_2b_{52}, & -k_4b_{52}, & -k_6b_{52}, & 0, & k_7z_5b_{52} - k_8z_5^2b_{52}, \\ 0, & 0, & 0, & 0, & 0, \end{bmatrix}$$

by means of which any other column of the matrix $[Q_{D1}, Q_{D2}]$ may be expressed as a linear combination.

It is obvious that $h_{12} < h_1 + h_2$. From the form of the matrix B in eqn. (11.37) it is clear that the two modes corresponding to the eigenvalues z_3 and z_4, respectively, are controlled by both the input variable u_1 and the input variable u_2. This is in agreement with Theorems 11.2 and 11.3. The difference $\pi_{12} = h_1 + h_2 - h_{12} = 2$. We calculate that $n - h_{12} = 6 - 5 = 1 > 0$ so that according to Theorem 11.5 there exists a subsystem (corresponding to the eigenvalue z_6) influenced by neither of the variables u_1 and u_2. From the form of the matrix B in eqn. (11.37) it follows that the mode corresponding to the eigenvalue z_6 is controlled by the input variable u_3. Since the vector b_3 of the matrix B has only one non-zero element b_{63}, $Q_{D3} = b_3$ and

$$h(Q_{D3}) = h[b_3] = h_3 = 1 \qquad (11.42)$$

It also holds that

$$h(Q_{D1}, Q_{D2}, Q_{D3}) = h_{123} = 6$$

Since the difference $n - h_{123} = 0$, then, according to Theorem 11.5, the input variables u_1, u_2 and u_3 just influence all modes of the system. It may be noted that u_1, u_2 and u_3 represent the smallest possible number of input variables guaranteeing the system to be reachable. If we omit any of the vectors b_1, b_2 or b_3 and the corresponding input variable, some modes of the system cease to be reachable.

It is apparent that $r_n = 3$, but $r_n < r$, since $\overset{\circ}{r} = 4$. According to Theorem 11.6 some of the modes of the system are influenced by a further input variable u_4. From the column b_4 of the matrix B of the given system it follows that this variable influences only the mode corresponding to the eigenvalue z_1.

Let us now neglect the variable u_4 and the corresponding column b_4, denote $B^* = [b_1, b_2, b_3]$ and calculate the index of reachability v for a minimally reachable system. According to formula (11.32) we have $2 \leq v \leq 3$. To determine the value of v indicated by the last inequality we arrange the matrix Q_D^* using as many successive columns of the matrices B^*, JB^*, J^2B^*, ... as many are needed in order that $h(Q_D^*) = n$.

$$\begin{bmatrix} b_{11}, & 0, & 0, & z_1b_{11}, & 0, & 0, & z_1^2b_{11} \\ b_{21}, & 0, & 0, & z_2b_{21}, & 0, & 0, & z_2^2b_{21} \\ b_{31}, & b_{32}, & 0, & z_3b_{31}, & z_3b_{32}, & 0, & z_3^2b_{31} \\ b_{41}, & b_{42}, & 0, & z_4b_{41}, & z_4b_{42}, & 0, & z_4^2b_{41} \\ 0, & b_{52}, & 0, & 0, & z_5b_{52}, & 0, & 0 \\ 0, & 0, & b_{63}, & 0, & 0, & z_6b_{63}, & 0 \end{bmatrix} \quad (11.43)$$

The last vector $J^2b_1 = J^{v-1}b_1$ is necessary to attain the rank $h = 6$ of the matrix (11.43) because the third and sixth column of the matrix (11.43) are linearly dependent. The index of reachability is, therefore, $v = 3$. Thus there exist three input variables of the vector $u(k)$ by which the given system is transferred from the initial state $x(0)$ to any desired state $x(3)$ in three steps. The input variables are calculated using eqn. (11.30) with $k = 0$, $N = 3$ and $B^* = [b_1, b_2, b_3]$.

$$[B^*, JB^*, J^2B^*] \begin{bmatrix} u(2) \\ u(1) \\ u(0) \end{bmatrix} = x(3) - A^3 x(0) \quad (11.44)$$

Equation (11.44) represents 6 linearly independent equations for 9 unknown components of the vectors $u(0)$, $u(1)$ and $u(2)$. Since the rank $(B^*, JB^*, J^2B^*) = 6$ and the rank of the extended matrix $[B^*, JB^*, J^2B^*, x(3) - A^3 x(0)]$ is also $h^+ = 6$ for any value of $x(3)$ and $h < n_u$, where n_u is the number if unknowns, the set of equations (11.44) has an infinite number of solutions. It should be noted that for $N = 2$ we obtain 6 equations for 6 unknowns but the rank $h(B^*, JB^*) = 5$. In order that this set of equations have a solution the rank of the extended matrix $[B^*, JB^*, x(2) - A^2 x(0)]$ must be $h^+ = 5$, too. This condition, however, restricts the choice of $x(2)$.

We can, therefore, calculate any number of the input vectors $u(0)$, $u(1)$ and $u(2)$ satisfying eqn. (11.44). We may, for instance, select in the matrix $[B^*, JB^*, J^2B^*]$ the following 6 linearly independent vectors:

$$\begin{bmatrix} 0, & z_1b_{11}, & 0, & z_1^2b_{11}, & 0, & 0 \\ 0, & z_2b_{21}, & 0, & z_2^2b_{21}, & 0, & 0 \\ b_{32}, & z_3b_{31}, & z_3b_{32}, & z_3^2b_{31}, & z_3^2b_{32}, & 0 \\ b_{42}, & z_4b_{41}, & z_4b_{42}, & z_4^2b_{41}, & z_4^2b_{42}, & 0 \\ b_{52}, & 0, & z_5b_{52}, & 0, & z_5^2b_{52}, & 0 \\ 0, & 0, & 0, & 0, & 0, & z_6^2b_{63} \end{bmatrix} = [b_2, Jb_1, Jb_2, J^2b_1, J^2b_2, J^2b_3] \quad (11.45)$$

The unknowns $u_1(2)$, $u_3(2)$ and $u_3(1)$ which are multiplied by linearly dependent vectors are set equal to zero. The remaining unknowns are calculated from the equation

$$\begin{bmatrix} u_2(2) \\ u_1(1) \\ u_2(1) \\ u_1(0) \\ u_2(0) \\ u_3(0) \end{bmatrix} = [b_2, Jb_1, Jb_2, J^2b_1, J^2b_2, J^2b_3]^{-1} [x(3) - A^3 x(0)] \qquad (11.46)$$

The reason for the use of this procedure lies in the fact that all the individual variables enter into action at the instant $k = 0$ and do not change in the subsequent steps after having reached a zero value.

Including the variable u_4 among the input variables we can easily ascertain that the rank $h(B, JB) = 6$. The index of reachability follows directly from the equality $JB = J^{\nu-1}B$ so that $\nu = 2$. By adding a further input variable, the index of reachability was reduced from 3 to 2 so that there are two input vectors, $u(0)$ and $u(1)$, whose action allows the system to be transferred from any initial state $x(0)$ to any desired state $x(2)$.

11.3 Canonical forms of multi-input/multi-output systems

11.3.1 Jordan's canonical form

The transformation of state equations into Jordan's canonical form for single-input/single-output systems was discussed in detail in Sec. 10.2 where the procedure was also shown how the simple forms of the matrices b_J and c_J^T may be determined. The results may easily be generalized to apply to multi-input/multi-output systems.

It should, however, be supplemented that for a larger number of input and output variables even the non-cyclic systems having Jordan's blocks with equal eigenvalues may be reachable and observable if the partial systems with equal eigenvalues are mutually independently reachable and observable. This possibility may be verified as follows [68.4]. In the Jordan form of the matrix of dynamics A_J we search out in all Jordan blocks with equal eigenvalues the last row and the corresponding rows of the matrix B_J. The partial systems considered are mutually independently reachable if the mentioned rows of the matrix B_J are mutually linearly independent.

A similar rule also applies to the observability of partial systems with equal eigenvalues except that in this case one must search out the first column of the appropriate Jordan blocks with equal eigenvalue and the corresponding columns of the matrix C_J. Then the partial systems are mutually independently observable if the mentioned columns of the matrix C_J are mutually linearly independent.

It can be said that in general it is possible to achieve that at least n elements of the matrix C_J or B_J will be equal to zero or one, where n is the dimension of the state space. If the system has m Jordan blocks ($m \leq n$), the minimum number of non-zero elements required for a unique description of the system is:

$\qquad m \qquad$ parameters of the matrix of dynamics A_J
$\qquad nr \qquad$ parameters of the input matrix B_J
$\qquad n(p-1) \qquad$ parameters of the output matrix C_J

$n(r + p - 1) + m =$ total number of parameters

If, however, $m = n$, such a system may be described by

$$L_J = n(r + p) \qquad (11.47)$$

coefficients. On the other hand, the largest number of parameters occurs if each input variable is acting on all modes of the system and if all modes are observable by all output variables.

The structure of reachability and observability directly follows from the Jordan canonical form of state equations. If, for instance, the coefficients $b_{v-s+1,i}$ through b_{vi} in the equation

$$x(k+1) = \begin{bmatrix} z_1, & 1, & 0, & \ldots, & 0 \\ 0, & z_1, & 1, & \ldots, & 0 \\ \multicolumn{5}{c}{\ldots\ldots\ldots\ldots} \\ 0, & 0, & 0, & \ldots, & z_1 \end{bmatrix} x(k) + \begin{bmatrix} b_{1i} \\ \vdots \\ b_{(v-s),i} \\ 0 \\ \vdots \\ 0 \end{bmatrix} u_i(k); \quad \begin{array}{l} s = 0, 1, \ldots, v-1 \\ i = 1, 2, \ldots, r \\ j = 1, 2, \ldots, p \end{array}$$

$$y_j(k) = [0, \ldots, 0, c_{v-s}, \ldots, c_v] \, x(k)$$

are equal to zero, then $h(Q_{DJ}) = v - s$, which means that the corresponding input variable is acting only on the partial system of the order $v - s$. A similar conclusion will also be arrived at in considering the system observability when the first elements of the row matrix c_j^T, inclusive of the element c_{v-s-j}^T, are equal to zero. On the other hand it may be stated that if the input variable u_i is acting only on the partial system of the order n_i, the maximum number of non-zero coefficients of the vector b_i will be n_i. If the output variable y_j permits an observability of a partial system of only the m_j-th order, the maximum number of non-zero coefficients of the row vector c_j^T will be m_j. If the Jordan matrix of the n-th order has a diagonal form corresponding to altogether different eigenvalues, then all modes of the system are reachable by the input variable u_i only if all elements of the column b_i are non-zero and are observable by the output variable y_j only if all elements of the row c_j^T are non-zero.

If in the case of a diagonal Jordan matrix the structure of reachability is known then, to describe the system,

$$L_D = np + \sum_{i=1}^{r} n_i \tag{11.48}$$

coefficients must be available in addition to the Jordan matrix eigenvalues where, for minimally reachable systems,

$$\sum_{i=1}^{r} n_i \leq nr - r(r-1) \tag{11.49}$$

since in r rows of the matrix B must always be at least one non-zero element.

Conversely, if in the same case the structure of observability is known, then, to describe the system,

$$L_P = nr + \sum_{j=1}^{p} m_j \tag{11.50}$$

coefficients must be known in addition to the Jordan matrix eigenvalues, where for minimally observable systems,

$$\sum_{j=1}^{p} m_j \leq np - p(p-1) \tag{11.51}$$

since in p columns of the matrix C must always be at least one non-zero element.

If the system is both minimally reachable and minimally observable, the number of parameters in addition to the Jordan matrix eigenvalues must be

$$L_{DP} \leq n(r+p) - p(p-1) - r(r-1) \tag{11.52}$$

The Jordan canonical form of state equations has no doubt certain advantages which make it useful, for instance, in mathematical proofs. For numerical calculations, however, this canonical form is disadvantageous since it requires to calculate the eigenvalues of the system matrix of dynamics, the corresponding eigenvectors and the generalized eigenvectors, besides additional calculation difficulties arising when the eigenvalues are complex conjugate. Therefore, as in the case of single-input/single-output systems, it is advisable to use for numerical calculations the canonical forms of state equations given in the subsequent sections.

Example 11.2: Consider a system described by the state equations

$$x(k+1) = \begin{bmatrix} 0, & 1, & 0, & 0, & 0 \\ -1, & -1, & 1, & 0, & 0 \\ 1, & 0, & -2, & 0, & 0 \\ -1, & 0, & 1, & -1, & 0 \\ 1, & 0, & -1, & 4, & 3 \end{bmatrix} x(k) + \begin{bmatrix} 1, & 2 \\ 1, & 1 \\ 2, & 1 \\ 2, & 1 \\ 1, & 3 \end{bmatrix} u(k)$$

$$y(k) = \begin{bmatrix} 3, & 1, & 1, & 1, & 2 \\ 1, & 2, & 2, & 1, & 1 \end{bmatrix} x(k) \tag{11.53}$$

MULTI-INPUT/MULTI-OUTPUT SYSTEMS

The system was obtained by adding further input and output variables to the system described in Example 10.3. Calculate the simplest form of the matrix C_J.

Solution: Using the transformation matrix $Q = {}^1Q^2Q$ of Example 10.4 (see page 125) we calculate

$$C_J = C {}^1Q {}^2Q = \begin{bmatrix} 4, & 0, & 1, & -1, & 2 \\ 1, & 0, & 2, & 0, & 1 \end{bmatrix} \begin{bmatrix} u_1, & u_2, & u_3, & u_4, & 0 \\ 0, & u_1, & u_2, & 0, & 0 \\ 0, & 0, & u_1, & 0, & 0 \\ 0, & v_2, & v_3, & v_4, & 0 \\ 0, & 0, & 0, & 0, & u_5 \end{bmatrix} =$$

$$= \begin{bmatrix} 4u_1, & 4u_2 - v_2, & 4u_3 + u_1 - v_3, & 4u_4 - v_4, & 2u_5 \\ u_1, & u_2, & u_3 + 2u_1, & u_4, & u_5 \end{bmatrix}$$

It is apparent that the system can be observable by both output variables y_1 and y_2. Choosing $u_1 = u_2 = 1$, $u_3 = -2$, $u_4 = u_5 = 1$, $v_2 = 4$, $v_3 = -7$ and $v_4 = 4$ we arrive at the simplest form of the matrix C_J:

$$C_J = \begin{bmatrix} 4, & 0, & 0, & 0, & 2 \\ 1, & 1, & 0, & 1, & 1 \end{bmatrix}$$

With this solution the matrix ${}^2Q$ is non-singular, the mode corresponding to the eigenvalue $z_2 = 3$ is observable by both output variables, the partial system corresponding to the distinct (non-multiple) eigenvalue $z_1 = -1$ is observable only by the variable y_2 and the partial system corresponding to the triple eigenvalue $z_1 = -1$ is again observable by both output variables.

In the original system with one input and one output variable, analysed in Example 10.3, both subsystems corresponding to the same eigenvalue z_1 are now, after introducing a second input and second output variable, observable and controllable as can easily be verified. The observability of both partial systems is confirmed by the fact that the determinant of the matrix compiled by means of the first and fourth column of the matrix C_J is $\det \begin{bmatrix} 4, & 0 \\ 1, & 1 \end{bmatrix} \neq 0$.

11.3.2 Canonical forms of reachability and observability with one-way coupling

We shall now discuss in more detail the canonical form of observability of multi-input/multi-output systems, specifically its special version designated as the one-way coupling case. This means that a multi-input/multi-output system may be arranged in such a way that its individual partial systems are elements of such a sequence that the state variables of any partial system can act only on the state variables of the succeeding partial systems of this sequence but cannot influence the state variables of the preceding partial systems. In other words, any partial system can only be influenced by the preceding partial systems of the sequence but not by the succeeding partial systems. This coupling between the individual partial systems is well apparent from the form of the individual matrices of the state equations.

In analogy with the questions of observability we could as well discuss the questions of reachability, controllability and reconstructability. Since, however, the reader will be able to derive these alternative canonical forms on the basis of the exemplary case of observability by himself, only some supplementary notes to the mentioned forms will be given here.

The *canonical form of observability* of multi-input/multi-output systems is

$$x_P(k+1) = \begin{bmatrix} 0, 1, \ldots 0 & & 0 & & 0 \\ \vdots & & & & \\ 0, 0, \ldots 1 & & & & \\ \alpha_{11}^T & & & & \\ \hline 0, \ldots & 0 \mid 0, 1, \ldots 0 & & 0 & \\ \vdots & \vdots & & & \\ 0, \ldots & 0 \mid 0, 0, \ldots 1 & & & \\ \alpha_{21}^T & \alpha_{22}^T & & & \\ \hline 0, \ldots & 0 \mid 0, \ldots & 0 \mid 0, 1, \ldots 0 \\ \vdots & \vdots & \vdots \\ 0, \ldots & 0 \mid 0, \ldots & 0 \mid 0, 0, \ldots 1 \\ \alpha_{m1}^T & \alpha_{m2}^T & \alpha_{mn}^T \end{bmatrix} x_P(k) + B_P u(k)$$

$$y(k) = \begin{bmatrix} 1, 0, \ldots 0 & 0, \ldots\ldots & 0 & 0, \ldots\ldots & 0 \\ 0, \ldots & 0 \mid 1, 0, \ldots 0 & \vdots & \vdots \\ \vdots & \vdots & & \vdots \\ 0, \ldots\ldots & 0 \mid 0, \ldots\ldots & 0 \mid 1, 0, \ldots 0 \\ \hline & & c_{P(m+1)}^T & & \\ & & \vdots & & \\ & & c_{P,p}^T & & \end{bmatrix}$$

(11.54)

The first m, $m \leq p$, output variables of the vector y correspond to the partial systems of the order $n_1, n_2, \ldots, n_m$ described by the appropriate blocks of the canonical form of observability (10.75) constituting the diagonal blocks of the matrix of dynamics. These m blocks correspond to the first m rows of the output equation matrix C_P. The triangular block arrangement of the matrix of dynamics is very convenient and enables us to calculate the characteristic polynomial of a multi-input/multi-output system as a product of the characteristic polynomials of the individual diagonal blocks. It should, however, be emphasized that this very convenient triangular block form can only be applied with a one-way partial system coupling as described at the beginning of this section. The degree of the resultant characteristic polynomial is

$$n = \sum_{i=1}^{m} n_i \qquad (11.55)$$

If the multi-input/multi-output system has additional output variables for which $n_i = 0$, $i > m$, then no canonical form of the matrix C_P exists for these variables. In such a case the matrix C_P may include additional rows $c_{P(m+1)}^T, c_{P(m+2)}^T, \ldots, c_{P,p}^T$ whose all elements may be non-zero.

The individual orders n_i and the number of partial systems m, depend on the way the partial system sequence is generated. This fact will be elucidated by an example of a system with three output variables shown symbolically in Fig. 11.2. Here the intersections of the partial systems S_1, S_2 and S_3 bounded by circles show their mutual interaction. In Fig. 11.2 the output variable y_1 has been assigned to the partial system of the highest order which, in the matrix of dynamics A_P, would correspond to the block A_{11}. The output variable y_2 has been assigned to the next partial system S_2 of the order $n_2 < n_1$, corresponding in the matrix A_P to the block A_{22} while the intersection $S_1 \cap S_2$ corresponds to the block A_{21}. The variable y_3 constitutes the output of the partial system S_3 of the lowest order $n_3 < n_2 < n_1$ corresponding in

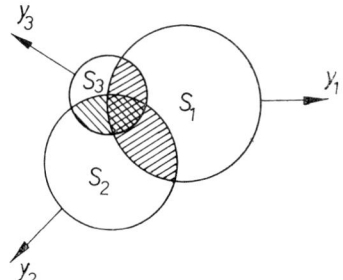
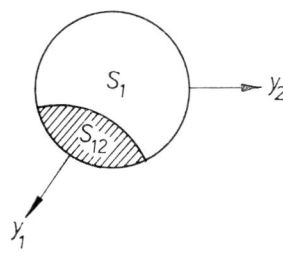

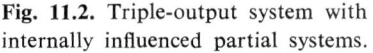

Fig. 11.2. Triple-output system with internally influenced partial systems.

Fig. 11.3. System with two outputs.

the matrix A_P to the block A_{33} while the intersections $S_1 \cap S_3$ and $S_2 \cap S_3$ correspond to the blocks A_{31} and A_{32}, respectively. The output variables may, however, be assigned to the individual partial systems in any other way. This will, of course, change the sequence and dimensions of the individual blocks of A_P.

If the system consisted of only the partial system S_1 and of $S_{12} = S_1 \cap S_2$ as shown in Fig. 11.3, then, for the output variable sequence y_1, y_2, the system would consist of two partial systems with a canonical form of the matrix C_P, while for the output sequence y_2, y_1 there would be only one system being completely observable by the variable y_2.

It is obvious that we always endeavour to describe the system by the smallest number of parameters. In eqns. (11.54) the number of parameters is

nr in the matrix B_P

$nm - n_2 - 2n_3 - \ldots - (m-1)n_m$ in the matrix A_P

$(p-m)n$ in the matrix C_P

The total number of parameters is

$$L_P = n(r+p) - n_2 - 2n_3 - \ldots - (m-1)n_m \tag{11.56}$$

The most unfavourable case occurs if the entire system is observable by all output variables. Then

$$L_P = n(r+p) \tag{11.57}$$

as for the Jordan canonical form.

The smallest total number of parameters given by eqn. (11.56) may be obtained by assigning the first output variable to the partial system of the lowest order, the second output variable to the partial system of the next higher order, etc. In practice, however, different aspects may prevail.

As in the single-variable case, the most important property of the canonical form of observability of multi-input/multi-output systems is the unit matrix of observability:

$$Q_{PF} = \begin{bmatrix} c_{P1}^T \\ c_{P1}^T A_P \\ \vdots \\ c_{P1}^T A_P^{n_1-1} \\ \hdashline c_{P2}^T \\ \vdots \\ \hdashline c_{Pm}^T A_P^{n_m-1} \end{bmatrix} = E \tag{11.58}$$

According to eqn. (10.19) with $^2Q_P = Q_{PF} = E$ we can determine the transformation matrix T which may be used to transform a given system described in a space spanned by an arbitrary basis into a canonical form of observability with the state vector

$$x_P = Tx$$

MULTI-INPUT/MULTI-OUTPUT SYSTEMS

where
$$T = \begin{bmatrix} c_1^T \\ c_1^T A \\ \vdots \\ c_1^T A^{n_1-1} \\ \hline c_2^T \\ \vdots \\ \hline \vdots \\ \hline c_m^T A^{n_m-1} \end{bmatrix} \quad (11.59)$$

The coefficients α_{ij} of the matrix of dynamics in eqns. (11.54) are calculated according to the relation

$$c_{Pi}^T A_P^{n_i} T = c_i^T A^{n_i}, \quad i = 1, 2, \ldots, m$$

$$[\alpha_{i1}^T, \alpha_{i2}^T, \ldots, \alpha_{ii}^T, 0, \ldots, 0] T = c_i^T A^{n_i}, \quad i = 1, 2, \ldots, m \quad (11.60)$$

The calculation proceeds, starting with $i = 1$, by a successive determination of the rows $c_1^T, c_1^T A, c_1^T A^2, \ldots$ until we find the first linearly dependent row given by the relation

$$c_1^T A^{n_1} = \alpha_{11}^T \begin{bmatrix} c_1^T \\ c_1^T A \\ \vdots \\ c_1^T A^{n_1-1} \end{bmatrix} \quad (11.61)$$

Then the first partial system is of the order n_1 and the coefficients of the characteristic polynomial are

$$\alpha_{11}^T = [-a_0, -a_1, \ldots, -a_{n_1-1}] \quad (11.62)$$

where $a_{n_1} = 1$.

Then we calculate the rows $c_2^T, c_2^T A, c_2^T A^2, \ldots$ until we find the first linearly dependent row given by the relation

$$c_2^T A^{n_2} = [\alpha_{21}^T, \alpha_{22}^T] \begin{bmatrix} c_1^T \\ \vdots \\ c_1^T A^{n_1-1} \\ \hline c_2^T \\ \vdots \\ c_2^T A^{n_2-1} \end{bmatrix} \quad (11.63)$$

The second partial system is of the order n_2 and the coefficients of its characteristic polynomial are elements of the row α_{22}^T. The row α_{21}^T represents the coefficients of the coupling between the first and the second partial system. Similarly we proceed further with the rows c_3^T through c_m^T. Thus we determine all linearly independent rows of the transformation matrix T. Now we can calculate $B_P = TB$. From the above procedure it is evident that the inverse of T need not be calculated.

If c_i^T is a linear combination of the previously selected rows $c_j^T A^{n_j}$, $j < i$, or if $c_i^T = 0$, no further partial system is observable by the output variable y_i and, therefore, the relevant order $n_i = 0$. The corresponding row of the canonical form of observability is calculated with the aid of the relation

$$c_{P_i}^T T = c_i^T$$

The *canonical form of reachability* may be calculated in a similar manner. This is a dual form of the canonical form of observability so that $A_D = A_P^T$, $B_D = C_P^T$, $C_D = B_P^T$. However, the orders of the individual partial systems depend now on the structure of reachability and the coefficients of the matrix of dynamics A_D follow from the linear dependence of the columns of the matrix of reachability Q_D.

The *canonical form of controllability* may be calculated by determining first the canonical form of reachability using the method described in the preceding paragraph and then transforming it in such a way that the matrix A assumes the form (11.54) and the matrix B has in the last row of the i-th column a 1 for all values of i denoting the i-th partial system.

A somewhat simpler procedure was described by Anderson and Luenberger [67.1] resulting in an upper triangular block matrix of dynamics.

The *canonical form of reconstructability* may be determined as the dual form of the canonical form of controllability.

In conclusion it should be noted that the canonical forms of multivariable systems, described in this section, are characterized by a one-way coupling between the individual partial systems. These canonical forms are obtained by assigning the first input or output variable to the complete partial system reachable or observable by this single variable. The second input or output variable is attached to the next complete partial system which, in addition, is reachable or observable, respectively, by this variable, etc. With this procedure the first variables are being assigned to the largest partial systems so that it may happen that no partial system will be left to the last variables. The advantage of one-way forms is the relatively small number of coupling blocks and elements and the small number of parameters required to describe the system. In addition, the coefficients of the characteristic polynomials of the individual partial systems may be read from the elements of diagonal blocks and the product of these characteristic polynomials gives the characteristic polynomial of the complete multivariable system.

11.3.3 Canonical forms with two-way coupling

In some cases the one-way coupling within a multi-input/multi-output system is not advantageous. It may, for instance, be desired that the multivariable system be divided into partial systems of approximately the same order, provided that such a division would be realizable by the actual structure of reachability or observability. This version requires that approximately equal tasks could be imposed on the individual input or output variables, respectively. Luenberger [66.5, 67.4] suggested general canonical forms of multivariable systems which have such properties. Mutual couplings arise here in both directions and the one-way forms may be regarded as special cases of the normal two-way canonical form.

The *canonical form of observability* with a two-way coupling is obtained by means of the transformation matrix

$$T = \begin{bmatrix} c_1^T \\ c_1^T A \\ \vdots \\ c_1^T A^{n_1-1} \\ \hline c_2^T \\ \vdots \\ \hline \vdots \\ c_p^T A^{n_p-1} \end{bmatrix} \tag{11.64}$$

where, in contrast with eqn. (11.59), the individual orders n_i are selected so as to have approximately equal values. All p rows fo the matrix C are involved in such a way that the matrix of observability is generated by the rows $c_1^T, c_2^T, \ldots, c_p^T, c_1^T A$, $c_2^T A, \ldots$. As soon as the first row, linearly dependent on the preceding rows, appears in this sequence, the transformation matrix T may be set up of the linearly independent rows. If $c_i^T A^{\mu-1}$ is the last linearly independent row, then μ is the *index of observability* in the sense of Definition 11.3 and at the same time the highest order of the partial system: $\mu = \max [n_i, i = 1, 2, \ldots, p]$.

A system described as indicated above is characterized by that a partial system is assigned to each output variable. The matrix of dynamics, however, is coupled in both directions according to the sequence of state variables.

$$x_P(k+1) = \begin{bmatrix} 0, 1, \ldots 0 & 0, \ldots & 0 & 0, \ldots & 0 \\ \cdots & \cdots & & \cdots & \\ 0, 0, \ldots 1 & 0, \ldots & 0 & 0, \ldots & 0 \\ \alpha_{11}^T & \alpha_{12}^T & & \alpha_{1p}^T & \\ \hline 0, \ldots & 0 & 0, 1, \ldots 0 & 0, \ldots & 0 \\ \cdots & & \cdots & \cdots & \\ 0, \ldots & 0 & 0, 0, \ldots 1 & 0, \ldots & 0 \\ \alpha_{21}^T & & \alpha_{22}^T & \alpha_{2p}^T & \\ \hline 0, \ldots & 0 & 0, \ldots & 0 & 0, 1, \ldots 0 \\ \cdots & & \cdots & & \cdots \\ 0, \ldots & 0 & 0, \ldots & 0 & 0, 0, \ldots 1 \\ \alpha_{p1}^T & & \alpha_{p2}^T & & \alpha_{pp}^T \end{bmatrix} x_P(k) + B_P u(k)$$

$$y(k) = \begin{bmatrix} 1, 0, & \ldots & 0 & 0, & \ldots & & 0 & 0, & \ldots & & 0 \\ 0, & \ldots & 0 & 1, 0, & \ldots & 0 & & \vdots & & & \vdots \\ \vdots & & \vdots & \vdots & & & \vdots & 0, & \ldots & & 0 \\ 0, & \ldots & 0 & 0, & \ldots & & 0 & 1, 0, & \ldots & 0 \end{bmatrix} x_P(k) = C_P x_P(k) = E_P x_P(k)$$

(11.65)

In analogy with eqn. (11.59) we have

$$[\alpha_{i1}^T, \alpha_{i2}^T, \ldots, \alpha_{ip}^T] T = c_i^T A^{n_i} \tag{11.66}$$

where A and C are the original, arbitrary given matrices. Equation (11.66) allows to calculate the transformation matrix T by the method described in Par. 11.3.2, which also allows to transform the given system to the canonical form of observability with two-way coupling.

The *canonical form of reachability* with two-way coupling is the dual form of the form of observability just described, so that $A_D = A_P^T$, $B_D = C_P^T$, $C_D = B_P^T$.

The *canonical form of controllability* has the form

$$x_R(k+1) = \begin{bmatrix} 0, & 1, & \ldots & 0 & 0, & \ldots & & 0 & 0, & \ldots & & 0 \\ \cdots & & & & \cdots & & & & \cdots & & & \\ 0, & 0, & \ldots & 1 & 0, & \ldots & 0 & 0, & \ldots & & 0 \\ \alpha_{11}, & \alpha_{12}, & \ldots & \alpha_{1n_1} & \cdots & & & & \cdots & & & \alpha_{1n} \\ \hline 0, & \ldots & & 0 & 0, 1, & \ldots & & 0, & \ldots & & 0 \\ \cdots & & & & \cdots & & & & \cdots & & & \\ 0, & \ldots & & 0 & 0, 0, & \ldots & 1 & 0, & \ldots & & 0 \\ \alpha_{21}, & \alpha_{22}, & \ldots & \alpha_{2n_1} & \cdots & & & & & \ldots & & \alpha_{2n} \\ \hline 0, & \ldots & & 0 & 0, & \ldots & & 0 & 0, 1, & \ldots & 0 \\ \cdots & & & & \cdots & & & & \cdots & & & \\ 0, & \ldots & & 0 & 0, & \ldots & & 0 & 0, 0, & \ldots & 1 \\ \alpha_{r1}, & \alpha_{r2}, & \ldots & \alpha_{rn_1} & \cdots & & & & & \ldots & & \alpha_{rn} \end{bmatrix} x_R(k) + B_R u(k)$$

(11.67)

where
$$B_R = \begin{bmatrix} 0, 0, & \ldots & 0 \\ \vdots & & \\ 0, 0, & \ldots & 0 \\ 1, 0, & \ldots & 0 \\ \hline 0, 0, & \ldots & 0 \\ \vdots & \vdots & \\ 0, 1, & \ldots & 0 \\ \hline 0, 0, & \ldots & 0 \\ \vdots & \vdots & \vdots \\ 0, 0, & \ldots & 1 \end{bmatrix} \begin{bmatrix} 1, m_{12}, m_{13}, & \ldots & m_{1r} \\ 0, 1, & m_{23}, & \ldots & m_{2r} \\ \cdots & & & \\ 0, & \ldots & & 1 \end{bmatrix} = E_R M$$

Any controllable system defined by the matrices (A, B, C) may be transformed by the matrix $Q = T^{-1}$ to the canonical form of controllability with the state vector $x_R = Tx$. The transformation matrix T is calculated as follows [66.5]:

(1) We calculate the matrix of reachability

$$Q_{D1} = [B, AB, \ldots, A^{n-1}B] \tag{11.68}$$

determine from the left the first n linearly independent columns and construct of them a matrix of the form

$$Q_{D2} = [b_1, Ab_1, \ldots, A^{n_1-1}b_1, b_2, \ldots, A^{n_r-1}b_r] \tag{11.69}$$

where

$$n = \sum_{i=1}^{r} n_i$$

is the system order. If $A^{\nu-1}b_i$ is the last linearly independent column and ν is equal to the highest order of the partial system, $\nu = \max[n_i, i = 1, 2, \ldots, r]$, then ν is the *index of reachability* in the sense of Definition 11.1.

(2) We denote the rows of the inverse matrix Q_{D2}^{-1} as follows:

$$Q_{D2}^{-1} = \begin{bmatrix} \eta_{11}^T \\ \vdots \\ \eta_{1n_1}^T \\ \hline \eta_{21}^T \\ \vdots \\ \hline \vdots \\ \eta_{rn_r}^T \end{bmatrix} \tag{11.70}$$

Then we denote each last row of the individual chains by

$$\eta_i^T = \eta_{in_i}^T, \quad i = 1, 2, \ldots, r$$

(3) The transformation matrix is now

$$T = \begin{bmatrix} \eta_1^T \\ \eta_1^T A \\ \vdots \\ \eta_1^T A^{n_1-1} \\ \hline \eta_2^T \\ \vdots \\ \hline \vdots \\ \eta_r^T A^{n_r-1} \end{bmatrix} \tag{11.71}$$

from which we may calculate

$$A_R = TAT^{-1}, \quad C_R = CT^{-1}$$

As in the case of systems with one-way coupling, where we have determined the elements of A_P as a linear combination of linearly independent rows of the matrix C, which in turn determines further linearly dependent rows, e.g. by (11.60), we may also in this case write that

$$N_R T = \begin{bmatrix} \eta_1^T A^{n_1} \\ \vdots \\ \eta_r^T A^{n_r} \end{bmatrix} \tag{11.72}$$

The elements of the matrix N_R are elements in the rows n_i, $i = 1, 2, \ldots, n_r$, of the matrix A_R in eqn. (11.67).

(4) Next we calculate

$$B_R = TB$$

and, using eqn. (11.67), we may calculate the matrix M. This matrix may, however, be also determined directly when checking for linearly independent vectors of the matrix of reachability by the method described in the literature [69.1].

If $A^{n_i-1} b_i$ is the end of the chain of columns corresponding to b_i in eqn. (11.69), then the further columns in $A^{n_i-1} b_{i+1}, A^{n_i-1} b_{i+2}, \ldots, A^{n_i-1} b_r$ are linearly dependent and it holds that

$$A^{n_i-1} b_j = m_{ij} A^{n_i-1} b_i + \sum_k c_k \beta_k, \quad j > i \tag{11.73}$$

where β_k are the remaining selected linearly independent columns of the matrix (11.69).

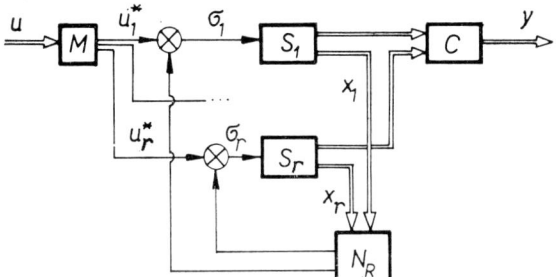

Fig. 11.4. Diagram of a multivariable system in the canonical form of controllability with two-way coupling.

The block diagram of a multi-input/multi-output system in the canonical form of controllability with two-way coupling is shown in Fig. 11.4. In this diagram the individual partial systems are denoted by S_1 through S_r and their input variables by σ_1 through σ_r.

The input vector u is transformed by the filter M into the vector u^* whose each component $u_1^*, u_2^*, \ldots, u_r^*$ is acting on only one partial system. The block arrangement of the individual partial systems is shown, for the system S_1, in Fig. 11.5. In this diagram α_1^T is the first row of the matrix N_R and the vector x_R is the state vector lying in the space spanned by the basis of controllability:

$$x_R^T = [x_1^T, x_2^T, \ldots, x_r^T], \quad \text{where} \quad x_i^T = [x_{11}, x_{12}, \ldots, x_{1n_i}], \quad i = 1, 2, \ldots, r$$

Fig. 11.5. Block arrangement of the partial systems S_i, $i = 1, 2, \ldots, r$.

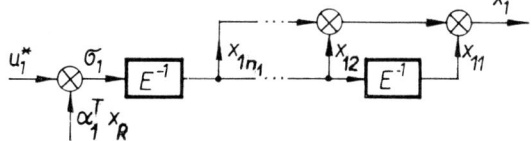

Example 11.3: Consider a system with two inputs and two outputs, whose matrices A, B, C are

$$A = \begin{bmatrix} 0.6, & 0.7, & 0 \\ 0, & 0.75, & 0 \\ 0, & 0, & 0.8 \end{bmatrix}, \quad B = \begin{bmatrix} 0.8, & 0.4 \\ 0, & 0.9 \\ 0.9, & 0.9 \end{bmatrix}, \quad C = \begin{bmatrix} 1, & 0, & 0 \\ 0, & 0, & 1 \end{bmatrix}$$

The corresponding block diagram of the system is shown in Fig. 11.6. Calculate the canonical form of controllability for this system.

Solution: According to eqn. (11.69) we have

$$Q_{D2} = [b_1, Ab_1, b_2] = \begin{bmatrix} 0.8, & 0.48, & 0.4 \\ 0, & 0, & 0.9 \\ 0.9, & 0.72, & 0.9 \end{bmatrix}$$

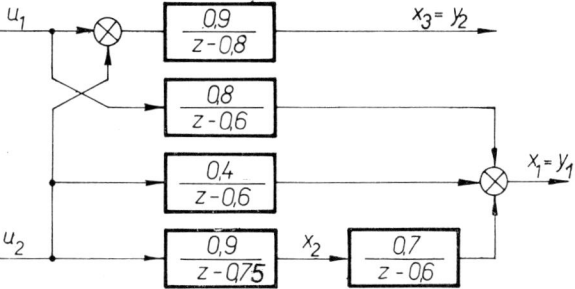

Fig. 11.6. Block diagram relating to Example 11.3.

The linear independence of the columns of this matrix may be verified, e.g. by calculating the determinant of the matrix Q_{D2}.

$$Q_{D2}^{-1} = \begin{bmatrix} 5, & 1.1111, & -3.3333 \\ -6.25, & -2.7778, & 5.5556 \\ 0, & 1.1111, & 0 \end{bmatrix}$$

Let us denote, according to eqn. (11.70), the second row of the matrix Q_{D2}^{-1} by $\eta_1^T = \eta_{12}^T$ and the third row by $\eta_2^T = \eta_{21}^T$. Using these rows we now calculate the transformation matrix (11.71)

$$T = \begin{bmatrix} -6.25, & -2.7778, & 5.5556 \\ -3.75, & -6.4584, & 4.4446 \\ 0, & 1.1111, & 0 \end{bmatrix}$$

and its inverse

$$T^{-1} = \begin{bmatrix} -0.64, & 0.8, & 3.05 \\ 0, & 0, & 0.9 \\ -0.54, & 0.9, & 3.8812 \end{bmatrix}$$

The matrix $A_R = TAT^{-1}$ is now

$$A_R = \begin{bmatrix} 0, & 1, & 0 \\ -0.48, & 1.4, & 0.2156 \\ \hline 0, & 0, & 0.75 \end{bmatrix}$$

The elements of the second and third row of A_R, in general the elements of the rows n_i, $i = 1, 2, \ldots, r$, may as well be determined according to the relation (11.72):

$$N_R = \begin{bmatrix} \eta_1^T A^2 \\ \eta_2^T A \end{bmatrix} T^{-1} = \begin{bmatrix} -2.25, & -7.4688, & 3.5556 \\ 0, & & 0.8333, & 0 \end{bmatrix} T^{-1}$$

$$N_R = \begin{bmatrix} -0.48, & 1.4, & 0.2156 \\ 0, & 0, & 0.75 \end{bmatrix}$$

The matrix $B_R = TB$ is

$$B_R = \begin{bmatrix} 0, & 0 \\ 1, & -3.3125 \\ 0, & 1 \end{bmatrix}$$

and can, by (11.67), also be written in the form

$$B_R = \begin{bmatrix} 0, & 0 \\ 1, & 0 \\ 0, & 1 \end{bmatrix} \begin{bmatrix} 1, & -3.3125 \\ 0, & 1 \end{bmatrix} = E_R M$$

The element $m_{12} = -3.3125$ of the matrix M may also be determined with the aid of the set of equations following from eqn. (11.73):

$$0.87 = 0.48 m_{12} + 0.8 c_1 + 0.4 c_2$$
$$0.675 = \phantom{0.48 m_{12} + 0.8 c_1 + {}} 0.9 c_2$$
$$0.72 = 0.72 m_{12} + 0.9 c_1 + 0.9 c_2$$

Solving this set of equations we obtain the previously given value of m_{12}.

Finally, the matrix $C_R = CT^{-1}$ is

$$C_R = \begin{bmatrix} -0.64, & 0.8, & 3.05 \\ -0.54, & 0.9, & 3.8812 \end{bmatrix}$$

The *canonical form of reconstructability* has the form

$$x_K(k+1) = \begin{bmatrix} 0, & \ldots & 0, & k_{11} & 0, & \ldots & 0, & k_{12} & 0, & \ldots & 0, & k_{1p} \\ 1, & \ldots & 0, & k_{21} & \vdots & & \vdots & \vdots & \vdots & & & \vdots \\ \ldots & \ldots & \ldots & \ldots & & & & & & & & \\ 0, & \ldots & 1, & k_{n_1 1} & 0, & \ldots & 0, & k_{n_1 2} & 0, & \ldots & 0, & k_{n_1 p} \\ \hline 0, & \ldots & 0, & \vdots & 0, & \ldots & 0, & \vdots & 0, & \ldots & 0, & \vdots \\ \vdots & & \vdots & \vdots & 1, & \ldots & 0, & \vdots & \vdots & & & \vdots \\ & & & & \ldots & \ldots & \ldots & & & & & \\ 0, & \ldots & 0, & \vdots & 0, & \ldots & 1, & \vdots & 0, & \ldots & 0, & \vdots \\ \hline 0, & \ldots & 0, & \vdots & 0, & \ldots & 0, & \vdots & 0, & \ldots & 0, & \vdots \\ \vdots & & \vdots & \vdots & \vdots & & \vdots & \vdots & 1, & \ldots & 0, & \vdots \\ \vdots & & \vdots & \vdots & \vdots & & \vdots & \vdots & & & & \vdots \\ 0, & \ldots & 0, & k_{n1} & 0, & \ldots & 0, & k_{n2} & 0, & \ldots & 1, & k_{np} \end{bmatrix} x_K(k) + B_K u(k)$$

$$y(k) = L \begin{bmatrix} 0, & \ldots & 0, & 1 & 0, & \ldots & & & & \ldots & 0 \\ 0, & \ldots & & & & \ldots & 0, & 1 & 0, & \ldots & \ldots & 0 \\ \vdots & & & & & & & & & & & \\ 0, & \ldots & & & & & & & & \ldots & 0, & 1 \end{bmatrix} c_K(k) = LE_K x_K(k)$$

(11.74)

where the matrix L is of the form

$$L = \begin{bmatrix} 1, & 0, & 0, & \ldots & 0 \\ l_{21}, & 1, & 0, & \ldots & 0 \\ l_{31}, & l_{32}, & 1, & \ldots & 0 \\ \vdots & & & & \\ l_{p1}, & l_{p2}, & \ldots & & 1 \end{bmatrix}$$

In many cases the matrix L is a unit matrix.

The matrix $Q = T^{-1}$, which transforms a given state vector x of the observable system (A, B, C) to the vector x_K lying in the space spanned by the basis of reconstructability according to the relation

$$x_K = Tx \quad \text{or} \quad Qx_K = x$$

is calculated as follows [66.5]:

(1) We determine n linearly independent rows of the matrix of observability

$$Q_{P1} = \begin{bmatrix} C \\ CA \\ \vdots \\ CA^{n-1} \end{bmatrix} \qquad (11.75)$$

If $c_i^T A^{\mu-1}$ is the last linearly independent row, then μ is the index of observability and at the same time the highest order of the partial system: $\mu = \max [n_i, i = 1, 2, ..., p]$.

(2) We arrange the linearly independent rows of the matrix (11.75) in the form

$$Q_{P2} = \begin{bmatrix} c_1^T \\ c_1^T A \\ \vdots \\ c_1^T A^{n_1-1} \\ \hdashline c_2^T \\ \vdots \\ \hdashline \vdots \\ c_p^T A^{n_p-1} \end{bmatrix} \qquad (11.76)$$

Now we denote the individual columns of the inverse matrix as follows:

$$Q_{P2}^{-1} = [\varphi_{11}, \varphi_{12}, ..., \varphi_{1n_1}, \varphi_{21}, ..., \varphi_{pn_p}] \qquad (11.77)$$

where each last column of the individual chains is denoted by

$$\varphi_i = \varphi_{in_i}, \quad i = 1, 2, ..., p$$

(3) The transformation matrix $Q = T^{-1}$ is

$$Q = [\varphi_1, A\varphi_1, ..., A^{n_1-1}\varphi_1, \varphi_2, ..., A^{n_p-1}\varphi_p] \qquad (11.78)$$

Now we can calculate

$$A_K = Q^{-1}AQ, \quad B_K = Q^{-1}B$$

As in the case of systems with one-way coupling, where we have determined the elements of the matrix A_P as a linear combination of linearly independent rows of the matrix C, which in turn determines further linearly dependent rows, e.g. by (11.60), we may also in this case write that

$$QN_K = [A^{n_1}\varphi_1, A^{n_2}\varphi_2, ..., A^{n_p}\varphi_p] \qquad (11.79)$$

The elements of the matrix N_K are the elements of the matrix A_K. When using this procedure the inversion of the matrix Q will be avoided.

(4) Next we calculate

$$C_K = CQ$$

and, if needed, we may calculate the matrix L by using eqns. (11.74). This matrix may, however, also be determined directly when checking for linear independence of the

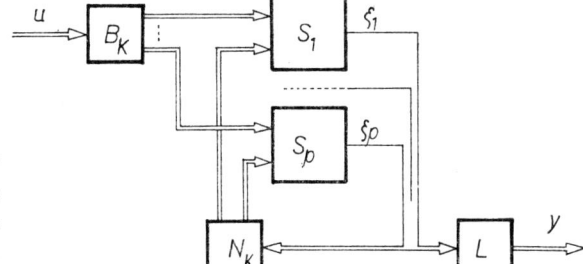

Fig. 11.7. Block diagram of a multivariable system in the canonical form of reconstructability with two-way coupling.

rows of the matrix of observability by the method described in the literature [69.1]. In analogy with eqn. (11.73) it holds that

$$c_j^T A^{n_i-1} = l_{ji} c_i^T A^{n_i-1} + \sum_k d_k \gamma_k^T \tag{11.80}$$

where γ_k^T are the remaining selected linearly independent rows of the matrix (11.76).

Figure 11.7 shows the block diagram of a multi-input/multi-output system in the canonical form of reconstructability with two-way couplings. In this diagram ζ_1

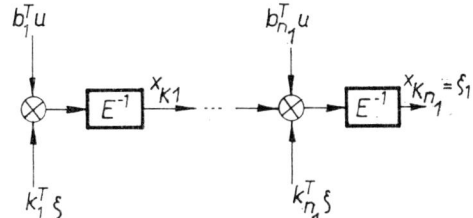

Fig. 11.8. Block arrangement of the partial systems S_i, $i = 1, 2, ..., p$.

through ζ_p denote the output variables of the individual partial systems S_1 through S_p, respectively. Since the matrix L is non-singular, it holds that

$$\zeta = L^{-1} y$$

The block diagram of the individual partial systems is shown in Fig. 11.8.

Example 11.4: Consider a system with two input and two output variables described by the matrices

$$A = \begin{bmatrix} 0.2, & 0.5, & 0 \\ 0, & 0.3, & 0 \\ 0, & 0, & 0.4 \end{bmatrix}, \quad B = \begin{bmatrix} 0.3, & 1 \\ 0, & 0.4 \\ 1, & 1 \end{bmatrix}, \quad C = \begin{bmatrix} 2, & 1, & -1 \\ -1, & 0.5, & 2 \end{bmatrix}$$

Calculate the canonical form of reconstructability for this system.

Solution: According to eqn. (10.76) we calculate the transformation matrix

$$Q_{P2} = \begin{bmatrix} c_1^T \\ c_1^T A \\ c_2^T \end{bmatrix} = \begin{bmatrix} 2, & 1, & -1 \\ 0.4, & 1.3, & -0.4 \\ -1, & 0.5, & 2 \end{bmatrix}$$

and its inverse

$$Q_{P2}^{-1} = \begin{bmatrix} 0.7568, & -0.6757, & 0.2432 \\ -0.1081, & 0.8108, & 0.1081 \\ 0.4054, & -0.5405, & 0.5946 \end{bmatrix} = [\varphi_{11}, \varphi_{12}, \varphi_{21}]$$

It should be noted that according to eqn. (11.64) the matrix Q_{P2} is identical with the matrix T and may, together with $Q_{P2}^{-1} = T^{-1}$, be used for the determination of the canonical form of observability.

In the example we must further calculate the transformation matrix Q according to eqn. (11.78)

$$Q = [\varphi_1, A\varphi_1, \varphi_2] = \begin{bmatrix} -0.6757, & 0.2703, & 0.2432 \\ 0.8108, & 0.2432, & 0.1081 \\ -0.5405, & -0.2162, & 0.5946 \end{bmatrix}$$

and its inverse

$$Q^{-1} = \begin{bmatrix} -0.6215, & 0.7891, & 0.1107 \\ 2, & 1, & -1 \\ 0.1622, & 1.0810, & 1.4108 \end{bmatrix}$$

Now we can calculate

$$A_K = Q^{-1} A Q = \begin{bmatrix} 0, & -0.0612, & -0.0119 \\ 1, & 0.5108, & 0 \\ 0, & -0.0146, & 0.3872 \end{bmatrix}$$

According to eqn. (11.79) we can calculate only the columns k_i, $i = 1, 2, \ldots, p$, of the matrix (11.74) set up of the coefficients of the characteristic polynomials of partial systems and of the elements of coupling members. We obtain

$$N_K = Q^{-1}[A^2\varphi_1, A\varphi_2] = Q^{-1}\begin{bmatrix} 0.1757, & 0.1027 \\ 0.0730, & 0.0324 \\ -0.0865, & 0.2378 \end{bmatrix}$$

$$N_K = \begin{bmatrix} -0.0612, & -0.0119 \\ 0.5108, & 0 \\ -0.0146, & 0.3872 \end{bmatrix}$$

The result agrees with the original solution.

The matrix $B_K = Q^{-1}B$ is

$$B_K = \begin{bmatrix} -0.0758, & -0.8264 \\ -0.4, & 1.4 \\ 1.4675, & 2.0134 \end{bmatrix}$$

Finally, the output matrix

$$C_K = CQ = \begin{bmatrix} 0, & 1, & 0 \\ 0, & -0.5811, & 1 \end{bmatrix} = \begin{bmatrix} 1, & 0 \\ -0.5811, & 1 \end{bmatrix}\begin{bmatrix} 0, & 1, & 0 \\ 0, & 0, & 1 \end{bmatrix}$$

$$C_K = LE_K$$

where $l_{21} = -0.5811$. The element l_{21} of the matrix L may also be determined by solving the set of equations resulting from eqn. (11.80).

$$-0.2 = 0.4l_{21} + 2d_1 - d_2$$
$$-0.35 = 1.3l_{21} + d_1 + 0.5d_2$$
$$0.8 = -0.4l_{21} - d_1 + 2d_2$$

By solving this set of equations we again calculate $l_{21} = -0.5811$.

11.4 Minimum description

Let us consider a multivariable system with r inputs and p outputs, whose dynamic properties are described by the matrix of transfer functions $G(z)$, having the dimension $(p; r)$. Each element of this matrix is a transfer function transforming the input variable u_i, $i = 1, 2, \ldots, r$, into the component of the corresponding output variable y_j, $j = 1, 2, \ldots, p$. If the system is described in state space by the matrices (A, B, C), the matrix of transfer functions $G(z)$ may be calculated according to formula (11.8). From this formula it is evident that the denominators of all elements of the matrix $G(z)$ may be identical polynomials in z, but the numerators may be different for each element. The number of these numerators is rp and, in general, each of them may be a polynomial with n parameters, where n is the order of the denominator and the number of parameters of the denominator. Thus, for this type of multivariable system description, the required number of parameters will be $n(rp + 1)$. In contrast with this we have shown in eqns. (11.47) and (11.57) that a multivariable system with r inputs and p outputs can be described by only $n(r + p)$ parameters. Both these formulae give the same number of parameters only for systems with one input and one output. If $r > 1$ and $p > 1$, the description by means of the matrix $G(z)$ has a redundant number of coefficients. Thus the question arises whether there exists a description by means of the matrix of transfer functions $G(z)$ which would manage with $n(r + p)$ parameters.

DEFINITION 11.5: *The description of a multivariable system with r inputs and p outputs is minimum if, for $r \geq 1$ and $p \geq 1$, the required number of parameters does not exceed $n(r + p)$, where n is the order of the system.*

The solution of this problem was described in the literature [71.1]. If the system is represented by state equations with the matrices (A, B, C), then, eliminating the state vector x in the way already used in Par. 7.7.4, we arrive at the set of equations

$$\begin{bmatrix} y(k) \\ y(k+1) \\ \vdots \\ y(k+n-1) \end{bmatrix} = \begin{bmatrix} C \\ CA \\ \vdots \\ CA^{n-1} \end{bmatrix} x(k) + \begin{bmatrix} 0, & 0, & \ldots & 0 \\ CB, & 0, & \ldots & 0 \\ CAB, & CB, & \ldots & 0 \\ \vdots & & & \\ CA^{n-2}B, & CA^{n-3}B, & \ldots, & CB \end{bmatrix} \begin{bmatrix} u(k) \\ u(k+1) \\ u(k+2) \\ \vdots \\ u(k+n-2) \end{bmatrix}$$
(11.81)

The matrix at the vector $x(k)$ is the matrix of observability of the system. We know from eqn. (11.58) that for the canonical form of observability with one-way coupling this is a unit matrix and it is, therefore, reasonable to transform first the system (A, B, C) to this form in order that the calculation of the vector x be as simple as possible. We, therefore, continue to operate with the matrices A_p, B_p and C_p. Assigning to the first m output variables y_i, $i = 1, 2, \ldots, m$, the partial systems of the order n_i, $i = 1, 2, \ldots, m$, where $n_1 + n_2 + \ldots + n_m = n$, we may express the state vector of the complete system as follows:

$$x(k) = \begin{bmatrix} y_1(k) \\ y_1(k+1) \\ \vdots \\ y_1(k+n_1-1) \\ \hline y_2(k) \\ y_2(k+1) \\ \vdots \\ y_2(k+n_2-1) \\ \hline \vdots \\ y_m(k+n_m-1) \end{bmatrix}$$

MULTI-INPUT/MULTI-OUTPUT SYSTEMS 183

$$-\begin{bmatrix} 0, & 0, & \cdots & & & 0 \\ b_1^T, & 0, & \cdots & & & 0 \\ \cdots\cdots\cdots\cdots\cdots\cdots\cdots\cdots\cdots\cdots\cdots\cdots \\ b_{n_1-1}^T, & \cdots & b_1^T, & 0, & \cdots & 0 \\ \hline 0, & 0, & \cdots & & & 0 \\ b_{n_1+1}^T, & 0, & \cdots & & & 0 \\ \cdots\cdots\cdots\cdots\cdots\cdots\cdots\cdots\cdots\cdots\cdots\cdots \\ b_{n_1+n_2-1}^T, & \cdots & b_{n_1+1}^T, & 0, & \cdots & 0 \\ \hline \vdots & & \vdots & & & \\ b_{n-1}^T, & \cdots & b_{n-n_m+1}^T, & 0, & \cdots & 0 \end{bmatrix} \begin{bmatrix} u(k) \\ u(k+1) \\ \vdots \\ u(k-2+\max_i n_i) \end{bmatrix}$$

$$x(k) = y^*(k) - B^* u^*(k) \tag{11.82}$$

where b_i^T, $i = 1, 2, \ldots, n$, are the relevant rows of the matrix B_P. Substituting eqn. (11.82) into eqn. (11.54) we eliminate the vectors $x(k)$ and $x(k+1)$. In the i-th partial system the rows 1 through $n_i - 1$ always yield a trivial solution but the last row represents the difference equation

$$y^*(k+1) = A_P y^*(k) + (B_P - A_P B^*) u^*(k) + B^* u^*(k+1)$$

$$y_i(k+n_i) - \alpha_{ii}^T y_i^{**}(k) = \sum_{j=0}^{n_i-1} \beta_{ij}^T u(k+j) + \sum_{j=1}^{i-1} \alpha_{ij}^T y_j^{**}(k) \tag{11.83}$$

where

$$y_i^{**}(k) = \begin{bmatrix} y_i(k) \\ y_i(k+1) \\ \vdots \\ y_i(k+n_i-1) \end{bmatrix}, \quad i = 1, 2, \ldots, m; \quad u(k+j) = \begin{bmatrix} u_1(k+j) \\ u_2(k+j) \\ \vdots \\ u_r(k+j) \end{bmatrix}$$

The subscript i in eqn. (11.83) denotes relevance to the last row of the matrix block of the i-th partial system and β_{ij}^T are the rows with the dimension $(1; r)$ of the individual blocks of the matrices B^* and $(B_P - A_P B^*)$ in eqn. (11.83) pertinent to the vectors $u(k+j)$, $j = 0, 1, \ldots, (n_i - 1)$. The further output variables y_{m+1} through y_p do not appear in the difference equations.

Example 11.5: A system with two input and two output variables is described by the Jordan form of state equations

$$x_J(k+1) = \begin{bmatrix} 0.2, & 0, & 0 \\ 0, & 0.4, & 0 \\ 0, & 0, & 0.6 \end{bmatrix} x_J(k) + \begin{bmatrix} 1, & 2 \\ 2, & 1 \\ 1, & 2 \end{bmatrix} u(k)$$

$$y(k) = \begin{bmatrix} 1, & 0, & 1 \\ 1, & 2, & 2 \end{bmatrix} x_J(k)$$

Determine the minimum description!

Solution: Since the state description of the system is in the Jordan form we can directly recognize that the output variable $y_1(t)$ observes the partial system of the second order with the eigenvalues 0.2 and 0.6 and the output variable $y_2(t)$ observes the complete system. To be able to compare the number of coefficients of the minimum description with the number of coefficients of the difference equations corresponding to the given problem we calculate first the matrix of transfer functions

$$G(z) = C(zE - A)^{-1} B =$$

$$= \begin{bmatrix} \dfrac{2z - 0.8}{z^2 - 0.8z + 0.12}, & \dfrac{4z - 1.6}{z^2 - 0.8z + 0.12} \\ \dfrac{7z^2 - 5.4z + 0.88}{z^3 - 1.2z^2 + 0.44z - 0.048}, & \dfrac{8z^2 - 6z + 1.04}{z^3 - 1.2z^2 + 0.44z - 0.048} \end{bmatrix}$$

With the aid of this matrix we can now easily write the pertinent difference equations:

$$y_1(k+2) - 0.8\, y_1(k+1) + 0.12\, y_1(k) =$$
$$= -0.8\, u_1(k) + 2\, u_1(k+1) - 1.6\, u_2(k) + 4\, u_2(k+1)$$

$$y_2(k+3) - 1.2\, y_2(k+2) + 0.44\, y_2(k+1) - 0.048\, y_2(k) =$$
$$= 0.88\, u_1(k) - 5.4\, u_1(k+1) + 7\, u_1(k+2) +$$
$$+ 1.04\, u_2(k) - 6\, u_2(k+1) + 8\, u_2(k+2)$$

To determine the minimum description we transform the given system, as already described in this paragraph, to the economical form of observability. The relevant transformation matrix is

$$T = \begin{bmatrix} c_{J1}^T \\ c_{J1}^T A_J \\ c_{J2}^T \end{bmatrix} = \begin{bmatrix} 1, & 0, & 1 \\ 0.2, & 0, & 0.6 \\ 1, & 2, & 2 \end{bmatrix} ; \quad T^{-1} = -\dfrac{1}{0.8} \begin{bmatrix} -1.2, & 2, & 0 \\ 0.2, & 1, & -0.4 \\ 0.4, & -2, & 0 \end{bmatrix}$$

The matrices of the canonical form of observability are

$$C_P = C_J T^{-1} = \begin{bmatrix} 1, & 0 & 0 \\ 0, & 0 & 1 \end{bmatrix}$$

$$A_P = T A_J T^{-1} = \begin{bmatrix} 0, & 1, & 0 \\ -0.12, & 0.8, & 0 \\ \hline -0.5, & 1.5, & 0.4 \end{bmatrix} ; \quad B_P = T B_J = \begin{bmatrix} 0, & 4 \\ 0.8, & 1.6 \\ \hline 7, & 8 \end{bmatrix}$$

MULTI-INPUT/MULTI-OUTPUT SYSTEMS

On comparison with (11.54) we easily determine

$$\alpha_{11} = [-0.12, \ 0.8]$$
$$\alpha_{21} = [-0.5, \ 1.5]$$
$$\alpha_{22} = 0.4$$

According to eqn. (11.82) the state vector

$$x(k) = \begin{bmatrix} y_1(k) \\ y_1(k+1) \\ y_2(k) \end{bmatrix} - \begin{bmatrix} 0, & 0 \\ 2, & 4 \\ 0, & 0 \end{bmatrix} \begin{bmatrix} u_1(k) \\ u_2(k) \end{bmatrix}$$

Substituting now this state vector into the equation

$$x_P(k+1) = A_P \, x_P(k) + B_P \, u(k)$$

we obtain

$$\begin{bmatrix} y_1(k+1) \\ y_1(k+2) \\ y_2(k+1) \end{bmatrix} = \begin{bmatrix} 0, & 1, & 0 \\ -0.12, & 0.8, & 0 \\ -0.5, & 1.5, & 0.4 \end{bmatrix} \begin{bmatrix} y_1(k) \\ y_1(k+1) \\ y_2(k) \end{bmatrix} + \begin{bmatrix} 0, & 0 \\ -0.8, & -1.6 \\ 4, & 2 \end{bmatrix} \begin{bmatrix} u_1(k) \\ u_2(k) \end{bmatrix} +$$

$$+ \begin{bmatrix} 0, & 0 \\ 2, & 4 \\ 0, & 0 \end{bmatrix} \begin{bmatrix} u_1(k+1) \\ u_2(k+1) \end{bmatrix}$$

The relevant difference equations are

$$y_1(k+2) - 0.8 \, y_1(k+1) + 0.12 \, y_1(k) =$$
$$= -0.8 \, u_1(k) + 2 \, u_1(k+1) - 1.6 \, u_2(k) + 4 \, u_2(k+1)$$
$$y_2(k+1) - 0.4 \, y_2(k) = 4 \, u_1(k) + 2 \, u_2(k) - 0.5 \, y_1(k) + 1.5 \, y_1(k+1)$$

It is evident that the first difference equation is the same as the one given. It describes the partial system of the second order with the eigenvalues 0.2 and 0.6 which is observable by the output variable y_1. The second difference equation of the resultant solution refers only to the system of the first order with the eigenvalue 0.4 which is observable by the output variable y_2. This means that the description of the second-order partial system does not repeat in the second difference equation.

The number of coefficients of the given description is 15 both in the Jordan form of state equations and the relevant difference equations. In this specific example the minimum description has 11 coefficients. According to eqn. (11.47) the maximum number of these coefficients must not exceed 12. This means that the system calculated in this example could have been described by a smaller number of coefficients than that corresponding to the upper limit for the minimum description.

12

Stability

12.1 Basic definitions

The stability of systems may be tested by many different methods. Our subsequent discussion will be restricted to certain methods suitable for the systems described by state equations and to the methods by which certain relationships may be elucidated.

Let us consider a discrete equation of dynamics of the form

$$x(t_{k+1}) = f[x(t_k), u(t_k), t_k] \qquad (12.1)$$

where $t_k < t_{k+1}$ for all values of k and $t_k \to \infty$ for $k \to \infty$.

DEFINITION 12.1: *The system described by eqn. (12.1) is called "free" if $u(t_k) \equiv 0$ for all values of t_k.*

The equation of a free system has, therefore, the general form

$$x(t_{k+1}) = f[x(t_k), t_k] \qquad (12.2)$$

where the zero vector at $u(t_k)$ in the function $f[\ldots]$ has, for simplicity, been omitted.

Let us denote the balanced or equilibrium state of the system by x_r.

DEFINITION 12.2: *A system is in the equilibrium state if $x_r = f(x_r, t_k)$ for all values of t_k.*

DEFINITION 12.3: *A system is linear if f in eqn. (12.1) is a linear function of x and u.*

DEFINTION 12.4: *A system is time-invariant if the function f does not depend on the time t_k, i.e. if $f(x, u, t_k) = f(x, u)$.*

In the subsequent discussion it will be assumed that for time-invariant systems $t_{k+1} - t_k = T$ for all values of k.

STABILITY

DEFINITION 12.5: *A system is called autonomous if it is free and time-invariant.*

Let us denote the solution of the difference equation (12.2) by $x(t_k)$. In linear and non-linear time-invariant systems, the solution $x(t_k)$ depends on the initial state $x(t_0)$ and on the initial instant t_0. In linear time-invariant systems the solution depends only on the initial state $x(t_0)$ and does not depend on the choice of t_0.

DEFINITION 12.6: *The equilibrium state x_r is stable in the Lyapunov sense if for a small deviation from the equilibrium state the solution remains in the neighbourhood of the equilibrium state.*

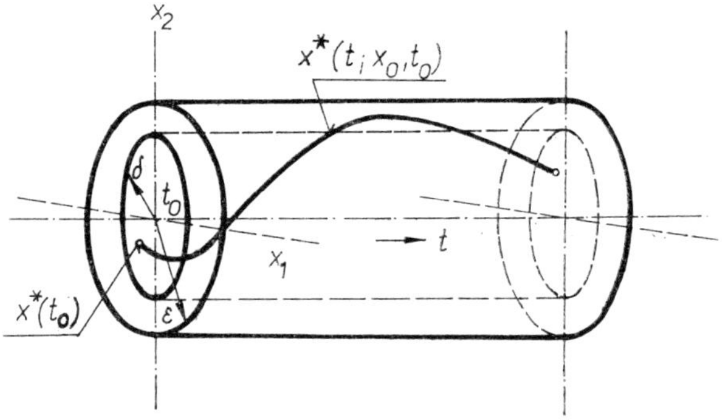

Fig. 12.1. Trajectory of a free system of the 2nd order stable in the Lyapunov sense.

Definition 12.6 may be expressed more precisely as follows: The equilibrium state of the free system given by eqn. (12.2) is stable in the Lyapunov sense or, which is equivalent, the system Σ is stable in the Lyapunov sense at zero input if there exists, for any value of t_0 and any value of $\varepsilon > 0$, such a number $\delta(\varepsilon, t_0)$ that for $\|x^*(t_0) - x_r\| \leq \delta$ the solution beginning at $x^*(t_0)$ remains in the ε-neighbourhood of x_r, which means that the norm $\|x^*(t_k) - x_r\| \leq \varepsilon$ for all values of $t_k > t_0$. For a system of the second order this concept of stability is illustrated in Fig. 12.1. The stability in the Lyapunov sense is uniform if δ does not depend on t_0 as, for instance, in the case of autonomous systems.

DEFINITION 12.7: *The equilibrium state is asymptotically stable if*

$$\lim_{k \to \infty} [x^*(t_k) - x_r] = 0.$$

This definition, in contrast with Definition 12.6, means that after the deviation from the equilibrium state the solution $x^*(t_k)$ returns to the equilibrium state x_r.

DEFINITION 12.8: *The region of asymptotic stability is a set of all initial states $x^*(t_0)$ at which the solutions $x^*(t_k)$ ending in the equilibrium state x_r begin.*

DEFINITION 12.9: *The equilibrium state is asymptotically stable in the large if there exists only one equilibrium state x_r and if the region of stability is the complete state space.*

If the equilibrium state of a system satisfies the definition 12.9, then such a system is called asymptotically stable in the large.

To the linear system the law of superposition applies. If $x_1^*(t)$ and $x_2^*(t)$ are partial solutions, then also $a\,x_1^*(t) + b\,x_2^*(t)$ is a solution where a and b are arbitrary real constants. For this reason finite regions of stability cannot occur for linear systems.

The following theorem holds:

THEOREM 12.1: *If the equilibrium state x_r of a linear system is asymptotically stable, then the system is asymptotically stable in the large, too.*

Example 12.1: Consider the equation of dynamics of a time-invariant linear system

$$x(k+1) = \begin{bmatrix} a, & 0, & 0 \\ 0, & \alpha e^{j\omega}, & 0 \\ 0, & 0, & \alpha e^{-j\omega} \end{bmatrix} x(k)$$

with the initial conditions for $k = 0$

$$x(0) = \begin{bmatrix} x_1(0) \\ \alpha_0 e^{j\omega_0} \\ \alpha_0 e^{-j\omega_0} \end{bmatrix}$$

Determine when the system is asymptotically stable and when it is stable in the Lyapunov sense.

The solution, according to eqn. (7.1), is

$$x(k) = A^k x(0) = \begin{bmatrix} a^k x_1(0) \\ \alpha^k \alpha_0 e^{j(\omega_0 + k\omega)} \\ \alpha^k \alpha_0 e^{-j(\omega_0 + k\omega)} \end{bmatrix}$$

It is evident that the system will be asymptotically stable if $|a| < 1$ and $|\alpha| < 1$. For $|a| = 1$ and $|\alpha| \leq 1$ as well as for $|a| \leq 1$ and $|\alpha| = 1$, the system will be stable in the Lyapunov sense.

A system stable in the Lyapunov sense but not asymptotically stable is referred to as a *system on the boundary of stability* or as a *neutrally stable system*.

In our example the values $|a|$ and $|\alpha|$ are obviously absolute values of the eigenvalues of the system matrix of dynamics. Let us remind that according to Note 10.2 these eigenvalues are invariant with respect to linear transformation.

The general solution of a linear autonomous discrete system described by the equation

$$x(k+1) = A\,x(k) \tag{12.3}$$

has the form

$$x(k) = A^k x(0) \tag{12.4}$$

STABILITY

The solution consists of components whose asymptotic behaviour depends on $|z_i|^k$, where z_i are eigenvalues of the matrix A. Evidently $\lim_{k\to\infty} x(k) = 0$ for $|z_i| < 1$, $i = 1, 2, \ldots, n$, where n is the order of the matrix A and $\lim_{k\to\infty} x(k) \to \infty$ if at least one eigenvalue $|z_i| > 1$.

According to this consideration we may state the following theorem:

THEOREM 12.2: *A linear autonomous discrete system is asymptotically stable if and only if the absolute values of all eigenvalues of the matrix of dynamics, A, are smaller than one.*

This statement follows from the properties of the Jordan matrix $A_J = Q^{-1}AQ$ provided that this matrix has distinct eigenvalues z_i, $i = 1, 2, \ldots, n$, since in that case

$$x(k) = A_J^k x(0) = \begin{bmatrix} z_1^k x_1(0) \\ z_2^k x_2(0) \\ \vdots \\ z_n^k x_n(0) \end{bmatrix} \tag{12.5}$$

To be able to prove Theorem 12.2 for multiple eigenvalues of the matrix A, let us consider a more general case where by substituting

$$x(k) = F \zeta(k) \tag{12.6}$$

where F is a non-singular linear matrix operator, we transform the matrix A into the upper triangular matrix

$$A^* = F^{-1}AF \tag{12.7}$$

Equation (12.3) then assumes the form

$$\begin{bmatrix} \zeta_1(k+1) \\ \vdots \\ \zeta_n(k+1) \end{bmatrix} = \begin{bmatrix} a_{11}, & a_{12}, & \ldots, & a_{1n} \\ 0, & a_{22}, & \ldots, & a_{2n} \\ \multicolumn{4}{c}{\dotfill} \\ 0, & 0, & \ldots, & a_{nn} \end{bmatrix} \begin{bmatrix} \zeta_1(k) \\ \vdots \\ \zeta_n(k) \end{bmatrix} \tag{12.8}$$

For the state variable ζ_n we now have

$$\zeta_n(k+1) = a_{nn} \zeta_n(k) \tag{12.9}$$

For an arbitrary $k \geq 0$ and for arbitrary initial conditions

$$\zeta(0) = F^{-1} x(0) \tag{12.10}$$

we have

$$\zeta_n(k) = a_{nn}^k \zeta_n(0) \tag{12.11}$$

From eqn. (12.11) it follows that $\lim_{k \to \infty} \zeta_n(k) = 0$ if and only if $|a_{nn}| < 1$. For the state variable ζ_{n-1}

$$\zeta_{n-1}(k + 1) = a_{n-1,n-1} \zeta_{n-1}(k) + a_{n-1,n} \zeta_n(k) \tag{12.12}$$

holds. Now $\lim_{k \to \infty} \zeta_{n-1}(k) = 0$ if and only if $|a_{n-1,n-1}| < 1$ and if at the same time $\lim_{k \to \infty} \zeta_n(k) = 0$. In a similar manner we may proceed up to the variable ζ_1. The conditions of stability may be stated as follows:

THEOREM 12.3: *A linear autonomous discrete system is asymptotically stable if and only if the absolute values of all elements of the main diagonal of the triangular matrix $A^* = F^{-1}AF$ are smaller than one, where A is the matrix of dynamics of the system and F is a non-singular transformation matrix.*

It can easily be seen that Theorem 12.3 defines the conditions of stability not only for the triangular matrices A^* but also for the Jordan matrices A_J with multiple eigenvalues.

The conditions of stability stated above may also be formulated by means of the characteristic polynomial

$$P(z) = \det(zE - A) = z^n + a_{n-1}z^{n-1} + \ldots + a_1 z + a_0 \tag{12.13}$$

THEOREM 12.4: *A linear autonomous discrete system is asymptotically stable if and only if the absolute values of all roots of the characteristic polynomial* (12.13) *are smaller than one, i.e. are within the unit circle in the z-plane.*

The above Theorems 12.2 through 12.4 represent a mathematical formulation of the conditions of stability but are not suited for practical testing. On the other hand, the rules based on matrix trace, given e.g. in Ref. [63.1], are very suitable. The starting point is the fact that the trace of the matrix A, i.e. the sum of the elements on the main diagonal of A, is equal to the sum of the eigenvalues of A. If the system is stable, then all eigenvalues of the matrix A are $|z_i| < 1$, $i = 1, 2, \ldots, n$, where n is the dimension of A. Then it holds that

$$|\mathrm{tr}\,[A]| < n \tag{12.14}$$

where

$$\mathrm{tr}\,[A] = \sum_{i=1}^{n} z_i \tag{12.15}$$

and also

$$|\mathrm{tr}\,[A^k]| < n, \quad k = 1, 2, 3, \ldots \tag{12.16}$$

If, however, at least one eigenvalue $|z_i| > 1$, the system will be unstable and

$$|\mathrm{tr}\,A^k| > n, \quad \sum_{i=1}^{n} z_i^k > n \tag{12.17}$$

STABILITY

for a sufficiently large number k. In practice, only a few values of k are usually enough to test the stability. The facts stated above may be summarized by the theorem:

THEOREM 12.5: *A linear autonomous discrete system is asymptotically stable if and only if the absolute value of the trace of the matrix A^k is smaller than the dimension of the matrix A for all values of natural k.*

This theorem also may be used for testing the stability of linear autonomous continuously working plants if in the characteristic equation

$$\det(sE - F) = 0 \tag{12.18}$$

of these plants we substitute for s the operator

$$s = \frac{z+1}{z-1} \tag{12.19}$$

by which the left half of the complex plane s is transformed into the region bounded in the complex plane z by a unit circle with the center at the origin.

Performing the mentioned substitution we obtain

$$\det(sE - F) = \det\left\{\frac{1}{1-z}[z(F - E) - (F + E)]\right\} \tag{12.20}$$

so that the characteristic equation corresponding to a conformal mapping of the plane s into the plane z by means of the operator (12.19) has the form

$$\det[z(F - E) - (F + E)] = 0 \tag{12.21}$$

or

$$\det[zE - (F - E)^{-1}(F + E)] = 0$$

It also holds that

$$(F - E)^{-1}(F + E) = E + 2(F - E)^{-1} \tag{12.22}$$

Similarly we may transform the characteristic equation

$$\det(zE - A) = 0 \tag{12.23}$$

of a discrete system by means of the operator

$$z = \frac{s+1}{s-1} \tag{12.24}$$

into the form

$$\det[s(A - E) - (A + E)] = 0 \tag{12.25}$$

and use for the stability test the criteria known from the theory of continuous systems.

Let the characteristic equation (12.23) have the general form

$$z^n + \alpha_{n-1} z^{n-1} + \ldots + \alpha_1 z + \alpha_0 = 0 \tag{12.26}$$

If $n > 3$, the current method of determination of the characteristic equation by resolving $\det [zE - A]$ into third-degree determinants and by using the Sarrus rule [63.7] is very inconvenient and unsuitable for numerical calculation on a digital computer. We shall, therefore, show another method by which the coefficients of the characteristic equation (12.26) may be determined without meeting with the mentioned disadvantages.

According to the Cayley-Hamilton theorem [47.1] the square matrix A must satisfy its own characteristic equation and, therefore, eqn. (12.26) may be rewritten in the form

$$A^n + \alpha_{n-1} A^{n-1} + \ldots + \alpha_1 A + \alpha_0 E = 0 \tag{12.27}$$

We now multiply this equation from the right by an arbitrary vector x_0 and determine n vectors x_i, $i = 1, 2, \ldots, n$, as follows:

$$x_1 = A x_0$$
$$x_2 = A x_1$$
$$\vdots$$
$$x_n = A x_{n-1} \tag{12.28}$$

With these vectors, eqn. (12.27) has now the form

$$x_n + \alpha_{n-1} x_{n-1} + \ldots + \alpha_1 x_1 + \alpha_0 x_0 = 0 \tag{12.29}$$

It represents the following set of non-homogeneous linear algebraic equations with unknown coefficients α_i, $i = 0, 1, 2, \ldots, n - 1$:

$$x_{n,1} + \alpha_{n-1} x_{n-1,1} + \ldots + \alpha_1 x_{1,1} + \alpha_0 x_{0,1} = 0$$
$$x_{n,2} + \alpha_{n-1} x_{n-1,2} + \ldots + \alpha_1 x_{1,2} + \alpha_0 x_{0,2} = 0$$
$$\vdots$$
$$x_{n,n} + \alpha_{n-1} x_{n-1,n} + \ldots + \alpha_1 x_{1,n} + \alpha_0 x_{0,n} = 0$$

Solving this set of equations we determine the coefficients of the characteristic equation (12.26) and this may then be subjected to a stability test using, for instance, the known Routh-Shur stability theorem. The first vector x_0 is selected naturally as simple as possible, for example $x_0 = [1, 0, \ldots, 0]^T$.

The coefficients of the characteristic equation may alternatively be calculated by Bôcher's formulae [47.1]. If we resolve the characteristic equation (12.26) into the product of root factors

$$(z - z_1)(z - z_2) \ldots (z - z_n) = 0 \tag{12.30}$$

the coefficient α_{n-1} in eqn. (12.26) is evidently

$$\alpha_{n-1} = -(z_1 + z_2 + \ldots + z_n) \tag{12.31}$$

As alreaay stated, the sum of the matrix eigenvalues is equal to the trace of the matrix A so that it also holds that

$$\alpha_{n-1} = -\text{tr}[A] = -S_1 \tag{12.32}$$

According to eqns. (10.24) and (10.51) we have

$$A^k = QA_J^k Q^{-1} \tag{12.33}$$

The trace of the Jordan matrix raised to the k-th power is obviously

$$\text{tr}[A_J^k] = z_1^k + z_2^k + \ldots + z_n^k \tag{12.34}$$

Since eqn. (12.33) applies, it also holds that

$$\text{tr}[A^k] = \text{tr}[A_J^k] = z_1^k + z_2^k + \ldots + z_n^k = S_k \tag{12.35}$$

Because integer powers of the matrix A can easily be calculated on a computer we can as well use the mentioned properties for the calculation of the coefficients of the characteristic equation (12.26). We can easily make sure of the validity of the Bôcher formulae

$$\alpha_{n-1} = -S_1$$
$$\alpha_{n-2} = -\tfrac{1}{2}(\alpha_{n-1}S_1 + S_2)$$
$$\alpha_{n-3} = -\tfrac{1}{3}(\alpha_{n-2}S_1 + \alpha_{n-1}S_2 + S_3)$$
$$\vdots$$
$$\alpha_0 = -\frac{1}{n}(\alpha_1 S_1 + \alpha_2 S_2 + \ldots + \alpha_{n-1}S_{n-1} + S_n) \tag{12.36}$$

These formulae also apply to multiple and complex eigenvalues.

Another useful definition of the stability of a linear continuous or discrete system refers to the behaviour of the system tested by the input and output variables.

DEFINITION 12.10: *A linear system is stable if, beginning with the instant t_0 when the system was in the equilibrium state, the response to each bounded input variable $|u(t)| \leq M_u < \infty$ is again a bounded output variable $|y(t)| = M_y < \infty$.*

The requirement imposed on stability in the sense of this definition is more severe than in the Lyapunov stability definition 12.6.

With discrete systems the independent time variable assumes only the discrete values $t = kT$, $k = 0, 1, 2, \ldots$.

It should be noted that the testing of a dynamic system according to its input and output is complete only for systems whose state is reachable and observable. In a system whose unreachable mode is unstable, the relevant initial value does not vanish and causes an unlimited rise of the corresponding output variable component. On the other hand, an unobservable unstable mode may cause the corresponding state variable to assume undesirable, dangerous values without a possibility of this being indicated by the output variable.

The latter case may occur even when the unstable output variable component, e.g. $\exp[kT]$, $k = 0, 1, 2, \ldots$, is being sampled with the period $T = \pi$. Sampling then takes place just at the instants when this component passes through zero value and cannot, therefore, be observed by the output variable.

For time-invariant and linear systems the stability expressed by Definition 12.10 is identical with the asymptotic stability of reachable and observable systems since the requirement of Definition 12.10 can be satisfied if all roots of the characteristic equation are within a unit circle of the z-plane.

For the sake of completeness it should be noted that for systems described by a bounded discrete impulse response, for which $s(k) = 0$ for $k < 0$ and $|s(k)| \leq \leq M_s < \infty$ for $k \geq 0$, and whose output is calculated by means of the convolution summation,

$$y(k) = \sum_{r=0}^{\infty} s(r) u(k - r), \quad k = 0, 1, 2, \ldots \quad (12.37)$$

the conditions of stability are defined as follows:

DEFINITION 12.11: *A linear discrete system to which the convolution summation* (12.37) *applies is stable if and only if*

$$\sum_{k=0}^{\infty} |s(k)| < \infty \quad (12.38)$$

From the stated relationships between the description by state equations, the transfer function and the impulse response, and from the stated interrelations and properties of the matrix of dynamics and the characteristic polynomial it obviously follows that the stability test of the individual dynamic parts as well as of the entire feedback system may be carried out using a great variety of stability theorems accepted from the theory of continuously working systems. Some of these theorems were also modified for discrete systems so that there exists the continuous as well as the discrete version. A thorough review of stability theorems may be found in the literature [71.25]. Our present discussion will, therefore, be confined to only two stability theorems which may be considered to be advantageous even in computer-aided calculations. This concerns the so-called Routh-Shur stability test and the Lyapunov stability theorem which will now be described.

STABILITY

12.2 Routh-Shur stability test

In this section we shall show that it is not necessary to transform the system characteristic equation by substituting the operator (12.24) with a following use of the stability theorems known from the theory of continuous control. Let us show here without proof, which was published for example in [65.3], the discrete version of the Routh-Shur test allowing to check the system stability directly in the z-plane according to the coefficients of the characteristic equation

$$A(z) = 0 \qquad (12.39)$$

where

$$A(z) = a_0 + a_1 z + \ldots + a_n z^n$$

As already stated, the necessary and sufficient condition of the stability of linear time-invariant discrete systems is that all roots z_v, $v = 1, 2, \ldots, n$, of the characteristic polynomial $A(z)$ lie in the region $|z| < 1$. The polynomial satisfying this condition will henceforth be referred to as the stable polynomial.

Let us calculate the polynomial difference

$$A_1(z) = \frac{1}{z}\left(\sum_{v=0}^{n} a_v z^v - \frac{a_0}{a_n} \sum_{v=0}^{n} a_{n-v} z^v\right) \qquad (12.40)$$

It is evident that the coefficients at the zero power of z cancel and we obtain a new polynomial of the $(n-1)$st degree whose coefficients are

$$a_v^{(1)} = a_{v+1} - \frac{a_0}{a_n} a_{n-v-1}, \quad v = 0, 1, \ldots, n-1 \qquad (12.41)$$

We may, therefore, write

$$A_1(z) = \sum_{v=0}^{n-1} a_v^{(1)} z^v \qquad (12.42)$$

Let us now introduce the general notation

$$A_j(z) = \sum_{v=0}^{n-j} a_v^{(j)} z^v$$

$$a_v^{(j+1)} = a_{v+1}^{(j)} - \frac{a_0^{(j)}}{a_{n-j}^{(j)}} a_{n-j-v-1}^{(j)}$$

$$a_v^{(0)} = a_v \qquad (12.43)$$

Using this notation we can formulate the following theorem:

THEOREM 12.6: *The polynomial $A(z)$ is stable if and only if $a^{(j)}_{n-j} > 0$ for $j = 1, 2, \ldots, n$.*

In this stability test, the polynomials $A(z), A_1(z), \ldots, A_n(z)$ have the degree n, $(n-1), \ldots, 0$ and, therefore, this procedure is designated as the *reduction of the characteristic polynomial* from the left.

We may, however, proceed by reducing successively the characteristic polynomial from the opposite side. Then

$$A_{j+1}(z) = \sum_{\nu=0}^{n-j} a^{(j)}_\nu z^\nu - \frac{a^{(j)}_{n-j}}{a^{(j)}_0} \sum_{\nu=0}^{n-j} a^{(j)}_{n-j-\nu} z^\nu \tag{12.44}$$

so that

$$a^{(j+1)}_\nu = a^{(j)}_\nu - \frac{a^{(j)}_{n-j}}{a^{(j)}_0} a^{(j)}_{n-j-\nu}$$

and the procedure is then designated as a reduction of the characteristic polynomial from the right. Both procedures are quite equivalent.

In the numerical calculation of $a^{(j)}_{n-j}$ it is reasonable to proceed as follows:

We set up the matrix

$$\begin{vmatrix} a_n, & a_{n-1}, & \ldots & a_2, & a_1, & a_0 \\ a_0, & a_1, & \ldots & a_{n-2}, & a_{n-1}, & a_n \\ a^{(1)}_{n-1}, & a^{(1)}_{n-2}, & \ldots & a^{(1)}_1, & a^{(1)}_0, & 0 \\ a^{(1)}_0, & a^{(1)}_1, & \ldots & a^{(1)}_{n-2}, & a^{(1)}_{n-1}, & 0 \\ a^{(2)}_{n-2}, & a^{(2)}_{n-3}, & \ldots & a^{(2)}_0, & 0, & 0 \\ \multicolumn{6}{c}{\ldots\ldots\ldots\ldots\ldots\ldots} \\ a^{(n)}_0, & 0, & \ldots & 0, & 0, & 0 \end{vmatrix} \tag{12.45}$$

This matrix has $2n + 1$ rows. Each even row is obtained by rearranging the preceding row in the way shown in matrix (12.45). Each odd row is then obtained from the preceding two rows as, for instance, the fifth row from the third and fourth row:

(a) We multiply the fourth row by the coefficient $a^{(1)}_0/a^{(1)}_{n-1}$.

(b) We subtract the fourth row from the third one and obtain the fifth row.

In order that the given polynomial be stable, the elements marked off in the matrix (12.45) must be positive.

Example 12.2: Test whether the polynomial $A(z) = z^4 - 2.75z^2 + 2.25z - 0.5$ is stable.

STABILITY

Solution: We write the relevant matrix

$$\begin{bmatrix} +1, & 0, & -2.75, & +2.25, & -0.5 \\ -0.5, & +2.25, & -2.75, & 0, & +1 \\ +0.75, & +1.13, & -4.13, & 2.25, & \\ +2.25, & -4.13, & +1.13, & 0.75, & \\ -4.50, & & & & \\ \cdots & \cdots & \cdots & \cdots & \cdots \end{bmatrix}$$

It can be seen that $a_n^{(2)} < 0$ and, therefore, according to Theorem 12.6, $A(z)$ is an unstable polynomial. It may be checked that $A(z)$ has an unstable root $z = 2$.

12.3 Lyapunov's stability theorem

The second Lyapunov theorem provides sufficient conditions of asymptotic stability for nonlinear dynamic systems and necessary as well as sufficient conditions of asymptotic stability for linear autonomous systems.

The second Lyapunov theorem [1892.1] modified for discrete systems [58.2, 60.1, 71.26] may be formulated as follows:

THEOREM 12.7: *Consider a discrete, free dynamic system described by the equation*

$$x(k + 1) = f[x(k), k]$$

where $f[0, k] = 0$ *for all values of k. Suppose that there exists such a scalar function* $V(x, k)$ *that* $V(0, k) = 0$ *for all values of k and that:*

(a) $V(x, k)$ *is positive definite, i.e. there exists such a continuous non-decreasing scalar function* α *that* $\alpha(0) = 0$ *and that*

$$0 < \alpha(\|x\|) \leq V(x, k)$$

for all values of k and all values of $x \neq 0$;

(b) *there exists such a continuous scalar function* γ *that* $\gamma(0) = 0$ *and*

$$\Delta V(x, k) = V[x^*(k + 1), k + 1] - V[x^*(k), k] \leq -\gamma(\|x\|) < 0$$

for all values of k and all values of $x \neq 0$;

(c) *there exists such a continuous non-decreasing scalar function* β *that* $\beta(0) = 0$ *and*

$$V(x, k) \leq \beta(\|x\|)$$

for all values of k and all values of $x \neq 0$;

(d) $\alpha(\|x\|) \to \infty$ *for* $\|x\| \to \infty$.

Then the equilibrium state $x_r = 0$ is asymptotically stable in the large and $V(x, k)$ is the Lyapunov function.

Theorem 12.7 evidently applies also to nonlinear, time-variant dynamic systems. Condition (a) requires the Lyapunov function to be positive definite and condition (b) calls for a negative definite first difference of the Lyapunov function. Condition (c) ensures that the Lyapunov function converges to zero for $\|x\| \to 0$ and, finally, condition (d) must be satisfied if asymptotic stability in the large is to apply.

Note 12.1: Instead of condition (b) it will suffice if $\Delta V(x^*, k)$ is negative semi-definite but not identically equal to zero. This means that $\Delta V(x^*, k) \leq 0$ for all values of x^*, but that there exists no value of K and no such value of x^* that $\Delta V(x^*, k) = 0$ for $k > K$.

Note 12.2: Theorem 12.7 introduces the symbol $\|x\|$ which denotes the Euclidean norm as in all cases henceforward.

For free time-invariant dynamic systems Theorem 12.7 may be simplified as follows:

THEOREM 12.8: *If the system is autonomous, V needs only be considered as a function of x such that $V(0) = 0$ and that:*

(a_1) $V(x) > 0$ *for* $x \neq 0$;

(b_1) $\Delta V(x) < 0$ *for* $x \neq 0$;

(c_1) $V(x)$ *is continuous in x*;

(d_1) $V(x) \to \infty$ *for* $\|x\| \to \infty$;

Note 12.3: The condition (b) in Theorem 12.8 may be replaced as follows:

(b_{11}) $\Delta V(x) < 0$ for all values of x;
(b_{12}) $\Delta V(x^*)$ is not identically equal to zero if $k \geq K$ for any value of K and any value of $x \neq 0$.

Note 12.3 states the conditions on which $\Delta V(x)$ may be negative semidefinite.

Note 12.4: Theorem 12.8 defines the necessary and sufficient conditions for linear time-invariant systems as the Lyapunov function may always be chosen in the form $V(x) = \|x\|$.

The practical consequences of the Lyapunov theorem will be elucidated by an example.

Example 12.3: A continuously working autonomous system of the second order with the eigenvalues $\lambda_{1,2} = \alpha \pm j\omega$ is described by the state equation in real form

$$x'(t) = F x(t) = \begin{bmatrix} \alpha & -\omega \\ \omega & \alpha \end{bmatrix} x(t) \tag{12.46}$$

STABILITY

The corresponding discrete state equation is

$$x(k+1) = A\,x(k) = e^{FT}\,x(k) = e^{\alpha T}\begin{bmatrix} \cos \omega T, & -\sin \omega T \\ \sin \omega T, & \cos \omega T \end{bmatrix} x(k) \quad (12.47)$$

Let us investigate the properties of this system as regards stability in the sense of the second Lyapunov theorem and calculate the Lyapunov function V for $\exp[\alpha T] = 0.5$, $\cos \omega T = 0.4$, $x(0) = [1, 0]$. From these values it follows that $\omega T \doteq 66.4°$ and $\sin \omega T = 0.916$.

Solution: The factor $\exp[\alpha T]$ determines the change in length of the vector x in the sense of the Euclidean norm $\|x\| = (x^T x)^{1/2} = (x_1^2 + x_2^2)^{1/2}$ within one sampling interval. The matrix factor turns the vector x in each interval through an angle ωT. For $\alpha < 0$, i.e. for $\exp[\alpha T] < 1$, all points $x(k)$ lie on a spiral tending toward the zero point. A suitable Lyapunov function is such a function V that the equation $V = \text{const.}$ is an equation of the circle. In other words, the Lyapunov function may be represented by the quadratic form

$$V = x^T x = x_1^2 + x_2^2 \quad (12.48)$$

The function V is a positive definite function of the components of the vector x and its first difference is

$$\Delta V(x^*) = V[x^*(k+1)] - V[x^*(k)] =$$
$$= x^{*T}(k+1)\,x^*(k+1) - x^{*T}(k)\,x^*(k) =$$
$$= x^{*T}(k)(A^T A - E)\,x^*(k)$$

Substituting for A from eqn. (12.47) we obtain

$$\Delta V(x^*) = (e^{2\alpha T} - 1)\,x^{*T}(k)\,x^*(k) \quad (12.49)$$

Here $\Delta V(x^*)$ is negative definite if $\alpha < 0$, i.e. if $e^{\alpha T} < 1$. This result confirms the previously stated condition that the absolute value of the eigenvalues of the matrix of dynamics must be smaller than one (Theorem 12.4).

Using the given numerical values we have

$$x(k+1) = \begin{bmatrix} 0.2, & -0.458 \\ 0.458, & 0.2 \end{bmatrix} x(k)$$

For the individual integer values of k we calculate:

k	$x_1(k)$	$x_2(k)$	$V(k)$
0	1	0	1
1	0.2	0.458	0.25
2	−0.170	0.183	0.062
3	−0.118	−0.0412	0.0156
4	−0.00473	−0.0623	0.00388

In Fig. 12.2 the Lyapunov function is represented in the (x_1, x_2)-plane. The figure also shows the circles corresponding to $V = $ const. The radius of these circles is $\sqrt{(V)} = (x_1^2 + x_2^2)^{1/2}$.

From Fig. 12.2 one may imagine that the chosen Lyapunov function (12.48) is not the only one possible. The curves for $V = $ const. could as well be, for instance, ellipses with a low value of longitudinal eccentricity so that all trajectories of $x^*(k)$

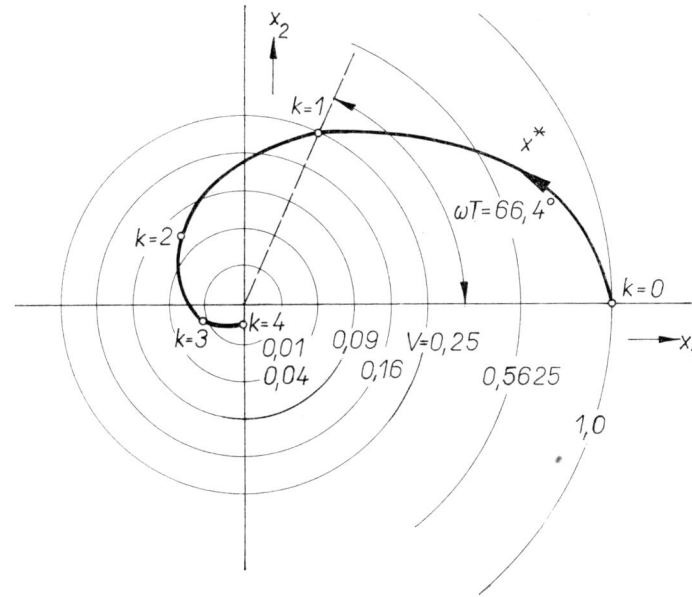

Fig. 12.2 Lyapunov function relating to Example 12.3.

would, with increasing k, intersect the curves for $V = $ const. from outside to inside. If, however, these ellipses had a great longitudinal eccentricity, they could also be intersected by the trajectories $x^*(k)$ in the opposite sense. In such a case ΔV is indefinite and the Lyapunov theorem yields no solution.

The example shows that Lyapunov functions are either suitable or unsuitable. For linear systems there always exists one most suitable, optimum Lyapunov function. In our example this optimum function was, with regard to the form of the matrix of dynamics (12.46), the function (12.48). In the solution of most problems one can manage with the suitable Lyapunov function and the optimum Lyapunov function need not be searched for. On the other hand, when determining a control algorithm with a bounded controlling variable, e.g. $|u| \leq k_u$, it is useful to know the optimum function V. Consider, for instance, the non-homogeneous state equation

$$x(k + 1) = A\,x(k) + b\,u(k) \qquad (12.50)$$

STABILITY

with A according to eqn. (12.47) and with $V = x^T x$. Then

$$\Delta V = (x^T A^T + u b^T)(A x + b u) - x^T x =$$
$$= x^T (A^T A - E) x + 2 u b^T A x + u^2 b^T b \qquad (12.51)$$

For high values of $|x|$, the quantity ΔV assumes the highest negative value if we choose the control law

$$u = -k_u \operatorname{sgn} [b^T A x] \qquad (12.52)$$

In this case V decreases in each step by a value corresponding to the boundary of the controlling variable.

This example was, of course, selected only for an illustration of the behaviour of the Lyapunov function and for the calculation of the control law (12.52). The stability may be tested directly with the aid of eqns. (12.46) and (12.47), whence it follows that the system is stable if $\alpha < 0$, i.e. if the real part of the eigenvalues of the system matrix of dynamics is negative, as known from the theory of linear continuous systems.

The optimum Lyapunov function may be determined when transforming the autonomous state equation

$$x(k+1) = A\, x(k) \qquad (12.53)$$

by substitution $x_J = Q_J x$ into the Jordan form; then the optimum Lyapunov function is

$$V = x_J^T x_J = x^T Q_J^T Q_J x = x^T P x \qquad (12.54)$$

where P is a symmetric and positive definite matrix.

This method of calculation of the Lyapunov function is very inconvenient for practical purposes since the transformation to the Jordan form requires to determine first the eigenvalues of the matrix A, and we want to avoid this step in stability checking. It shows, however, how the Lyapunov function may be selected as a function of the vector x.

Calculating ΔV with respect to the trajectory of x^* we obtain

$$\Delta V(x^*) = x^{*T}(k+1) P\, x^*(k+1) - x^{*T}(k) P\, x^*(k) =$$
$$= x^{*T}(k)(A^T P A - P) x^*(k) =$$
$$= -x^{*T}(k)\, Q\, x^*(k) \qquad (12.55)$$

where Q is again a symmetric positive definite matrix:

$$-Q = A^T P A - P \qquad (12.56)$$

The condition of stability in the sense of Lyapunov's theorem may now be formulated as follows:

THEOREM 12.9: *The autonomous discrete system* (12.53) *is stable if there exist positive definite symmetric matrices Q and P satisfying eqn.* (12.56).

Equation (12.56) is designated as the *Lyapunov equation*. It also represents the linear algebraic part of the matrix Riccati equation.

In testing stability according to Theorem 12.9 we content ourselves with only a suitable Lyapunov function. We have here the choice of two possibilities:

(1) We choose the positive definite symmetric matrix P, calculate the matrix Q of eqn. (12.56) and check whether it is positive definite. In most attempts of this kind the P chosen is unsuitable, the result being a negative definite matrix Q even if the system is stable. Therefore, this procedure gives no answer.

(2) We choose the positive definite matrix Q, calculate the matrix P of eqn. (12.56) and check whether it is positive definite. This procedure always gives a unique decision on the system stability.

Consider the positive definite symmetric matrix Q. Then, according to eqn. (12.55), we may write that

$$\Delta V[x^*(k)] = -x^{*T}(k) Q x^*(k) = V[x^*(k+1)] - V[x^*(k)] \qquad (12.57)$$

for all values of x^*. In eqn. (12.57) we have the solution for x from the instant k to the instant $k+1$; we determine how much the value of the Lyapunov function has decreased and can reconstruct the Lyapunov function $V(x)$ so it gives the corresponding fall $\Delta V[x^*(k)]$. From eqn. (12.57) if follows that

$$V[x^*(k+1)] = V[x^*(k)] - x^{*T}(k) Q x^*(k)$$
$$V[x^*(k+2)] = V[x^*(k+1)] - x^{*T}(k+1) Q x^*(k+1) =$$
$$= V[x^*(k)] - x^{*T}(k) Q x^*(k) - x^{*T}(k+1) Q x^*(k+1)$$
$$\vdots$$
$$V[x^*(k+N)] = V[x^*(k)] - \sum_{m=k}^{N-1} x^{*T}(m) Q x^*(m) \qquad (12.58)$$

If the system is asymptotically stable, then

$$\lim_{N \to \infty} V[x^*(k+N)] = V(0) = 0 \qquad (12.59)$$

Equation (12.58) then gives the sought Lyapunov function

$$V[x^*(k)] = \sum_{m=k}^{\infty} x^{*T}(m) Q x^*(m) \qquad (12.60)$$

The Lyapunov function is, therefore, the negative sum of all values of ΔV along the trajectory corresponding to the future values of the solution of x^*, i.e. $x^*(k)$, $x^*(k+1)$, ...

If the function

$$V[x^*(k)] = \sum_{m=k}^{\infty} x^{*T}(k)(A^T)^{m-k} Q A^{m-k} x^*(k) = x^{*T}(k) P x^*(k) \qquad (12.61)$$

according to eqns. (12.60) and (12.54) is positive definite and the series

$$P = \sum_{m=k}^{\infty} (A^T)^{m-k} Q A^{m-k} \qquad (12.62)$$

is converging, then the function in question is a Lyapunov function and the condition of Theorem 12.7 stated for the function β is also satisfied.

The necessary and sufficient conditions for the matrix P to be positive definite are the following:

(1) a non-singular matrix Q exists such that $P = Q^T Q$, or

(2) all eigenvalues of the matrix P are real and positive, or

(3) all main subdeterminants of the matrix P are positive, i.e.

$$p_{11} > 0, \quad \begin{vmatrix} p_{11}, & p_{12} \\ p_{21}, & p_{22} \end{vmatrix} > 0, \quad \begin{vmatrix} p_{11}, & p_{12}, & p_{13} \\ p_{21}, & p_{22}, & p_{23} \\ p_{31}, & p_{32}, & p_{33} \end{vmatrix} > 0, \ldots, \det P > 0$$

or

(4) the inequality $x^T P x > 0$ for any vector $x \neq 0$ holds.

In practical cases the check for positive definiteness is carried out according to condition (3) or (4).

Example 12.4: Check in the sense of Lyapunov's theorem the stability of the system

$$x(k+1) = \begin{bmatrix} 0.8, & -0.4 \\ 1.2, & 0.2 \end{bmatrix} x(k)$$

Solution: We choose $Q = E$. According to eqn. (12.56) we may write

$$-\begin{bmatrix} 1, & 0 \\ 0, & 1 \end{bmatrix} = \begin{bmatrix} 0.8, & 1.2 \\ -0.4, & 0.2 \end{bmatrix} \begin{bmatrix} p_{11}, & p_{12} \\ p_{21}, & p_{22} \end{bmatrix} \begin{bmatrix} 0.8, & -0.4 \\ 1.2, & 0.2 \end{bmatrix} - \begin{bmatrix} p_{11}, & p_{12} \\ p_{21}, & p_{22} \end{bmatrix}$$

This matrix equation determines a set of four linear algebraic equations which may be written as follows:

$$p_{12} = p_{21}$$

$$\begin{bmatrix} -0.36, & 1.92, & 1.44 \\ -0.32, & -1.32, & 0.24 \\ 0.16, & -0.16, & -0.96 \end{bmatrix} \begin{bmatrix} p_{11} \\ p_{12} \\ p_{22} \end{bmatrix} = \begin{bmatrix} -1 \\ 0 \\ -1 \end{bmatrix}$$

Solving this set of equations we obtain

$$P = \begin{bmatrix} 3.904, & -1.057 \\ -1.057, & 2.223 \end{bmatrix}$$

Since $p_{11} > 0$ and $\det P > 0$, the matrix P is, according to condition (3), positive definite and, therefore, the system is stable.

If it is desired to perform the test according to condition (4) we choose an arbitrary vector x, say, $x = [1, 2]^T$ and calculate

$$x^T P x = 9.568 > 0$$

This result leads to the same conclusion as for the stability test based on condition (3).

13

Identification of Model Parameters

The object of control plant identification is the determination of a mathematical model of the plant, describing its dynamic properties in a manner suitable for a given purpose. If the final objective is to determine optimum control, it will suffice if, for a given input, the output of the model is equivalent to the output of the actual system in a chosen sense. In this case it is, therefore, not necessary that the structure and parameters of the model be identical with the structure and parameters of the physical system.

In general, the determination of the mathematical model means to determine the plant model structure, the parameter values and, if needed, the values of dependent variables, e.g. state variables.

Sometimes the structure determination is considered as a separate task sharply segregated from parameter determination. For instance, L. A. Zadeh [62.2] designates structure determination as "classification". In some recent works, however, the structure and parameters are estimated simultaneously, i.e. by means of a single program. For example, M. Kárný [76.11] determines the order of the linear plant model and its parameters with the aid of a common algorithm.

The mathematical plant model was defined by P. Eykhoff [74.7] as a mathematical representation of the essential aspects of an existing or to be constructed plant, describing the knowledge of that plant in a usable form.

This definition is very general. It allows to determine the mathematical model in any manner. It emphasizes the model usability and the determination of only the essential properties of the actual plant. This is an important aspect since the model is expected to yield information for further decisions. A model which is too complex may not be usable for this purpose. However, a model with a limited complexity can describe reality only approximately. In many cases identification methods do not even permit, for a given limited measurement accuracy, to establish a complex model equivalent in structure and parameters to the actual plant. This fact, however, should not be prejudicial to a further utilization of the model provided that the model represents the essential properties of the plant. On the contrary, owing to its simplicity it may be well available.

The identification may be performed either by the method of *system analysis* or by the method of *experimental system identification*.

In system analysis we start from the plant design data and from the mathematical description of the elementary events taking place in the plant. We thus obtain a set of algebraic and differential equations containing partly input and output variables and partly state variables. These equations sometimes also include redundant internal plant variables which may be eliminated. If the plant description by system analysis is complete, the individual equations of the description may be arranged in the form of state equations. In this case the state vector components have a specific and simple physical meaning. The input variables may be both useful variables, i.e. controlling variables, and disturbing variables. They may be either deterministic variables or random processes. Simple examples of system analysis were given in Sec. 3.1.

In system identification we usually determine the mathematical model of a stable plant using the data measured at its inputs and outputs. A great many methods serving the purpose of system identification have been elaborated. A very good survey may be found in Refs. [71.21, 74.7]. Theoretically well founded procedures characterized by relatively simple and well converging algorithms are described in Ref. [73.18]. We shall here confine ourselves to only a few selected methods closely related to state space system analysis and synthesis.

System identification methods may in general be divided into deterministic and stochastic methods. In the deterministic methods one usually assumes the system to have a certain initial state, for instance $x(0) = 0$, and a certain relatively simple input signal, e.g. a unit rectangular impulse of unit area, a unit-step signal, a sinusoidal signal, etc.

In the stochastic methods of system identification an arbitrary, not pre-determined initial state and an arbitrary input signal may be used. Besides the useful input signal, corrupting noise, whose statistical properties may not be known in advance, is acting on the plant. In constrast to deterministic methods, stochastic methods allow to express the quality of estimates in the statistical sense, e.g. by dispersion, covariance matrix, etc.

A special case of stochastic estimates is the state vector estimate. It is assumed that the matrices A, B, C, D of the mathematical model in state space are known but the state vector components are not measurable. They must, therefore, be determined indirectly by calculation from the data measured at the plant input and output. It is also assumed that corrupting noise exists at the plant input and output.

The methods of optimum state vector estimate in the presence of corrupting noise acting on the plant rank, therefore, among stochastic system identification methods.

The individual methods of system identification differ not only in dependence on whether they are intended for linear or nonlinear plants, whether noise is present or not, whether noise can be measured or not, whether the order or structure of the

IDENTIFICATION OF MODEL PARAMETERS

model is known or not, but in the first place also on the type of mathematical model and on the chosen objective function testing the quality of the model. Although it is beyond the scope of this book to go into the details dealt with in specialized literature, it should yet be noted that the methods of identification used for a direct determination of the state equation matrices A, B, C, D are being developed to a relatively small extent in comparison with those used for the estimate of the "classical" models such as impulse response, transfer function, etc. This fact, however, is not detrimental since in Chap. 6 and 7 we became familiar with the necessary relations which elucidate sufficiently the relations between the "classical" models and the state space system description. However, any of the methods of system identification permits to determine only the model of the observable and reachable part of the plant since only the data measured at the input and output of the plant can be used for the calculation.

Of the stochastic methods of system identification the discussion will be restricted to the least squares method and the maximum likelihood method. In both cases main consideration will be devoted to such a formulation of the problem and resultant algorithms which, on the one hand, permits to update the identified plant parameters even with a continuously increaing number of measured samples and, on the other hand, ensures a stable numerical solution even with a large number of parameters being sought.

13.1 Deterministic methods

13.1.1 Minimum realization

The problem which will be discussed in this paragraph is frequently referred to as the *problem of minimum realization* of the impulse (weighting) or unit-step response. The question is to determine the coefficients $a_i, b_i, i = 0, 1, \ldots, n - 1$, of the transfer function (10.62) or of the corresponding minimum-order difference equation if we know the "exact" impulse response $s(k), k = 0, 1, \ldots$. By the word "exact" we mean that the errors of the impulse response ordinates are negligible and not considered.

The problem is the same even if we know the unit-step response since the relation between the two responses is a simple one. Denoting the discrete ordinates of the step response by $y_p(k)$ we may write

$$s(k) = y_p(k) - y_p(k - 1) \tag{13.1}$$

$$y_p(k) = \sum_{i=0}^{k} s(i) \tag{13.2}$$

Knowing the unit-step response we may, therefore, by means of eqn. (13.1) determine the impulse response and then proceed according to the original formulation of the problem.

For the sake of completeness it should be noted that the impulse response may be determined as a response to the unit impulse

$$u(\varkappa) = 1 \quad \text{for} \quad \varkappa = 0 \tag{13.3}$$

$$u(\varkappa) = 0 \quad \text{for} \quad \varkappa \neq 0 \tag{13.4}$$

if the initial state of the plant is $x(0) = 0$.

The reason why the minimum-order model should be determined is not the simplicity of the resultant model but the fact that the measured input and output data allow to describe only the reachable and observable part of the plant with a corresponding minimum characteristic polynomial and, therefore, also the mathematical model must be of the minimum order. It should be noted that it is not necessary to determine the impulse or step response directly by measurement. In most practical cases this is even impossible. It is supposed, however, that either of the two responses will be determined by calculation using some method of identification based on the measured values of the plant input and output variables. The calculation of these responses is, in contrast with the calculation of the coefficients of, say, a difference equation, easier and represents, from the point of view of mathematics, a linear problem.

Let us consider first the determination of the minimum realization of a single-input/single-output plant. In Sec. 10.3 and 10.5 we have shown that the parameters of the canonical forms of reachability and observability are directly the coefficients a_i, $i = 0, 1, \ldots, n - 1$, and the ordinates $s(k)$ of the discrete impulse response. According to eqn. (10.90)

$$s(k) = -[s(k-n), s(k-n+1), \ldots, s(k-1)] \begin{bmatrix} a_0 \\ a_1 \\ \vdots \\ a_{n-1} \end{bmatrix} \tag{13.5}$$

for $k > n$.

For $k = n + 1, n + 2, \ldots, 2n$ we obtain a set of linear algebraic equations which may be written in the form

$$\begin{bmatrix} s(n+1) \\ s(n+2) \\ \vdots \\ s(2n) \end{bmatrix} = -\begin{bmatrix} s(1), s(2), & \ldots, s(n) \\ s(2), s(3), & \ldots, s(n+1) \\ \cdots\cdots\cdots\cdots\cdots\cdots\cdots \\ s(n), s(n+1), \ldots, s(2n-1) \end{bmatrix} \begin{bmatrix} a_0 \\ a_1 \\ \vdots \\ a_{n-1} \end{bmatrix} = -S_n \begin{bmatrix} a_0 \\ a_1 \\ \vdots \\ a_{n-1} \end{bmatrix} \tag{13.6}$$

The minimum order of this set of equations is the minimum integer n for which $\det [S_n] = 0$. This means practically that $2n$ ordinates of a discrete impulse response determine n equations constituting the set (13.6) thus enabling us to calculate the coefficients a_i, $i = 0, 1, \ldots, n - 1$. Having these coefficients and the ordinates

$s(1), s(2), \ldots, s(n)$, the canonical form of reachability and observability is fully defined.

If it is desired to determine the coefficients $b_0, b_1, \ldots, b_{n-1}$ of the z-transfer function numerator, the solution (10.89) may be used:

$$b_{n-k} = s(k) + a_{n-1} s(k-1) + \ldots + a_0 s(k-n), \quad n \geq k$$

Let us now proceed to the determination of the minimum realization of a multi-input/multi-output system. The solution of this problem was originally published in Ref. [66.3] where no canonical form was introduced in advance and it was, therefore, necessary to determine $n(n + r + p)$ coefficients. More convenient is the procedure proposed in Ref. [71.2] where in the minimum realization approach a canonical form of observability is chosen in advance. Thus the solution is simplified and only $n(r + p)$ coefficients need to be calculated. We shall, therefore, devote attention to this procedure.

According to eqn. (11.9) $D = S(0)$ so that the problem of minimum realization reduces to the determination of the matrices A, B, C which, for an arbitrary k, would satisfy the equation

$$S(k) = CA^{k-1}B, \quad k = 1, 2, \ldots \tag{13.7}$$

Here the order n of the square matrix A should be as low as possible. The element $s_{ij}(k)$ of the matrix $S(k)$ denotes the value of the i-th output at the instant kT, due to the action of a unit impulse at the j-th input at the instant zero.

Let us establish, according to eqn. (13.7), the symmetric matrix

$$\begin{bmatrix} S(1), & S(2), & \ldots, & S(n) \\ S(2), & S(3), & \ldots, & S(n+1) \\ \ldots & \ldots & \ldots & \ldots \\ S(n), & S(n+1), & \ldots, & S(2n-1) \end{bmatrix} = \begin{bmatrix} CB, & CAB, & \ldots, & CA^{n-1}B \\ CAB, & CA^2B, & \ldots, & CA^nB \\ \ldots & \ldots & \ldots & \ldots \\ CA^{n-1}B, & CA^nB, & \ldots, & CA^{2n-2}B \end{bmatrix} \tag{13.8}$$

We can easily see that the matrix on the right-hand side of eqn. (13.8) may be written as the product of the matrix of observability and the matrix of reachability. Denoting in addition the matrix on the left-hand side of eqn. (13.8) by $\mathscr{S}_n$ we may write

$$\mathscr{S}_n = \begin{bmatrix} C \\ CA \\ \vdots \\ CA^{n-1} \end{bmatrix} [B, AB, \ldots, A^{n-1}B] \tag{13.9}$$

The matrix of reachability as well as the matrix of observability must have the rank n since they refer to that part of the system which is reachable and observable by means of the input and output variables. Since it holds that the rank of the product of two matrices, each of which having the rank $h = n$, is $h \leq n$, the order of the system will be

$$n = \max_i h[\mathscr{S}_i] \tag{13.10}$$

In Par. 11.3.2 we have learned how to determine the structure of reachability and observability according to the linear dependence of the columns and rows of the matrix of reachability and of the matrix of observability. The same linear dependences occur for the columns and rows of the matrix $\mathscr{S}_n$ so that the measured values of the plant input and output variables may be utilized for the determination of the plant structure.

In the following calculation of minimum realization we use preferably the structure of reachability, since with actual plants we can separately observe the individual output variables while the input variables are, as a rule, acting simultaneously and their influence cannot be separately observed.

To determine minimum realization we use the canonical form of observability (11.54) whose main properties are expressed in the relations (11.56), (11.58) and (11.61). Generalizing eqn. (11.61) to apply to the i-th partial system we may write

$$c_i^T A^{n_i} = \sum_{j=1}^{i} \alpha_{ij}^T \begin{bmatrix} c_j^T \\ c_j^T A \\ \vdots \\ c_j^T A^{n_j-1} \end{bmatrix} \tag{13.11}$$

Since the same linear relations must also apply to the rows of $\mathscr{S}$ we have

$$s_i^T(n_i + 1) = \sum_{j=1}^{i} \alpha_{ij}^T \begin{bmatrix} s_j^T(1) \\ s_j^T(2) \\ \vdots \\ s_j^T(n_j) \end{bmatrix} \tag{13.12}$$

According to this equation we construct first the rows $s_1^T(k)$, $s_1^T(k + 1)$, ... for the first output variable ($j = 1$) and check after adding each new row whether all rows are linearly independent. As soon as we find the first linear dependent row we obtain

$$s_1^T(n_1 + 1) = \alpha_{11}^T \begin{bmatrix} s_1^T(1) \\ s_1^T(2) \\ \vdots \\ s_1^T(n_1) \end{bmatrix} \tag{13.13}$$

In this way we determine the order n_1 of the first partial system observable by the output variable y_1. At the same time we determine the coefficients α_{11}^T of the characteristic equation of this partial system using eqn. (13.13). Now we add the rows $s_2^T(1)$, $s_2^T(2)$, ... whose elements are the measured values of the output variable y_2.

Again, as soon as we find the first dependent row we obtain

$$s_2^T(n_2 + 1) = [\alpha_{21}^T, \alpha_{22}^T] \begin{bmatrix} s_1^T(1) \\ \vdots \\ s_1^T(n_1) \\ \hline s_2^T(1) \\ \vdots \\ s_2^T(n_2) \end{bmatrix} \quad (13.14)$$

In this way we determine the order n_2 of the second partial system which is observable if we take up the variable y_2 in addition to the output variable y_1. At the same time we calculate from eqn. (13.14) the coefficients α_{22}^T of the characteristic equation of this partial system as well as the coefficients α_{21}^T of the coupling member. In a similar way we proceed by successively taking up further output variables until they are all used up. In this way we successively determine the entire matrix A_P.

The determination of the matrix B_P is facilitated by the fact that in the canonical form of observability $Q_{PF} = E$. With this relation eqn. (13.9) assumes the form

$$\begin{bmatrix} s_1^T(1) \\ \vdots \\ s_1^T(n_1) \\ \hline s_2^T(1) \\ \vdots \\ \hline \vdots \\ s_m^T(n_m) \end{bmatrix} = [B_P, A_P B_P, \ldots, A_P^{n-1} B_P] \quad (13.15)$$

From eqn. (13.15) it is evident that the matrix B_P is equal to the first r columns of the matrix on the left-hand side of this equation.

In the determination of the matrix C_P we use the fact that, according to eqn. (11.54), the first m rows have the form

$$c_i^T = [0, \ldots, 0, 1, 0, \ldots, 0] \quad (13.16)$$

where 1 represents the element $n_1 + n_2 + \ldots + n_{i-1} + 1$ if $n_i > 0$. For $n_i = 0$ we calculate c_i^T with the aid of eqn. (13.8), i.e. according to the relation

$$c_i^T [B_P, A_P B_P, \ldots, A_P^{n-1} B_P] = s_i^T(1) \quad (11.17)$$

To calculate the matrices A_P, B_P, C_P one must know the values of $S(k)$ for $k = 1, 2, \ldots, N$, where $N = n + \max n_i$.

13.1.2 Determination of the difference equation coefficients

In contrast with the task formulated in the preceding paragraph let us now be concerned with a considerably more difficult problem consisting in the determination of the difference equation coefficients, assuming that the "exact" values of the plant input and output variable at the instant of sampling are known. The difficulty of the solution lies primarily in the numerical processing of data by a deterministic procedure where even small deviations from the exact values may cause the results to be quite incorrect. In addition, the determination of the coefficients of a difference equation from data measured at the plant input and output is basically a nonlinear problem. It can, however, be converted into a linear problem by introducing the set of values of the system output variable as state variables. In this manner the given system may be described by a set of difference equations.

If the system has a single input and single output the relevant ordinary difference equation has the form

$$y(k + n) + a_{n-1} y(k + n - 1) + \ldots + a_1 y(k + 1) + a_0 y(k) =$$
$$= b_0 u(k) + b_1 u(k + 1) + \ldots + b_{n-1} u(k + n - 1) \qquad (13.18)$$

The number of unknown coefficients of eqn. (13.18) is $2n$. We therefore use in the deterministic case as many measured pairs of values of $u(i)$ and $y(i)$ as many there are necessary to set up $2n$ equations enabling us to determine the coefficients a_i and b_i, $i = 0, 1, \ldots, n - 1$. It holds that

$$\begin{bmatrix} y(n) \\ y(n+1) \\ \vdots \\ y(2n-1) \end{bmatrix} =$$

$$= \begin{bmatrix} y(0), & \ldots, & y(n-1) & u(0), & \ldots, & u(n-1) \\ y(1), & \ldots, & y(n-2) & u(1), & \ldots, & u(n-2) \\ \vdots & & \vdots & \vdots & & \vdots \\ y(2n-1), & \ldots, & y(3n-2) & u(2n-1), & \ldots, & u(3n-2) \end{bmatrix} \begin{bmatrix} -a_0 \\ -a_1 \\ \vdots \\ -a_{n-1} \\ \hline b_0 \\ b_1 \\ \vdots \\ b_{n-1} \end{bmatrix}$$

$$\qquad (13.19)$$

$$y = Z_{1n} \vartheta \qquad (13.20)$$

The matrix Z_{1n} in eqn. (13.20) is singular only in very special situations, e.g. with a permanently constant input signal; such situations may easily be eliminated ac-

IDENTIFICATION OF MODEL PARAMETERS

cording to the physical intuition. If, however, the signal excites sufficiently the system and is, therefore, suitable for the identification of the sought coefficients, the matrix in eqn. (13.20) is invertible (non-singular). The order of the system, n, is determined as the greatest number for which this matrix has the rank $h(Z_{1n}) = 2n$. In uncertain cases it should be verified whether the individual data sets in eqn. (13.19) satisfy the difference equation (13.18). If the order of the system is determined correctly, the vector of the sought parameters will be

$$\vartheta = Z_{1n}^{-1} y \tag{13.21}$$

In the analysis of the dynamic properties of multivariable systems with r inputs and p outputs we determine the order, the observability structure and the parameter values by the following procedure [47.1]:

Using the measured components of the vector of input variables, u, and the first output variable y_1 we construct the matrices

$$Z_{1m} = \begin{bmatrix} y_1(0), & \cdots & y_1(m-1) \\ \vdots & & \vdots \\ y_1[(1+r)m-1], & \cdots & y_1[(2+r)m-2] \\ u^T(0), & \cdots & u^T(m-1) \\ \vdots & & \vdots \\ u^T[(1+r)m-1], & \cdots & u^T[(2+r)m-2] \end{bmatrix} \tag{13.22}$$

for $m = 1, 2, \ldots, n_1$, until we find a linearly dependent column, i.e. until

$$\begin{bmatrix} y_1(n_1) \\ \vdots \\ y_1[(2+r)n_1-1] \end{bmatrix} = Z_{1n_1} \begin{bmatrix} \alpha_{11} \\ \beta_0^{(1)} \\ \vdots \\ \beta_{n_1-1}^{(1)} \end{bmatrix} \tag{13.23}$$

$$y = Z_{1n_1} \vartheta \tag{13.24}$$

In eqn. (13.23) $\alpha_{11}^T(1; n_1)$ is the row matrix of the coefficients of the partial system observable by the output variable y_1 and $\beta_0^{T(1)}, \ldots, \beta_{n_1-1}^{T(1)}$ are the rows of the matrix B_P in eqn. (11.54). The total number of unknown coefficients is, therefore, $(1+r)n_1$ and it can be checked that the order of the matrix Z_{1n_1} is $(1+r)n_1$ as well. For suitable input sequences of the vector u the matrix Z_{1n_1} is non-singular and, therefore, its inverse does exist. Consequently, we may calculate from eqn. (13.24) the column matrix ϑ whose elements are the sought coefficients. Then the first difference equation corresponding to the output variable y_1 is

$$y_1(k+n_1) - \alpha_{11}^T \begin{bmatrix} y_1(k) \\ \vdots \\ y_1(k+n_1-1) \end{bmatrix} = \sum_{j=0}^{n_1-1} \beta_j^{T(1)} u(k+j) \tag{13.25}$$

We now proceed to the identification of the next part of the system in that we use, in addition to the output variable y_1, the measured discrete values of the output variable y_2 and set up the matrix

$$Z_{2m} = [Y_{2m}, U_{2m}] \tag{13.26}$$

where

$$Y_{2m} = \begin{bmatrix} y_1(0), & \cdots & y_1(n_1 - 1) \\ \vdots & & \vdots \\ y_1[(1 + r)m - 1 + n_1], & \cdots, & y_1[(1 + r)m - 2 + 2n_1] \\ y_2(0), & \cdots & y_2(m - 1) \\ \vdots & & \vdots \\ y_2[(1 + r)m - 1 + n_1], & \cdots, & y_2[(2 + r)m - 2 + n_1] \end{bmatrix} \tag{13.27}$$

$$U_{2m} = \begin{bmatrix} u^T(0), & \cdots, & u^T(m - 1) \\ \vdots & & \vdots \\ u^T[(1 + r)m - 1 + n_1], & \cdots, & u^T[(2 + r)m - 2 + n_1] \end{bmatrix} \tag{13.28}$$

for $m = 1, 2, \ldots, n_2$ until we find a linearly dependent column, i.e. until

$$\begin{bmatrix} y_2(n_2) \\ \vdots \\ y_2[(2 + r)n_2 - 1 + n_1] \end{bmatrix} = Z_{2m} \begin{bmatrix} \alpha_{21} \\ \alpha_{22} \\ \beta_0^{(2)} \\ \vdots \\ \beta_{n_2-1}^{(2)} \end{bmatrix} \tag{13.29}$$

The second difference equation is then

$$y_2(k + n_2) - \alpha_{22}^T \begin{bmatrix} y_2(k) \\ \vdots \\ y_2(k + n_2 - 1) \end{bmatrix} - \alpha_{21}^T \begin{bmatrix} y_1(k) \\ \vdots \\ y_1(k + n_1 - 1) \end{bmatrix} =$$

$$= \sum_{j=0}^{n_2-1} \beta_j^{T(2)} u(k + j) \tag{13.30}$$

In like manner we proceed with the further output variables of the system. If, with the i-th output variable, we arrive at the order $n_i = 0$, the dynamics of the corresponding partial system will already be described by using the output variables $y_1, y_2, \ldots, y_{i-1}$. Instead of a difference equation we only obtain an algebraic linear equation determining the relation between $y_i(k)$ and the sets $u(k), y_1(k), y_1(k + 1), \ldots, y_{i-1}(k), y_{i-1}(k + 1), \ldots, y_{i-1}(k + n_{i-1} - 1)$.

13.2 The least squares method

In the majority of actual cases the measured values of input and output variables are not exact enough for us to manage with the deterministic methods of solution.

IDENTIFICATION OF MODEL PARAMETERS 215

On the contrary, in addition to the useful input signals u the plant input is influenced by unmeasurable disturbing signals w and the output variables are not measured exactly but with errors v. For such a case the state equations may be written in the form

$$x(k + 1) = A\,x(k) + B\,u(k) + w(k) \tag{13.31}$$

$$y(k) = C\,x(k) + D\,u(k) + v(k) \tag{13.32}$$

where $v(k)$ and $w(k)$ are random variables of which we usually know either nothing or merely some of their statistical properties. Depending on our a priori information on the random variables v and w we can formulate the problem of identification and propose various solution procedures. In this section we shall be concerned with the least squares method which belongs under the oldest, now already classical statistical methods.

Since the time it was first suggested and used by K. F. Gauss in 1795 for the calculation of the motion of "heavenly bodies" [1809.1] this method found application in a great many different fields thus facilitating to judge a number of problems from the same point of view. In this section the least squares method will be used for the calculation of the parameters of a controlled plant model or, in other words, for the identification of the dynamic properties of an open-loop system.

13.2.1 Mathematical controlled plant models

Before proceeding to the actual solution it appears reasonable to note that the identification of the dynamic properties of a controlled plant will be considerably facilitated if we choose for the calculation such a mathematical plant model which will enable us to describe the plant properties by means of the least number of parameters. Besides, in evaluating the data measured at the plant input and output, only the observable and reachable part of the system can be involved in the calculation. According to these considerations, therefore, the system description by state equations is not the most suitable one. We shall not discuss here the variety of further possible mathematical models since these questions are treated in the specialized literature [71.21, 73.18, 74.7]. We shall restrict the discussion to the linear regression model with finite memory introduced for multivariable systems by V. Peterka. In general, the regression model may be expressed by the relation

$$y(k) = r(k, y_{(0)}^{(k-1)}, u_{(0)}^{(k)}, v_{(0)}^{(k-1)}) + e(k) \tag{13.33}$$

where $y_0^{(k-1)}$ denotes the sequence $y(0), y(1), \ldots, y(k-1)$ and similar notation is applied to other variables, where $r(k)$ is a conditional mean value expressed as

$$r(k) = \mathscr{E}(y(k) \mid y_{(0)}^{(k-1)}, u_{(0)}^{(k)}, v_{(0)}^{(k-1)}) = \int y(k)\, p(y(k) \mid y_{(0)}^{(k-1)}, u_{(0)}^{(k)}, v_{(0)}^{(k-1)})\, \mathrm{d}y(k)$$

and where the vector $e(k)$ is a random component in the transformation into the present output vector $y(k)$, expressed by the *regression function* $r(k)$ of the past history of input and output signals.

Among input variables belong both the controlling variables, i.e. components of the vector u, and the independent input variables, e.g. disturbing variables, being components of the vector v. For the sake of clarity and simplicity of further expressions the input variables v may be omitted without prejudice of solution generality.

The random variable $e(k)$ has the following important properties:

(a) The zero mean value

$$\mathscr{E}[e(k)] = 0 \tag{13.34}$$

(b) The variable $e(k)$ is independent of the instantaneous input value and of the past input and output values, so that

$$\mathscr{E}[e(k) u^T(k - \varkappa)] = 0 \quad \text{for} \quad \varkappa = 0, 1, 2, ..., k$$
$$\mathscr{E}[e(k) y^T(k - \varkappa)] = 0 \quad \text{for} \quad \varkappa = 1, 2, ..., k \tag{13.35}$$

(c) The sequence $e(k)$, $k = 0, 1, 2, ...$ is a sequence of mutually non-correlated random variables.

$$\mathscr{E}(e(k) e^T(k - \varkappa)) = 0 \quad \text{for} \quad \varkappa = 1, 2, ..., k \tag{13.36}$$

In the case of a *generalized linear time-invariant regression model* with finite memory eqn. (13.33) may be given the form

$$y(k) = \sum_{i=1}^{N} Q_i\, y(k - i) + \sum_{i=0}^{N} P_i\, u(k - i) + e(k) \tag{13.37}$$

provided that all variables are measured from their long-term mean values and that the covariance matrix

$$R = \mathscr{E}[e(k) e^T(k)] \tag{13.38}$$

corresponds to the equally distributed random vectors $e(k)$ and Q_i and P_i are, in general, matrices of regression coefficients.

The generalized regression model (13.37) is a convenient model of a linear dynamic system with measurable input and output variables. At this point the simpler linear regression models which may be deduced from the generalized verison (13.37) are worth mentioning. Such models are, for instance, the *ordinary linear regression model*

$$y(k) = \sum_{i=0}^{N} P_i\, u(k - i) + e(k)$$

and the *linear autoregression model*

$$y(k) = \sum_{i=1}^{N} Q_i y(k - i) + e(k)$$

with lagged output values $y(k - i)$.

In the case of an open loop system the input values $u(k - i)$ may be considered as constants. On the other hand, in closed-loop systems the values of $u(k)$ are calculated by means of the measured past outputs. Consequently, the values of $u(k)$ depend on the past values of $y(k)$ which, owing to the noise v, are actually observed values of the random variable y. Accentuating this fact it can be stated that all variables of the closed-loop system, described by the generalized regression model (13.37) or its simpler versions, are random variables.

Equation (13.37) may formally be rewritten in the simple form

$$y(k) = \Theta^T z(k) + e(k), \quad k = 1, 2, \ldots \tag{13.39}$$

where let

$$\Theta^T = [P_N, Q_N, \ldots, P_1, Q_1, P_0]$$
$$z^T(k) = [u^T(k - N), y^T(k - N), \ldots, u^T(k - 1), y^T(k - 1), u^T(k)] \tag{13.40}$$

With r inputs and p outputs the blocks of the matrix Θ^T, that is P_i, $i = 0, 1, \ldots, N$, and Q_j, $j = 1, 2, \ldots, N$, have, in general, the dimensions $(p; r)$ and $(p; p)$, respectively, and the matrix Θ^T has, therefore, the dimension $(p; v)$ where $v = N(p + r) + r$. The matrices Θ^T and $z(k)$ may, of course, have even a different internal arrangement, e.g. according to the principle applied in eqn. (13.19). Let us mention additionally the relationships to some other mathematical plant models.

Model (13.37) is a special case of the regression model with infinite memory, where $N \to \infty$. In reducing the parameters to a finite number we now assume that the matrices Q_i and P_i are independent only for $i \leq n$ while for all values of $i > n$ they are linearly dependent according to the following relations:

$$Q_i = -\sum_{j=1}^{n} C_j Q_{i-j}$$

$$P_i = -\sum_{j=1}^{n} C_j P_{i-1} \tag{13.41}$$

If we express the *regression model with infinite memory* using the argument $k - j$ instead of k and with $Q_0 = -E$ as

$$-\sum_{i=0}^{\infty} Q_i y(k - j - i) = \sum_{i=0}^{\infty} P_i u(k - j - i) + e(k - j)$$

we obtain, for $j + i = l$, the relation

$$-\sum_{l=j}^{\infty} Q_{l-j}\, y(k - l) = \sum_{l=j}^{\infty} P_{l-j}\, u(k - l) + e(k - j)$$

Multiplying now this relation from the left by the matrix of coefficients, C_j, and adding up the individual expressions for $j = 0, 1, 2, \ldots, n$, then, on interchanging the sequence of additions and using relations (13.41), we obtain

$$\sum_{l=0}^{n} A_l\, y(k - l) = \sum_{l=0}^{n} B_l\, u(k - l) + \sum_{l=0}^{n} C_l\, e(k - l) \tag{13.42}$$

where

$$A_0 = C_0 = E$$

$$A_l = -\sum_{j=0}^{l} C_j Q_{l-j}; \quad B_l = \sum_{j=0}^{l} C_j P_{l-j}$$

Equation (13.42) is a *stochastic difference equation*.

Let us introduce the shift operator

$$y(k - l) = E^{-l}\, y(k) \tag{13.43}$$

The individual polynomials may then be written in the form

$$Q(E) = E - Q_1 E^{-1} - Q_2 E^{-2} - \ldots - Q_N E^{-N}$$

$$P(E) = P_0 + P_1 E^{-1} + P_2 E^{-2} + \ldots + P_N E^{-N} \tag{13.44}$$

$$A(E) = E + A_1 E^{-1} + A_2 E^{-2} + \ldots + A_n E^{-n}$$

$$B(E) = B_0 + B_1 E^{-1} + B_2 E^{-2} + \ldots + B_n E^{-n}$$

$$C(E) = E + C_1 E^{-1} + C_2 E^{-2} + \ldots + C_n E^{-n} \tag{13.45}$$

Let us introduce also for the impulse model the relations

$$S(E) = S_0 + S_1 E^{-1} + S_2 E^{-2} + \ldots + S_M E^{-M}$$

$$F(E) = E + F_1 E^{-1} + F_2 E^{-2} + \ldots + F_M E^{-M} \tag{13.46}$$

where $M \to \infty$ for a model with infinite memory.

Using this notation we can easily show the mutual relationship between the individual models and their physical interpretation.

The regression model equation (13.37) assumes the form

$$Q(E)\, y(k) = P(E)\, u(k) + e(k) \tag{13.47}$$

IDENTIFICATION OF MODEL PARAMETERS

The stochastic difference equation may now be expressed as

$$A(E) y(k) = B(E) u(k) + C(E) e(k) \tag{13.48}$$

For the sake of completeness let us also write the impulse model equation

$$y(k) = S(E) u(k) + F(E) e(k) \tag{13.49}$$

Considering the linear relations between the individual time sequences we may imagine the plant output as a linear superposition of the ideal output y^* and the corrupting noise e^*, that is

$$y(k) = y^*(k) + e^*(k) \tag{13.50}$$

This conception is best satisfied by the impulse model (13.49) where

$$y^*(k) = S(E) u(k) \tag{13.51}$$

and

$$e^*(k) = F(E) e(k)$$

The corresponding block diagram is shown in Fig. 13.1.

If it is desirable to interpret the regression model by means of the diagram in Fig. 13.1, then

$$S(E) = Q^{-1}(E) P(E)$$
$$F(E) = Q^{-1}(E) \tag{13.52}$$

For the stochastic difference equation we obtain analogously

$$S(E) = A^{-1}(E) B(E)$$
$$F(E) = A^{-1}(E) C(E) \tag{13.53}$$

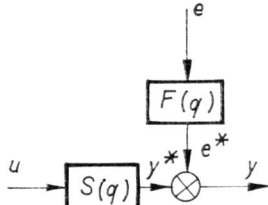

Fig. 13.1. Impulse model of a plant.

The inverse operators in eqns. (13.52) and (13.53) are calculated by comparing the matrix coefficients at equal powers of E. For instance, for the first of eqns. (13.53) we obtain

$$A(E) S(E) = B(E)$$

$$S_i = -\sum_{j=1}^{i} A_j S_{i-j} + B_i, \quad i = 1, 2, \ldots \tag{13.54}$$

13.2.2 Least sum of squares

For the calculation of the plant coefficients we shall use model (13.39) since any of the plant models in Par. 13.2.1 may be converted into this form.

Let us write eqn. (13.39) in the transposed form

$$y^T(k) = z^T(k) \Theta + e^T(k) \tag{13.55}$$

where the components of the vectors $y^T(k)$ and $e^T(k)$ correspond to the individual output variables $y_i(k)$, $i = 1, 2, \ldots, p$, of a multi-input/multi-output plant.

Now we construct a set of linear equations by writing eqn. (13.55) for $k = 1, 2, \ldots, K$, where $K \geq v$. We obtain

$$Y = Z\Theta + E_e \tag{13.56}$$

where

$$Y = \begin{bmatrix} y_1(1), & y_2(1), & \ldots, & y_p(1) \\ \vdots & & & \vdots \\ y_1(K), & y_2(K), & \ldots, & y_p(K) \end{bmatrix}$$

$$Z = \begin{bmatrix} u_1(1-N), & \ldots, & u_r(1-N), & y_1(1-N), & \ldots \\ \vdots & & & & \\ u_1(K-N), & \ldots, & u_r(K-N), & y_1(K-N), & \ldots \end{bmatrix}$$

$$\begin{bmatrix} \ldots, & y_p(1-N), & \ldots, & u_1(1), & \ldots, & u_r(1) \\ & \vdots & & & & \vdots \\ \ldots, & y_p(K-N), & \ldots, & u_1(K), & \ldots, & u_r(K) \end{bmatrix}$$

etc. The dimensions of the matrices Y, Z, Θ and E_e are $(K; p)$, $(K; v)$, $(v; p)$ and $(K; p)$, respectively, where $v = N(p + r) + r$. From eqn. (13.56) we easily write the simpler form, corresponding to a single-input/single-output plant,

$$y = Z\vartheta + e \tag{13.57}$$

where, in agreement with the earlier notation, lower-case letters have again been used for column matrices and vectors. Since the identification procedure is essentially the same for both single-input/single-output and multi-input/multi-output plants we shall henceforth assume without prejudice to solution generality that the plant is a single-input/single-output one.

The sum of squares of the components of the random vector estimate $e(k)$ of eqn. (13.57) is

$$J = (y - Z\vartheta)^T (y - Z\vartheta) \tag{13.58}$$

IDENTIFICATION OF MODEL PARAMETERS

Assuming that $Z^T Z$ is a non-singular matrix we calculate the vector of the sought parameters according to the condition

$$\frac{\partial J}{\partial \vartheta} = 0$$

At the same time it must hold that

$$\frac{\partial J}{\partial \vartheta} \left[\frac{\partial J}{\partial \vartheta} \right]^T$$

is positive definite. We obtain

$$\hat{\vartheta} = (Z^T Z)^{-1} Z^T y \qquad (13.59)$$

where $\hat{\vartheta}$ is the parameter estimate in the sense of the chosen cost function. Solution (13.59) represents the classical formula given already in the early literature, for example in [58.1]. For a digital solution this formula is not suitable, however. In Appendix D a different approach will be shown, suitable particularly for computer-aided calculations.

13.2.3 Geometrical interpretation

Let us denote the individual columns of the matrix Z in eqn. (13.57) by ζ_i, $i = 1, 2, \ldots, v$, where $v = 2N + 1$, and the elements of the column matrix ϑ by π_i, $i = 1, 2, \ldots, v$. Using this notation we may rewrite eqn. (13.57) in the form

$$y = \sum_{i=1}^{v} \zeta_i \pi_i + e \qquad (13.60)$$

Let us now regard y and ζ_i, $i = 1, 2, \ldots, v$, as vectors of the K-dimensional Euclidean space. The least squares problem may now be formulated as a problem of approximation of the vector y by a linear combination of the vectors ζ_i, $i = 1, 2, \ldots, v$, such that the value of $\|e\|^2$ be a minimum. As we know, the solution of this problem is obtained by an orthogonal projection of the vector y onto the individual vectors ζ_i, $i = 1, 2, \ldots, v$, of the Euclidean space. If y^* is the orthogonal projection of y, then $y - y^*$ is a vector normal to all vectors ζ_i, $i = 1, 2, \ldots, v$ and, therefore, the pertinent scalar products are equal to zero:

$$(y - y^*)^T \zeta_i = 0, \quad i = 1, 2, \ldots, v \qquad (13.61)$$

Introducing

$$y^* = \sum_{i=1}^{v} \zeta_i \pi_i \qquad (13.62)$$

the set of equations (13.61) may be rewritten in the form

$$y^T[\zeta_1, \zeta_2, \ldots, \zeta_\nu] = [\pi_1, \pi_2, \ldots, \pi_\nu]\begin{bmatrix} \zeta_1^T\zeta_1, & \zeta_1^T\zeta_2, & \ldots, & \zeta_1^T\zeta_\nu \\ \zeta_2^T\zeta_1, & \zeta_2^T\zeta_2, & \ldots, & \zeta_2^T\zeta_\nu \\ \vdots & & & \\ \zeta_\nu^T\zeta_1, & \zeta_\nu^T\zeta_2, & \ldots, & \zeta_\nu^T\zeta_\nu \end{bmatrix}$$

$$y^T Z = \vartheta^T Z^T Z \qquad (13.63)$$

After a transposition of this equation we may determine the sought matrix of coefficients, $\hat{\vartheta}$. The result will be the same as in eqn. (13.59). If the vectors ζ_i are linearly dependent, the solution will not be unique. A unique solution may, however, be reached by reducing the number of parameters so as to make it equal to the number of linearly independent vectors ζ_i.

13.2.4 Increasing the number of parameters

In many identification problems the plant order, and consequently the corresponding number of coefficients of the regression model, is not known. We therefore commence the calculation with a small number of coefficients and then increase their number if this reduces the value of the cost function (13.58). It would be unreasonable to repeat the calculation with a new number of coefficients regardless of what we already have calculated. The extent to which previous results may be utilized will be shown in the present paragraph [68.2].

Suppose that the matrix Z in eqn. (13.57) has n_1 columns and we increase the number to $n_2 > n_1$ in order that we can increase the number of the regression model coefficients from n_1 to n_2. The matrix Z may then be written as a matrix with two blocks

$$Z = [Z_1, Z_2] \qquad (13.64)$$

and similarly also the vector ϑ

$$\vartheta_{12} = \begin{bmatrix} \vartheta_1 \\ \vartheta_2 \end{bmatrix} \qquad (13.65)$$

In eqn. (13.64) the block Z_1 has n_1 columns and the block Z_2 has $n_2 - n_1$ columns. The dimension of Z is $(K; n_2)$ where $K \geq n_2$. Analogously, in eqn. (13.65) ϑ_1 has n_1 elements and ϑ_2 has $n_2 - n_1$ elements.

Equation (13.59) may now be rewritten as follows:

$$\begin{bmatrix} Z_1^T Z_1 & Z_1^T Z_2 \\ \hline Z_2^T Z_1 & Z_2^T Z_2 \end{bmatrix} \begin{bmatrix} \hat{\vartheta}_1 \\ \hat{\vartheta}_2 \end{bmatrix} = \begin{bmatrix} Z_1^T \\ Z_2^T \end{bmatrix} y \qquad (13.66)$$

IDENTIFICATION OF MODEL PARAMETERS

or

$$Z_1^T Z_1 \hat{\vartheta}_1 + Z_1^T Z_2 \hat{\vartheta}_2 = Z_1^T y$$
$$Z_2^T Z_1 \hat{\vartheta}_1 + Z_2^T Z_2 \hat{\vartheta}_2 = Z_2^T y \tag{13.67}$$

Solving this set of equations we obtain

$$\hat{\vartheta}_1 = \hat{\vartheta} + P_3 Z_2^T (Z_1 \hat{\vartheta} - y)$$
$$\hat{\vartheta}_2 = P_2 Z_2^T (Z_1 \hat{\vartheta} - y) \tag{13.68}$$

We may as well determine the matrix inverse

$$\begin{bmatrix} Z_1^T Z_1 & Z_1^T Z_2 \\ \hline Z_2^T Z_1 & Z_2^T Z_2 \end{bmatrix}^{-1} = \begin{bmatrix} P_1 - P_3 Z_2^T Z_1 P_1 & P_3 \\ \hline P_3^T & P_2 \end{bmatrix} \tag{13.69}$$

In eqns. (13.68) and (13.69) we have introduced

$$\hat{\vartheta} = (Z_1^T Z_1)^{-1} Z_1^T y$$
$$P_1 = (Z_1^T Z_1)^{-1}$$
$$P_2 = [Z_2^T Z_2 - Z_2^T Z_1 (Z_1^T Z_1)^{-1} Z_1^T Z_2]^{-1}$$
$$P_3 = (Z_1^T Z_1)^{-1} Z_1^T Z_2 P_2 \tag{13.70}$$

From the above relations it follows that to determine the increased number of parameters it will be sufficient to perform the inversion of matrices of only the $(n_2 - n_1)$th order. It should, however, be noted that the relations given above are rather complex so that their usefulness may manifest itself only if n_1 is sufficiently large and the difference $n_2 - n_1$ is relatively small.

13.2.5 Increasing the number of samples

When solving practical problems it frequently occurs that the number of rows, K, of the matrix Z in eqn. (13.56) is successively supplemented. Since the number K may be quite large it would be very uneconomical to repeat the calculation with all the previously measured data stored in the matrix Z if the number of its rows were increased by, say, only one to give $K + 1$. Such a situation occurs when the dynamic properties of the plant are identified according to a measurement just being run or, in other words, in real time identification.

If, for the same reason as in Par. 13.2.2, we restrict our discussion to single-input/single-output plants we may introduce for the recursive method of identification the notation

$$y_{K+1} = \begin{bmatrix} y_K \\ y \end{bmatrix} \qquad Z_{K+1} = \begin{bmatrix} Z \\ z^T \end{bmatrix} \tag{13.71}$$

where
$$z^T = [u(K + 1 - N), y(K + 1 - N), \ldots, u(K), y(K), u(K + 1)]$$

The parameter estimate according to the least sum of square errors is, by (13.59),

$$\hat{\vartheta}_K = (Z^T Z)^{-1} Z^T y_K \tag{13.72}$$

$$\hat{\vartheta}_{K+1} = [Z_{K+1}^T Z_{K+1}]^{-1} Z_{K+1}^T y_{K+1} = (Z^T Z + zz^T)^{-1} (Z^T y_K + zy) \tag{13.73}$$

Using the matrix inversion lemma (Appendix A), the inverse of the matrix on the right-hand side of eqn. (13.73) may be written as follows:

$$(Z^T Z + zz^T)^{-1} = (Z^T Z)^{-1} - (Z^T Z)^{-1} z[1 + z^T(Z^T Z)^{-1} z]^{-1} z^T(Z^T Z)^{-1} \tag{13.74}$$

so that

$$\hat{\vartheta}_{K+1} = (Z^T Z)^{-1} Z^T y_K - (Z^T Z)^{-1} z[1 + z^T(Z^T Z)^{-1} z]^{-1} z^T(Z^T Z)^{-1} Z^T y_K +$$
$$+ (Z^T Z)^{-1} zy - (Z^T Z)^{-1} z[1 + z^T(Z^T Z)^{-1} z]^{-1} z^T(Z^T Z)^{-1} zy \tag{13.75}$$

Let us introduce

$$M(K) = (Z^T Z)^{-1} z[1 + z^T(Z^T Z)^{-1} z]^{-1} \tag{13.76}$$

and arrange the sum of the last two terms in eqn. (13.75) in such a way that

$$\{(Z^T Z)^{-1} z - (Z^T Z)^{-1} z[1 + z^T(Z^T Z)^{-1} z]^{-1} z^T(Z^T Z)^{-1} z\} y =$$
$$= (Z^T Z)^{-1} z[1 + z^T(Z^T Z)^{-1} z]^{-1} y$$

Using these relations we may rewrite eqn. (13.75) in the form

$$\hat{\vartheta}_{K+1} = \hat{\vartheta}_K + M(K)(y - z^T \hat{\vartheta}_K) \tag{13.77}$$

From this expression it follows that the estimate from $K + 1$ samples is calculated by adding to the estimate from K samples a correction proportional to $(y - z^T \hat{\vartheta}_K)$. The second term of this expression may be regarded as a prediction of the value y, expressed in terms of the model parameters $\hat{\vartheta}_K$ and of the set of measured values in z. The predicted value can be equal directly to y only in the case of an absolutely exact plant model and in absence of corrupting noise. In such a case the correction would be zero.

The elements of the matrix $M(K)$ are weighting coefficients. In order that we can calculate $M(K)$ also recursively, let us introduce the relation

$$P(K) = \alpha(Z^T Z)^{-1} \tag{13.78}$$

where α is a positive constant.
Then

$$M(K) = P(K) z[\alpha + z^T P(K) z]^{-1} \tag{13.79}$$

Substituting the matrix $P(K)$ defined in eqn. (13.78) also into eqn. (13.74), we have

$$P(K + 1) = P(K) - P(K) z [\alpha + z^T P(K) z]^{-1} z^T P(K) =$$
$$= P(K) - M(K) z^T P(K)$$
$$= [E - M(K) z^T(K + 1)] P(K) \qquad (13.80)$$

We have thus arrived at the conclusion that the estimate of plant parameters by the least squares method may be calculated by means of the following recursive formulae:

$$M(K) = P(K) z(K + 1) [\alpha + z^T(K + 1) P(K) z(K + 1)]^{-1} \qquad (13.81)$$

$$P(K + 1) = [E - M(K) z^T(K + 1)] P(K) \qquad (13.82)$$

$$\hat{\vartheta}(K + 1) = \hat{\vartheta}(K) + M(K) [y(K + 1) - z^T(K + 1) \hat{\vartheta}(K)] \qquad (13.83)$$

These formulae may easily be generalized for the case of a multivariable plant by using the original regression model (13.56). In Par. 14.2.1 we shall show the relationship between the derived recursive formulae and the recursive formulae following from Kalman's filtering.

According to the definition of $P(K)$ given by eqn. (13.78) it is necessary that $(Z^T Z)$ be a non-singular matrix. Denoting the individual rows of the matrix Z by $z^T(k)$, then

$$Z^T Z = \sum_{k=1}^{K} z(k) z^T(k) \qquad (13.84)$$

From this expression it follows that $(Z^T Z)$ can never be non-singular for $K < 2N + 1$. We must, therefore, demand that the number of measurements K be at least equal to the number of sought parameters. If we want to use the recursive formulae (13.81) through (13.83) even at the beginning of the calculation, when $K < 2N + 1$, we can proceed by choosing a $z(K_0)$ having as many elements as many there are sought parameters. Then the matrix $Z^T(K_0) Z(K_0)$ will be non-singular and we may determine the first estimate

$$P(K_0) = \alpha [Z^T(K_0) Z(K_0)]^{-1} \qquad (13.85)$$

$$\hat{\vartheta}(K_0) = [Z^T(K_0) Z(K_0)]^{-1} Z^T(K_0) y(K_0) \qquad (13.86)$$

If we want to use the recursive formulae (13.81) through (13.83) starting with the step $k = 1$ we may choose $z(0)$ and calculate $P(k)$ according to eqn. (13.78):

$$P^{-1}(k) = \frac{1}{\alpha} [z(0), Z^T(k)] \begin{bmatrix} z^T(0) \\ Z(k) \end{bmatrix} = \frac{1}{\alpha} [z(0) z^T(0) + Z^T(k) Z(k)] =$$

$$= P^{-1}(0) + \frac{1}{\alpha} Z^T(k) Z(k) \qquad (13.87)$$

for $k = 1, 2, \ldots$. We may, of course, choose directly

$$P(0) = \frac{1}{\varepsilon} E$$

where $\varepsilon > 0$ is a sufficiently small number so that for $k > K$ the matrix $P^{-1}(k)$ approaches, depending on the chosen value of ε, to the matrix $[Z^T(k) Z(k)]/\alpha$.

Some additional modifications of the least squares method, elaborated for the estimate of controlled plant parameters are discussed in a reviewing article published by V. Strejc [77.28].

13.2.6 Properties of least squares estimates

In Sec. 13.2 we have given the solution of the following problem:

Let (13.57) be the linear model of a plant characterized by the parameters of the matrix ϑ and let $\{u(i - N), y(i - N), i = 1, 2, \ldots, K\}$ be the set of measured values of the input and output variable. Find a matrix of parameters, ϑ, such that the cost function (13.58) assumes a minimum value.

We shall now show some important properties of estimates determined by the least squares method. For this purpose we write the mathematical model of the plant simply in the form

$$y = Z\vartheta + e \tag{13.88}$$

where the errors (residuals) $e(i - N)$, $i = 1, 2, \ldots, K$, are random variables. Let us again distinguish the matrix of actual parameters, ϑ, from the matrix of their estimates, $\hat{\vartheta}$.

We can now state the following theorem:

THEOREM 13.1: *The least squares estimate of the matrix of parameters, ϑ, is unbiased if the mean value of the components of the vector e are zero and if Z and e are mutually independent.*

Proof: The estimate $\hat{\vartheta}$ is, according to eqn. (13.59),

$$\hat{\vartheta} = (Z^T Z)^{-1} Z^T y$$

Substituting for y from eqn. (13.88) we have

$$\hat{\vartheta} = (Z^T Z)^{-1} Z^T (Z\vartheta + e) = \vartheta + (Z^T Z)^{-1} Z^T e \tag{13.89}$$

If Z and e are mutually independent we obtain

$$\mathscr{E}\hat{\vartheta} = \vartheta + \mathscr{E}[(Z^T Z)^{-1} Z^T e] =$$
$$= \vartheta + (Z^T Z)^{-1} Z^T \mathscr{E}e = \vartheta \quad \text{Q. E. D.}$$

IDENTIFICATION OF MODEL PARAMETERS

THEOREM 13.2: *If Z and e are mutually independent, $\mathscr{E}e(i) = 0$, $\mathscr{E}e^2(i) = \sigma^2$ for $i = 1, 2, \ldots, K$, it holds that*

$$\mathscr{E}[(\hat{\vartheta} - \vartheta)(\hat{\vartheta} - \vartheta)^T] = \sigma^2 (Z^T Z)^{-1} \tag{13.90}$$

Proof: According to eqn. (13.59) we may write

$$\mathscr{E}[(\hat{\vartheta} - \vartheta)(\hat{\vartheta} - \vartheta)^T] = \mathscr{E}[\hat{\vartheta}\hat{\vartheta}^T - \vartheta\vartheta^T] =$$

$$= \mathscr{E}[(Z^T Z)^{-1} Z^T y y^T Z (Z^T Z)^{-1} -$$

$$- (Z^T Z)^{-1} (Z^T Z) \vartheta\vartheta^T (Z^T Z)(Z^T Z)^{-1}] =$$

$$= (Z^T Z)^{-1} Z^T \mathscr{E}[(y - Z\vartheta)(y - Z\vartheta)^T] Z (Z^T Z)^{-1} =$$

$$= (Z^T Z)^{-1} Z^T \mathscr{E}[ee^T] Z (Z^T Z)^{-1} = \sigma^2 (Z^T Z)^{-1} \quad \text{Q.E.D.}$$

In this proof we have made use of the fact that the components of the random vector e have equal dispersion so that

$$\mathscr{E}[ee^T] = \sigma^2 \mathrm{E} \tag{13.91}$$

Analogously, we could derive for a multivariable case that

$$\mathscr{E}[(\hat{\Theta} - \Theta)_j (\hat{\Theta} - \Theta)_s^T] = \sigma^2 (Z^T Z)^{-1} \tag{13.92}$$

where $(\hat{\Theta} - \Theta)_j$ and $(\hat{\Theta} - \Theta)_s$ are the j-th and s-th columns of the matrix $(\hat{\Theta} - \Theta)$, $j, s = 1, 2, \ldots, p$, respectively. A more complex version is given in Par. 13.3.2.

COROLLARY 13.1: *The matrix P defined by eqn. (13.78) is an error covariance matrix of the parameter estimates provided that $\mathscr{E}[ee^T] = \sigma^2 \mathrm{E}$.*
The proof follows directly from a comparison of (13.78) with (13.90).

Note 12.1: The assumptions expressed in Theorems 13.1 and 13.2 restrict actually the validity of the results to open-loop systems described by the ordinary regression model derived from the generalized version (13.37). In this case the matrix Z contains only the input values which are independent of the output values. However, the properties of the least squares estimates are different for closed-loop systems described by the generalized regression model (13.37) where $y(k - i)$ and $u(k - i)$, $i = 0, 1, \ldots, N$, are observed values of mutually dependent random variables.

Durbin [60.2] summarized the respective properties as follows: Let $M = \mathscr{E}[(1/K) Z^T Z]$ and suppose that M^{-1} converges to a finite positive definite matrix V as $K \to \infty$. Suppose also that $M^{-1}[(1/K) Z^T Z]$ converges stochastically to the unit matrix. Then asymptotically $\sqrt{(K)}(\hat{\vartheta} - \vartheta)$ has a zero mean and variance matrix $\sigma_n^2 V$.

If we do not know the dispersion σ^2 of the components of the random vector e we may estimate it according to the formula

$$\hat{\sigma}^2 = \frac{1}{K-N}(y - Z\hat{\vartheta})^T(y - Z\hat{\vartheta}) \tag{13.93}$$

The proof of this statement may be found in the literature [68.2].

From Theorem 13.2 it follows that with an unbiased estimate of parameters the estimate of the error covariance matrix $\sigma^2(Z^TZ)^{-1}$ offers a possibility to judge the accuracy of the parameter values calculated by the least squares method.

13.3 Method of maximum likelihood

In constrast with the least squares method which we have applied to the open-loop case we shall now be concerned with the more general case of closed-loop identification. For this purpose we shall use the method of maximum likelihood based on the likelihood function L, which in this case is defined as the joint probability density function

$$L\big(\Theta, R; y^{(K)}_{(-N+1)}, u^{(K)}_{(-N+1)}\big) = f\big(y^{(K)}_{(-N+1)}, u^{(K)}_{(-N+1)}; \Theta, R\big) =$$

$$= \prod_{k=-N+1}^{K} f\big(y(k) \big| y^{(k-1)}_{(-N+1)}, u^{(k)}_{(-N+1)}; \Theta, R\big) \prod_{k=-N+1}^{K} f\big(u(k) \big| y^{(k-1)}_{(-N+1)}, u^{(k-1)}_{(-N+1)}\big) \tag{13.94}$$

where $[\Theta, R] = [P_N, Q_N, \ldots, P_1, Q_1, P_0, R] = \Theta^*$ are, in general, the parameter matrices of the plant regression model (13.39) and R is the covariance matrix (13.38). The object of the maximum likelihood method is to determine the estimate $\hat{\Theta}^*$ for Θ^* in such a way that the likelihood function L reaches its maximum. Since $\ln L$ is an increasing function which attains its maximum when the value of L is highest, the estimate $\hat{\Theta}^*$ is usually calculated using $\ln L$. That is to say, if the likelihood function is differentiable with respect to Θ^*, it is simpler to calculate the estimate $\hat{\Theta}^*$ from the condition

$$\frac{\partial}{\partial \Theta^*} \ln L = 0 \tag{13.95}$$

then from the condition

$$\frac{\partial L}{\partial \Theta^*} = 0 \tag{13.96}$$

Any solution of $\hat{\Theta}^*$ that satisfies eqn. (13.95) is called the *maximum likelihood estimate* Θ^* and eqn. (13.95) is referred to as the *likelihood equation*.

IDENTIFICATION OF MODEL PARAMETERS

The significance of the likelihood function and the application of its logarithm were known to K. F. Gauss already at the time he proposed the least squares method. However, it was not till 1912 that R. A. Fisher [12.1] submitted the generalized and exact formulation of the maximum likelihod method.

In the sum of the logarithms of joint probability density functions in eqn. (13.95) it only makes sense to consider that probability density which contains unknown plant parameters. This requirement is satisfied by the product

$$\prod_{k=-N+1}^{K} f(y(k)|y_{(-N+1)}^{(k-1)}, u_{(-N+1)}^{(k)}) \tag{13.97}$$

whose all factors for $k \geq 1$ may be determined by means of the regression model (13.37). If the random variable $e(k)$ of this model has a normal distribution and does not depend on the variables $y(k)$ and $u(k)$, it holds that

$$f[e(k)|y_{(-N+1)}^{(k-1)}, u_{(-N+1)}^{(k)}] =$$
$$= f[e(k)] = (2\pi)^{-p/2} |R^{-1}|^{1/2} \exp\{-\tfrac{1}{2} e^T(k) R^{-1} e(k)\} \tag{13.98}$$

where p is the number of plant outputs.

Expressing now the random variable $e(k)$ by means of the regression model (13.39), the likelihood function

$$L(\Theta, R) = \prod_{k=1}^{K} f[y(k)|y_{(-N+1)}^{(k-1)}, u_{(-N+1)}^{(k)}; \Theta, R] =$$
$$= (2\pi)^{-pK/2} |R^{-1}|^{K/2} \exp\{-\tfrac{1}{2} \sum_{k=1}^{K} [y(k) - \Theta^T z(k)]^T R^{-1} [y(k) - \Theta^T z(k)]\}$$
$$\tag{13.99}$$

It should, however, be noted that the product (13.97) does not define the probability density functions for $k < 1$, which must be given as initial conditions.

The regression model (13.39) corresponds to a feedback loop since the matrix $z(k)$ and, consequently, the matrix Z whose rows are $z^T(k)$, written for $k = 1, 2, ..., K$, contain variables depending on past values of the plant output and, therefore, on the past random variables $e(k - i)$, $i = 1, 2, ...$ In addition, the inputs u, being also elements of $z(k)$, depend on the past output values as well.

13.3.1 Calculation of maximum likelihood estimates

The general solution of the problem, i.e. the determination of the parameters Θ of the likelihood function (13.99) in such a way that this function attains its maximum, is well known and was published in the literature, for example in [58.1]. Here we shall concentrate ourselves particularly on the numerical solution of the problem which, as we shall show, can be applied even to the least squares method and to the real-time identification.

The natural logarithm of the likelihood function (13.99) is

$$\ln L(\Theta, R) = -\frac{pK}{2} \ln (2\pi) + \frac{K}{2} \ln |R^{-1}| -$$

$$-\tfrac{1}{2} \sum_{k=1}^{K} [y(k) - \Theta^T z(k)]^T R^{-1} [y(k) - \Theta^T z(k)] \qquad (13.100)$$

The sought parameters are determined by means of the last term on the right-hand side of eqn. (13.100)

First of all it should be noted that the covariance matrix R is necessarily positive definite since none of the random variables of the vector $e(k)$ has zero dispersion. Let us factorize this matrix so that

$$R = (R^{1/2})^T R^{1/2} \qquad (13.101)$$

where $R^{1/2}$ is an upper triangular matrix, i.e. the Cholesky square root of the matrix R, and introduce the notation

$$e^*(k) = (R^{-1/2})^T [y(k) - \Theta^T z(k)] = (R^{-1/2})^T e(k) \qquad (13.102)$$

The last term in eqn. (13.100) now assumes the form

$$\tfrac{1}{2} \sum_{k=1}^{K} e^{*T}(k) e^*(k) = \tfrac{1}{2} \sum_{k=1}^{K} \sum_{i=1}^{p} e_i^{*2}(k) \qquad (13.103)$$

where $e_i^*(k)$ are the individual components of the column vector $e^*(k)$.

Transposing now eqn. (13.102) and writing it for all values of $k = 1, 2, ..., K$ we obtain a set of linear algebraic equations which may be written in the matrix form

$$E_e^* = [Y - Z\Theta] R^{-1/2} \qquad (13.104)$$

The dimension of the individual matrices is the same as in eqn. (13.56). Now we modify eqn. (13.104) to the form

$$E_e^* = [Z, Y] \begin{bmatrix} -\Theta \\ E \end{bmatrix} R^{-1/2} = D\tilde{\Theta} R^{-1/2} \qquad (13.105)$$

where the dimensions of the individual matrices are $E_e^*(K; p)$, $D(K; v + p)$, $\tilde{\Theta}(v + p; p)$, $R(p; p)$. The unit matrix E has the same dimension as the matrix R.

Note that the sum in the last expression of eqn. (13.100) may now be expressed as the square of the norm of E_e^*, i.e.

$$\sum_{k=1}^{K} [y(k) - \Theta^T z(k)]^T R^{-1} [y(k) - \Theta^T z(k)] = \|E_e^*\|^2 \qquad (13.106)$$

IDENTIFICATION OF MODEL PARAMETERS

Now we make use of the fact that orthogonal transformation does not change the matrix norm, so that

$$\|E_e^*\| = \|TE_e^*\| \qquad (13.107)$$

if

$$T^T T = E$$

We choose the transformation matrix so as to transform the matrix D to the upper triangular form (see Appendix D), i.e. that

$$TD = \begin{bmatrix} D_{\bar{v}} \\ 0 \end{bmatrix} \qquad (13.108)$$

Thus all data of the matrix D are collected into the upper triangular matrix $D_{\bar{v}}$ whose dimension is $(v + p; v + p)$, and which, therefore, does not depend on K and contains all information required for the determination of the estimates $\hat{\Theta}$ and $\hat{R}$.

Since with a growing number of samples of the measured variables being successively transformed into the matrix $D_{\bar{v}}$ the values of the elements of this matrix are increasing, we store in the computer memory the so-called *information matrix* F defined by the relation

$$F = \frac{1}{\sqrt{K}} D_{\bar{v}} \qquad (13.109)$$

Let us partition this matrix into four blocks

$$F = \begin{bmatrix} F_{\Theta\Theta}, & F_{\Theta R} \\ 0, & F_{RR} \end{bmatrix} \qquad (13.110)$$

where $F_{\Theta\Theta}$ and F_{RR} are again upper triangular matrices of the dimensions $(v; v)$ and $(p; p)$, respectively. By means of the information matrix partitioned in this way and using eqn. (13.105) we may write

$$\|E_e^*\|^2 = \left\| TD \begin{bmatrix} -\Theta \\ E \end{bmatrix} R^{-1/2} \right\|^2 =$$

$$= K \left\| F \begin{bmatrix} -\Theta \\ E \end{bmatrix} R^{-1/2} \right\|^2 =$$

$$= K \|[F_{\Theta R} - F_{\Theta\Theta}\Theta] R^{-1/2}\|^2 + K \|F_{RR} R^{-1/2}\|^2 \qquad (13.111)$$

To attain the maximum of the likelihood function we may nullify the first term of the resultant expression by choosing Θ since in the logarithm of the likelihood function this term appears with a negative sign. Hence follows the relation for the matrix of parameters

$$F_{\Theta R} - F_{\Theta\Theta}\Theta = 0$$

$$\Theta = F_{\Theta\Theta}^{-1} F_{\Theta R} \qquad (13.112)$$

With this result the logarithm of the likelihood function reduces to the form

$$\ln L(\Theta, R) = -\frac{pK}{2} \ln(2\pi) + \frac{K}{2} \ln|R^{-1}| - \frac{K}{2} \|F_{RR}R^{-1/2}\|^2 \quad (13.113)$$

It has been proved (see e.g. Ref. [58.1, 73.18]) that the logarithm of the likelihood function reaches its maximum for

$$F_{RR}R^{-1/2} = E$$

$$F_{RR} = R^{1/2} \quad (13.114)$$

so that the norm in eqn. (13.113) is $\|E\|^2 = p$ and F_{RR} is directly the estimate of the right Cholesky square root of the covariance matrix R. The likelihood function assumes the value

$$L(\Theta, R) = \{(2\pi e)^{p/2} \prod_{i=1}^{p} [\hat{R}^{1/2}]_{ii}\}^{-K} \quad (13.115)$$

where $[\hat{R}^{1/2}]_{ii}$ are elements on the main diagonal of the matrix $\hat{R}^{1/2}$.

On the basis of this information we can summarize: If $K \geq pv + (p + p^2)/2$, where pv is the number of parameters of the plant regression model, $(p + p^2)/2$ the number of unknown elements of the covariance matrix R, p the number of linearly independent solution sets and v the number of parameters in one solution set, and if the matrix D has a full rank and the covariance matrix R is positive definite, then the maximum likelihood estimates of the matrix Θ and of the right Cholesky square root of the covariance matrix R of the regression model (13.104) are defined by eqns. (13.112) and (13.114), where $F_{\Theta\Theta}$, $F_{\Theta R}$ and F_{RR} are submatrices of the information matrix F defined by eqns. (13.108) and (13.109) and partitioned according to eqn. (13.110).

Using eqns. (13.108) and (13.109) we may write that

$$F^T F = \frac{1}{K} D^T D \quad (13.116)$$

Applying the substitution

$$D = [Z, Y]$$

as introduced in eqn. (13.105) and substituting further for F according to eqn. (13.110) we easily derive for the estimates $\hat{\Theta}$ and $\hat{R}$ the repeatedly published, now already classical formulae

$$\hat{\Theta} = F_{\Theta\Theta}^{-1} F_{\Theta R} = (Z^T Z)^{-1} Z^T Y \quad (13.117)$$

$$\hat{R} = F_{RR}^T F_{RR} = \frac{1}{K} [Y^T Y - Y^T Z (Z^T Z)^{-1} Z^T Y] \quad (13.118)$$

IDENTIFICATION OF MODEL PARAMETERS

It can also be shown that the calculated maximum likelihood estimates (13.112) and (13.114) or (13.117) and (13.118) derived on the assumption of a normal distribution of the random vector $e(k)$ in the regression model (13.39) have a natural physical meaning even if this random vector is not normal and if the matrix R in eqn. (13.111) is, for instance, an arbitrary weighting positive semidefinite matrix W.

Comparing the resultant expression (13.117) with the result (13.59) it is apparent that both estimates are identical. This identity can be reached only in the case of linear systems. We may, therefore, summarize that for linear systems with Gaussian noise the least squares and the maximum likelihood estimates of the parameters of the regression model (13.39) are identical.

13.3.2 Properties of maximum likelihood estimates

In specifying the statistical properties of estimates obtained by the maximum likelihood method we shall confine ourselves, as in the case of the least squares method, to the first general and the second central moment, i.e. to the mean value of the estimate and the covariance matrix of the estimate. We shall give without proofs, which may be found in the literature [73.18], only the significant results.

First of all it should be noted that it is useful to distinguish the properties of maximum likelihood estimates separately for open control loops and for closed control loops. Let the pertinent regression model be

$$Y = Z\Theta + E_e \qquad (13.119)$$

Substituting for Y in eqn. (13.117) the expression of eqn. (13.119) we obtain

$$\Delta\hat{\Theta} = \hat{\Theta} - \Theta = (Z^T Z)^{-1} Z^T E_e = \mathscr{L}^T E_e \qquad (13.120)$$

In an open control loop the random variable $e(k)$ and, consequently, the elements of the matrix E_e do not depend on the element values of the matrix Z and, therefore,

$$\mathscr{E}[E_e|Z] = \mathscr{E}[E_e] = 0 \qquad (13.121)$$

and we determine from eqn. (13.120) that

$$\mathscr{E}\,\Delta\Theta = \mathscr{E}(\hat{\Theta} - \Theta) = 0 \qquad (13.122)$$

Expressed in words, the main error value of the parameter estimates is zero.

For the covariance of the estimate $\hat{\Theta}$ we have the relation

$$\mathscr{E}\,\Delta\hat{\Theta}_{ij}\,\Delta\hat{\Theta}_{rs} = \mathscr{E}\left\{\sum_{k=1}^{K}\mathscr{L}_{ki}\,e_j(k)\sum_{l=1}^{K}\mathscr{L}_{lr}\,e_s(l)\right\} =$$

$$= \mathscr{E}\left\{\sum_{k=1}^{K}\sum_{l=1}^{K}\mathscr{L}_{ki}\mathscr{L}_{lr}\,\mathscr{E}[e_j(k)\,e_s(l)\,|\,Z]\right\}, \quad \begin{array}{l} i, r = 1, 2, \ldots, v \\ j, s = 1, 2, \ldots, p \end{array} \qquad (13.123)$$

where the subscripts k and l refer to the time instants and the subscripts j and s refer to the individual outputs. Since, as already stated, the random variables do not depend on the element values of the matrix Z we have

$$\mathscr{E}[e_j(k)\, e_s(l)] = \delta_{kl} R_{js} \qquad (13.124)$$

where δ_{kl} is the Kronecker symbol and R_{js} the element of the noise covariance matrix.

After several rearrangements eqn. (13.123) assumes the form

$$\mathscr{E}[\Delta\hat{\Theta}_j\, \Delta\hat{\Theta}_s^T] = R_{js}\, \mathscr{E}[(Z^T Z)^{-1}] = R_{js}\, \mathscr{E}\left[\frac{1}{K} F_{\Theta\Theta}^{-1} F_{\Theta\Theta}^{-T}\right] \qquad (13.125)$$

From the results it appears that $\mathscr{E}[(Z^T Z)^{-1}]$ is, except for the coefficient R_{js}, the covariance matrix of the j-th and s-th column of the matrix $\hat{\Theta}$.

When calculating the mean of the estimate $\hat{R}$ we again use the regression model (13.119) and the expression (13.118) for the estimate R. We obtain

$$\mathscr{E}\hat{R} = \frac{K - v}{K} R \qquad (13.126)$$

where v is the number of parameters corresponding to one output component of the vector $y(k)$. From the result it follows that the estimate R is not unbiased; it may, however, be corrected by the application of an appropriate modifying factor shown in the relations

$$\hat{R} = \frac{K}{K - v} F_{RR}^T F_{RR} \qquad (13.127)$$

or

$$\hat{R}^{1/2} = \sqrt{\left(\frac{K}{K - v}\right)} F_{RR}$$

making the estimate unbiased. For a numerical calculation this means that it is expedient that the information matrix (13.109) should be modified to the form

$$F = \frac{1}{\sqrt{(K - v)}} D_v \qquad (13.128)$$

in order that the unbiased estimate $\hat{R}^{1/2} = F_{RR}$ can be calculated all the time.

Let us now turn our attention to the closed control loop. Although the estimates of the parameter matrices Θ and R are the same as for the open control loop, their statistical properties are different. If, however,

(a) the matrix M^{-1}, where $M = \mathcal{E}[(1/K) Z^T Z] = \mathcal{E}[F_{\theta\theta}^T F_{\theta\theta}]$, converges as $K \to \infty$ to a positive definite matrix and

(b) $M^{-1}[(1/K) Z^T Z) = M^{-1} F_{\theta\theta}^T F_{\theta\theta}$ converges in probability to a unit matrix,

it can be proved that eqns. (13.122), (13.125) and (13.126) hold asymptotically.

13.3.3 Determination of state model parameters

In the preceding paragraphs of Sec. 13.3 the controlled plant parameter identification has been referred to regression models. The same procedure may, however, be used for the calculation of the estimate of the state model parameters of a controlled plant. It should be kept in view that a state model has, in general, a larger number of parameters than the regression model and that in their determination it is hardly possible to manage with only the information carried by the plant input and output variables. Besides these variables one must also make use of the measured state variables to facilitate the determination of state model parameters. A simultaneous estimation of the parameters and state variables of a plant model is a very difficult task. One of the possible methods of solution was described by L. Ljung [77.18].

In the following we shall show a procedure for the determination of the state model parameters of a controlled plant. In this procedure it is assumed that the state variables are measurable in addition to the input and output variables. Consider again the mathematical model (13.31) and (13.32). First we determine the parameters of the equation of dynamics (13.31). Let the random variable $w(k)$ have the following properties

$$\mathcal{E}[w(k)] = 0$$
$$\mathcal{E}[w(k) w^T(k - \varkappa)] = 0; \quad \varkappa = 1, 2, \ldots, k$$
$$\mathcal{E}[w(k) u^T(k - \varkappa)] = 0; \quad \varkappa = 1, 2, \ldots, k$$
$$\mathcal{E}[w(k) y^T(k - \varkappa)] = 0; \quad \varkappa = 1, 2, \ldots, k \qquad (13.129)$$

We rewrite eqn. (13.31) in the form

$$x(k + 1) = [A, B] z(k) + w(k) \qquad (13.130)$$

where

$$z(k) = \begin{bmatrix} x(k) \\ u(k) \end{bmatrix}$$

After a transposition of the last equation and for $k = 1, 2, \ldots, K$, where K is larger than the number of sought parameters, we obtain

$$X = Z\Theta + E_w \qquad (13.131)$$

where

$$\Theta^T = [A, B]$$

$$X = \begin{bmatrix} x_1(2), & x_2(2), & \ldots, & x_n(2) \\ \ldots\ldots\ldots\ldots\ldots\ldots\ldots\ldots\ldots\ldots \\ x_1(K+1), & x_2(K+1), & \ldots, & x_n(K+1) \end{bmatrix}$$

$$Z = \begin{bmatrix} x_1(1), & \ldots, & x_n(1), & u_1(1), & \ldots, & u_r(1) \\ \ldots\ldots\ldots\ldots\ldots\ldots\ldots\ldots\ldots\ldots\ldots\ldots \\ x_1(K), & \ldots, & x_n(K), & u_1(K), & \ldots, & u_r(K) \end{bmatrix}$$

$$E_w = \begin{bmatrix} w_1(1), & \ldots, & w_r(1) \\ \ldots\ldots\ldots\ldots\ldots\ldots \\ w_1(K), & \ldots, & w_r(K) \end{bmatrix}$$

Equation (13.131) is formally identical with eqn. (13.56) so that all relations derived for the solution of eqn. (13.56), after execution of the appropriate modifications, equally apply to the solution of eqn. (13.131).

It is obvious that the parameters may also be calculated using the square-root filtering described in Par. 13.3.1. For this purpose we further rewrite eqn. (13.131) in the form

$$E_w^* = [X - Z\Theta] R^{-1/2} \tag{13.132}$$

$$E_w^* = [Z, X] \begin{bmatrix} -\Theta \\ E \end{bmatrix} R^{-1/2} \tag{13.133}$$

These two equations are formally identical with eqns. (13.104) and (13.105) and the meaning of the matrix R as well as E_w^* directly follows from a comparison. The further solution procedure may, therefore, be taken over from Par. 13.3.1.

In complete analogy we may calculate the parameters of the matrices C and D of eqn. (13.32).

14

System State Vector Estimation

When solving practical problems by means of the state space theory of control we frequently meet with the case where either all state variables or some state vector components are non-measurable. This usually concerns plants with distributed parameters and complex plants such as heat exchangers, distillation columns, chemical reactors, driers, etc. In order to be able to use the state space theory for the control of these plants one must calculate the estimate of the non-measurable plant state by means of a mesurable plant input and output. Problems where we do not know and cannot directly measure all state vector components belong in the category of problems concerning the control or compensation of disturbing variables with incomplete information. As in the case of identification methods of plant model parameters we shall distinguish between the deterministic and the statistical methods of state vector estimation.

14.1 Deterministic state estimator

14.1.1 Estimator of the order n

In the case of discrete control loops where controller operation is effected by a computer, it is possible to design a discrete-type estimator which, by means of measured plant inputs and outputs, reconstructs the plant state vector provided the plant is observable.

Let us consider a multi-input/multi-output plant described in state space by the equations

$$x(k+1) = A\,x(k) + B\,u(k)$$
$$y(k) = C\,x(k) \qquad (14.1)$$

The plant has $p \leq n$ outputs, where n is the plant order. The plant output variables are assumed to be linearly independent so that the rank of the matrix C is p. The components of the input vector u may also be measurable disturbing variables.

The estimator which will now be described was suggested by D. G. Luenberger [66.5]. The principal idea is based on the assumption that the matrices A, B, C in eqn. (14.1) are known so that the plant whose model is a part of the estimator can be simulated. On the plant model we measure the state estimate $\hat{x}(k)$ and calculate the vector $u(k) = -K\hat{x}(k)$, where K is the matrix of the controller by which we are acting on both the real plant and its model. The estimator is described by the equation

$$\hat{x}(k+1) = A_E \hat{x}(k) + B_E u(k) + H_E y(k) \tag{14.2}$$

The task is to determine the elements of the matrices A_E, B_E, H_E in such a way that $\hat{x}(k)$ approaches asymptotically to $x(k)$.

Subtracting eqn. (14.2) from eqn. (14.1) we obtain for the error vector

$$\Delta x(k) = x(k) - \hat{x}(k)$$
$$\Delta x(k+1) = A_E \Delta x(k) + (A - A_E - H_E C) x(k) + (B - B_E) u(k) \tag{14.3}$$

In order that the error vector $\Delta x(k)$ of the estimates $\hat{x}(k)$ be asymptotically stable the following three conditions must be satisfied:

$$A_E = A - H_E C \tag{14.4}$$
$$B_E = B \tag{14.5}$$
$$\Delta x(k+1) = (A - H_E C) \Delta x(k) \tag{14.6}$$

The system described by the last equation must be a stable system.

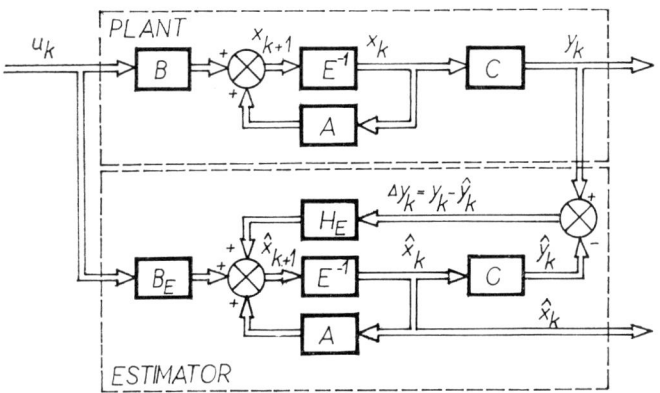

Fig. 14.1. Controlled plant with a system state estimator.

It should be noted that the condition (14.5) ensures that the vector Δx is uncontrollable so that the controlling variables or disturbances of the vector u cannot influence the error vector Δx. The matrix H_E of the system (14.6) is chosen such that

$$\det(zE - A + H_E C) = \prod_{i=1}^{n}(z - z_i), \quad |z_i| < 1 \tag{14.7}$$

SYSTEM STATE VECTOR ESTIMATION

The system (14.6) is then stable and the error vector $\Delta x(k)$ approaches to zero for $k \to \infty$. Additional more detailed determination possibilities of the matrix H_E are given by some methods of synthesis.

On substitution of eqns. (14.4) and (14.5) into eqn. (14.2) and after rearranging the terms, the equation of the estimator takes the form

$$\hat{x}(k+1) = A\hat{x}(k) + Bu(k) + H_E C \Delta x(k) \tag{14.8}$$

It can be seen that the estimator is a plant model acted upon by the vector $\Delta x(k)$. The block diagram is shown in Fig. 14.1.

14.1.2 Estimator of the reduced order

The estimator of the n-th order derived in the preceding paragraph may be simplified considering that a part of the system state vector is observable by the output variables. If the vector y has a total of p linearly independent components, the order of the estimator may be reduced to the value $n - p$. The estimator reduced in this way was also suggested by D. G. Luenberger [66.5]. The same subject is treated in the work [72.2].

To derive the equations of the reduced-order estimator we decompose the state vector x into the part x_a with $n - p$ components and the part x_b with p components. The partition must be such that the matrix $C = [C_1, C_2]$ has a non-singular block $C_2(p; p)$ of rank p. In order to satisfy this requirement, the components of the vector x must be suitably renumbered. Now the system state equations may be rewritten in the form

$$\begin{bmatrix} x_a(k+1) \\ x_b(k+1) \end{bmatrix} = \begin{bmatrix} A_{11}, A_{12} \\ A_{21}, A_{22} \end{bmatrix} \begin{bmatrix} x_a(k) \\ x_b(k) \end{bmatrix} + \begin{bmatrix} B_1 \\ B_2 \end{bmatrix} u(k)$$

$$y(k) = [C_1, C_2] \begin{bmatrix} x_a(k) \\ x_b(k) \end{bmatrix} \tag{14.9}$$

We calculate from the output equation the vector x_b and substitute it into the equation of dynamics (14.9). Thus the vector x_b is replaced by the vector y and the output variables become components of the state vector. This substitution corresponds to the linear transformation

$$w = T_1 x \tag{14.10}$$

where

$$T_1 = \begin{bmatrix} E_{n-p}, & 0 \\ C_1, & C_2 \end{bmatrix}; \quad T_1^{-1} = \begin{bmatrix} E_{n-p}, & 0 \\ -C_2^{-1}C_1, & C_2^{-1} \end{bmatrix} \tag{14.11}$$

Equations (14.9) with the transformation (14.10) assumes the form

$$w(k+1) = T_1 A T_1^{-1} w(k) + T_1 B u(k) = \begin{bmatrix} P, & Q \\ R, & S \end{bmatrix} w(k) + \begin{bmatrix} B_1 \\ B_3 \end{bmatrix} u(k)$$

$$y(k) = [C_1, C_2] T_1^{-1} w(k) = [0, E_p] w(k) \qquad (14.12)$$

where

$$P = A_{11} - A_{12} C_2^{-1} C_1, \quad Q = A_{12} C_2^{-1},$$
$$R = C_1 A_{11} + C_2 A_{21} - (C_1 A_{12} + C_2 A_{22}) C_2^{-1} C_1,$$
$$S = (C_1 A_{12} + C_2 A_{22}) C_2^{-1}, \quad B_3 = C_1 B_1 + C_2 B_2$$

The dimensions of the individual matrices are $P[(n-p);(n-p)]$, $Q[(n-p);p]$, $R[p;(n-p)]$, $S(p;p)$, $C_1[p;(n-p)]$, $C_2(p;p)$.

The estimator matrix $H(n-p;p)$ is introduced by a further transformation in that we replace the vector $w_a = x_a$ by a new vector

$$v = x_a - Hy \qquad (14.13)$$

We denote the resultant vector as

$$v^* = \begin{bmatrix} v \\ y \end{bmatrix} = T_2 w \qquad (14.14)$$

where the transformation matrix

$$T_2 = \begin{bmatrix} E_{n-p}, & -H \\ 0, & E_p \end{bmatrix}; \quad T_2^{-1} = \begin{bmatrix} E_{n-p}, & H \\ 0, & E_p \end{bmatrix} \qquad (14.15)$$

With the transformations (14.10) and (14.14), eqns. (14.9) assume the form

$$v^*(k+1) = T_2 T_1 A T_1^{-1} T_2^{-1} v^*(k) + T_2 T_1 B u(k)$$

$$v^*(k+1) = \begin{bmatrix} P - HR, & PH - HRH + Q - HS \\ R, & RH + S \end{bmatrix} v^*(k) + \begin{bmatrix} B_1 - HB_3 \\ B_3 \end{bmatrix} u(k)$$

$$y(k) = [0, E_p] v^*(k) \qquad (14.16)$$

Because y is a part of the state vector v^*, it is sufficient to design the estimator for an estimate of only the state vector v. From eqns. (14.16) it follows that

$$v(k+1) = A_E v(k) + H_E y(k) + B_E u(k) \qquad (14.17)$$

where

$$A_E = P - HR$$
$$H_E = PH - HRH + Q - HS$$
$$B_E = B_1 - HB_3$$

SYSTEM STATE VECTOR ESTIMATION

For this set of relations the estimator may be designed as in eqn. (14.2):

$$\hat{v}(k+1) = A_E \hat{v}(k) + H_E y(k) + B_E u(k) \qquad (14.18)$$

Subtracting eqn. (14.18) from eqn. (14.17) we obtain for the error vector $\Delta v = v - \hat{v}$ the equation

$$\Delta v(k+1) = A_E \Delta v(k) \qquad (14.19)$$

By this we have in substance converted the solution into the problem described in the preceding paragraph except that now the estimator is of the order $n - p$.

14.2 Statistical and probabilistic state estimation

14.2.1 State estimation in the Kalman sense

In this section we shall deal with the state estimation of a system acted upon by random input signals. A problem of this type is usually designated as the problem of filtering which falls within a larger category of interpolation, filtering and prediction problems. All of these three problems are closely associated and may be solved by the same mathematical procedure. In the following we shall show only the principal relationship of these problems and restrict the discussion to the problem of filtering in the Kalman sense, i.e. with the minimum error variance of the system state estimate.

The discrete version of Kalman's filtering was first published in 1960 [61.2]. That is a solution in the time domain in the state space. Without detriment to generality it is assumed that the system is acted upon by white noise. That is to say, if the actual random input is a rational spectrum, it can always be expressed as the output of a linear filter whose input is acted upon by white noise.

The problem of system state estimation in the Kalman sense may be solved using various approaches, and the solutions were repeatedly published. At random we mention, for instance, the works [64.2, 68.9, 69.6, 70.1]. Of the many possibilities we shall now describe the recursive procedure for the derivation of the Kalman filter since this procedure is extraordinarily simple and represents the natural approach for deriving the resultant Kalman's recursive relations.

Consider a dynamic system described by the equations

$$\begin{aligned} x(k+1) &= A\,x(k) + B\,w(k) \\ y(k) &= C^T x(k) + v(k) \end{aligned} \qquad (14.20)$$

where $w(k)$ and $v(k)$ are mutually independent Gaussian random sequences with zero means and with the covariance matrices Q and R, respectively.

$$\begin{aligned} \mathscr{E}[w(k)] &= 0 \\ \mathscr{E}[w(j)\,w^T(k)] &= Q(k)\,\delta_{jk} \end{aligned} \qquad (14.21)$$

for all values of $j, k = 1, 2, \ldots$, where $Q(k)$ is a positive semidefinite matrix of the type $(r; r)$.

$$\mathscr{E}[v(k)] = 0$$
$$\mathscr{E}[v(j) v^T(k)] = R(k) \delta_{jk} \qquad (14.22)$$

for all values of $j, k = 1, 2, \ldots$, where $R(k)$ is a positive semidefinite matrix of the type $(p; p)$.

$$\mathscr{E}[v(j) w^T(k)] = 0 \qquad (14.23)$$

for all values of $j = 1, 2, \ldots$ and $k = 0, 1, 2, \ldots$.

The initial state $x(0)$ is a Gaussian random vector with the mean value

$$\mathscr{E}[x(0)] = 0$$

and with the covariance matrix of the type $(n; n)$

$$\mathscr{E}[x(0) x^T(0)] = P(0) \qquad (14.24)$$

It is assumed that the state $x(0)$ is independent of $w(k)$, $k = 0, 1, 2, \ldots$ and $v(k + 1)$, $k = 0, 1, 2, \ldots$, so that

$$\mathscr{E}[x(0) w^T(k)] = 0$$
$$\mathscr{E}[x(0) v^T(k + 1)] = 0 \qquad (14.25)$$

for all values of $k = 0, 1, 2, \ldots$.

The model (14.20) has the following properties:

$$\mathscr{E}[x(j) w^T(k)] = 0, \quad k \geq j, \; j = 0, 1, 2, \ldots$$
$$\mathscr{E}[x(j) v^T(k)] = 0 \qquad (14.26)$$

for an arbitrary value of j and k, where $j = 0, 1, \ldots$ and $k = 1, 2, \ldots$,

$$\mathscr{E}[y(j) w^T(k)] = 0, \quad k \geq j, \; j = 1, 2, \ldots$$
$$\mathscr{E}[y(j) v^T(k)] = 0, \quad k > j, \; j, k = 1, 2, \ldots$$

The matrices A, B, C can be time-variable.

The problem may now be formulated as follows:

Provided that $y(k)$, $k = 1, 2, \ldots$, and A, B, C^T are known and that v and w are Gaussian white noises, determine the optimum estimate of the states $x(i)$, $i = 0, 1, 2, \ldots, n$, in the given sense.

It should be noted that for $i < n$ the problem in question becomes an interpolation problem, for $i = n$ a filtering problem and for $i > n$ a prediction problem.

SYSTEM STATE VECTOR ESTIMATION

Let us denote, in general, the state estimate by $\hat{x}(i \mid r)$. The quality of estimates may be considered according to various cost functions. In the case of Kalman's filtering the estimates fulfil the requirement of a minimum error variance of an arbitrary linear state function or, in other words, the problem involves the least squares estimation. For this reason the results arrived at in the present paragraph may be considered as a mere modification of those given in Par. 13.2.2 and 13.2.5.

The system state estimate in the Kalman sense may be summarized by the following two Theorems (14.1 and 14.2).

THEOREM 14.1: *If the estimate* $\hat{x}(j \mid j) = \mathscr{E}[x(j) \mid y(1), y(2), \ldots, y(j)]$ *of the state* $x(j)$ *and the covariance matrix* $P(j \mid j)$ *of the corresponding estimate error* $\Delta \hat{x}(j \mid j) = x(j) - \hat{x}(j \mid j)$ *are known for some value of* $j = 0, 1, \ldots$, *then, for all values of* $k > j$,

(a) *the optimum predicted estimate* $\hat{x}(k \mid j)$, $k > j$, *may, for all admissible cost functions of the quality of estimates, be expressed by the relation*

$$\hat{x}(k \mid j) = A^{k-j} \hat{x}(j \mid j) \tag{14.27}$$

(b) *the stochastic process* $\{\Delta \hat{x}(k \mid j), k = j+1, j+2, \ldots\}$, *where* $\Delta \hat{x}(k \mid j) = x(k) - \hat{x}(k \mid j)$ *is a Gauss-Markov random sequence with a zero mean and with the covariance matrix* $P(k \mid j)$ *given by the relation*

$$P(k \mid j) = A^{k-j} P(j \mid j) A^{(k-j)T} + \sum_{i=j+1}^{k} A^{k-i} B Q_{i-1} B^T A^{(k-i)T} \tag{14.28}$$

Proof: According to eqn. (7.2) we have

$$x(k \mid j) = A^{k-j} x(j) + \sum_{i=j+1}^{k} A^{k-i} B w(i-1) \tag{14.29}$$

for $k \geq j+1$. Therefore,

$$\hat{x}(k \mid j) = \mathscr{E}\left[A^{k-j} x(j) + \sum_{i=j+1}^{k} A^{k-i} B w(i-1) \mid y(1), \ldots, y(j)\right] =$$

$$= A^{k-j} \mathscr{E}[x(j) \mid y(1), \ldots, y(j)] + \sum_{i=j+1}^{k} A^{k-i} \mathscr{E}[w(i-1) \mid y(1), \ldots, y(j)] \tag{14.30}$$

Since the random vectors $\{w(i-1), i = j+1, j+2, \ldots, k\}$ and $\{y(1), \ldots, y(j)\}$ are not mutually dependent and since eqn. (14.21) applies, eqn. (14.30) assumes the form

$$\hat{x}(k \mid j) = A^{k-j} \hat{x}(j \mid j) \tag{14.31}$$

Thus we have proved Theorem 14.1a. Theorem 14.1b will be proved in the following steps.

$$\Delta \hat{x}(k \mid j) = x(k) - \hat{x}(k \mid j) =$$

$$= A^{k-j} x(j) + \sum_{i=j+1}^{k} A^{k-i} B w(i-1) - A^{k-j} \hat{x}(j \mid j) =$$

$$= A^{k-j} \Delta \hat{x}(j \mid j) + \sum_{i=j+1}^{k} A^{k-i} B w(i-1) \qquad (14.32)$$

From eqn. (14.32) it is apparent that $\{\Delta \hat{x}(k \mid j), k = j+1, j+2, \ldots\}$ is a Gaussian discrete process with a zero mean since for $w(i-1)$ eqn. (14.21) holds.

From eqn. (14.32) also follows the relation

$$\Delta \hat{x}(k \mid j) = A \, \Delta \hat{x}(k-1 \mid j) + B w(k-1) \qquad (14.33)$$

from which it is apparent that the process considered is a Markov process.

Finally we derive the relation for $P(k \mid j)$

$$P(k \mid j) = \mathscr{E}[\Delta \hat{x}(k \mid j) \, \Delta \hat{x}^T(k \mid j)] \qquad (14.34)$$

Substituting for $\Delta \hat{x}(k \mid j)$ from eqn. (14.32), the covariance matrix

$$\mathscr{E}[\Delta \hat{x}(j \mid j) w^T(i-1)] = \mathscr{E}[x(j) - \hat{x}(j \mid j)] w^T(i-1), \quad i > j \qquad (14.35)$$

will appear in the resultant expression.

We know from eqn. (14.26) that $\mathscr{E}[x(j) w^T(i-1)] = 0$; further it holds that

$$\mathscr{E}[\hat{x}(j \mid j) w^T(i-1)] = \sum_{l=1}^{j} a(l) \, \mathscr{E}[y(l) w^T(i-1)] = 0$$

where $\hat{x}(j \mid j)$ has been expressed as a linear combination of output variables. Therefore, also the covariance matrix (14.35), where $\Delta \hat{x}(j \mid j) = x(j) - \hat{x}(j \mid j)$, is equal to zero in the stochastic model considered. We may, therefore, write that

$$P(k \mid j) = A^{k-j} \mathscr{E}[\Delta \hat{x}(j \mid j) \, \Delta \hat{x}^T(j \mid j)] A^{(k-j)T} +$$

$$+ \sum_{i=j+1}^{k} A^{k-i} B \, \mathscr{E}[w(i-1) w^T(i-1)] B^T A^{k-i} \qquad (14.36)$$

where use has been made of the fact that $\mathscr{E}[w(j) w^T(k)] = 0$ for $j \neq k$. Equation (14.36) may now be given the final form

$$P(k \mid j) = A^{k-j} P(j \mid j) A^{(k-j)T} + \sum_{i=j+1}^{k} A^{k-i} B Q_{i-1} B^T A^{(k-i)T} \quad \text{Q.E.D.} \qquad (14.37)$$

COROLLARY 14.1: *If the optimum estimate $\hat{x}(k \mid k)$ and the covariance matrix $P(k \mid k)$ of the corresponding estimate error $\Delta \hat{x}(k \mid k) = x(k) - \hat{x}(k \mid k)$ are known for $k = 0, 1, \ldots$ it holds that*

(a) *the single-stage optimum predicted system state estimate for all admissible cost functions is given by the expression*

$$\hat{x}(k+1 \mid k) = A\, \hat{x}(k \mid k) \tag{14.38}$$

(b) *the stochastic process $\{\Delta \hat{x}(k+1 \mid k), k = 0, 1, \ldots$ where $\Delta \hat{x}(k+1 \mid k) = x(k+1) - \hat{x}(k+1 \mid k)$ is a zero mean Gauss-Markov sequence whose covariance matrix is*

$$P(k+1 \mid k) = A\, P(k \mid k)\, A^T + B Q_k B^T \tag{14.39}$$

Proof: The corollary 14.1 follows directly from Theorem 14.1 if $k+1$ is substituted for k and k for j.

THEOREM 14.2:

(a) *The optimum system state estimate in the Kalman sense is expressed by the recursive relation*

$$\hat{x}(k+1 \mid k+1) = A\, \hat{x}(k \mid k) + K(k+1)\left[y(k+1) - C^T A\, \hat{x}(k \mid k)\right] \tag{14.40}$$

for $k = 0, 1, \ldots$, where $\hat{x}(0 \mid 0) = 0$.

(b) *$K(k+1)$ is a gain or weighting matrix of the dimension $(n; p)$ determined by the following relations:*

$$K(k+1) = P(k+1 \mid k)\, C\left[C^T P(k+1 \mid k)\, C + R_{k+1}\right]^{-1} \tag{14.41}$$

$$P(k+1 \mid k) = A\, P(k \mid k)\, A^T + B Q_k B^T \tag{14.42}$$

$$P(k+1 \mid k+1) = \left[E - K(k+1)\, C^T\right] P(k+1 \mid k) \tag{14.43}$$

for $k = 0, 1, \ldots$, where E is an identity matrix of the dimension $(n; n)$ and $P(0 \mid 0) = P(0)$ is the initial condition for eqn. (14.42).

(c) *The stochastic process $\{\Delta \hat{x}(k+1 \mid k+1), k = 0, 1, \ldots\}$, where $\Delta \hat{x}(k+1 \mid k+1) = x(k+1) - \hat{x}(k+1 \mid k+1), k = 0, 1, \ldots$ is a zero mean Gauss-Markov sequence whose covariance matrix is given by eqn. (14.43).*

Proof: To simplify somewhat the proof we shall carry it out in full on the simplifying assumption $w(k) = 0, k = 0, 1, \ldots$. In conclusion of the solution the proof procedure is also shown for $w(k) \neq 0, k = 0, 1, \ldots$. Using the state equations (14.20) we may write the set of equations

$$\begin{bmatrix} v(1) \\ v(2) \\ \vdots \\ v(k) \\ v(k+1) \end{bmatrix} = \begin{bmatrix} y(1) \\ y(2) \\ \vdots \\ y(k) \\ y(k+1) \end{bmatrix} + \begin{bmatrix} C^T A^{-1} B, & C^T A^{-2} B, & \ldots & C^T A^{-k} B, & 0 \\ 0, & C^T A^{-1} B, & \ldots & C^T A^{-k+1} B, & 0 \\ \cdots & \cdots & \cdots & \cdots & \cdots \\ 0, & 0, & \ldots & C^T A^{-1} B, & 0 \\ 0, & 0, & \ldots & 0, & 0 \end{bmatrix} \begin{bmatrix} w(1) \\ w(2) \\ \vdots \\ w(k) \\ w(k+1) \end{bmatrix} -$$

$$- \begin{bmatrix} C^T A^{-k+1} A^{-1} \\ \vdots \\ C^T A^{-1} \\ C^T \end{bmatrix} x(k+1) \tag{14.44}$$

This equation may formally be rewritten in the form

$$\begin{bmatrix} V_k \\ v(k+1) \end{bmatrix} = \begin{bmatrix} Y_k \\ y(k+1) \end{bmatrix} + \begin{bmatrix} \Gamma, & 0 \\ 0, & 0 \end{bmatrix} \begin{bmatrix} W_k \\ w(k+1) \end{bmatrix} - \begin{bmatrix} \Phi A^{-1} \\ C^T \end{bmatrix} x(k+1) =$$

$$= \begin{bmatrix} Y_k + \Gamma W_k \\ y(k+1) \end{bmatrix} - \begin{bmatrix} \Phi A^{-1} \\ C^T \end{bmatrix} x(k+1) =$$

$$= \begin{bmatrix} Y_k^* \\ y(k+1) \end{bmatrix} - \begin{bmatrix} Z_k \\ z(k+1) \end{bmatrix} x(k+1) \tag{14.45}$$

Let us remind the dimensions of the individual matrices and vectors: $V_k(kp; 1)$, $Y_k(kp; 1)$, $\Gamma(kp; kr)$, $W_k(kr; 1)$, $\Phi(kp; n)$, $A(n; n)$, $B(n; r)$, $C^T(p; n)$, $Y^*(kp; 1)$, $Z_k(kp; n)$, $v_{k+1}(p; 1)$, $y_{k+1}(p; 1)$, $w_{k+1}(r; 1)$, $z_{k+1}(p; n)$, $x_{k+1}(n; 1)$.

With k values of input and output variables and on the assumption $w(k) = 0$, $k = 1, 2, \ldots$, we obtain the best estimate $\hat{x}(k \mid k)$ in the sense of the least squares when the quadratic cost function

$$J_k = [Y_k - \Phi \hat{x}(k \mid k)]^T R_k^{-1} [Y_k - \Phi \hat{x}(k \mid k)] \tag{14.46}$$

reaches its minimum value. According to the condition

$$\frac{\partial J_k}{\partial \hat{x}(k \mid k)} = 0 \tag{14.47}$$

we calculate

$$\hat{x}(k \mid k) = [\Phi^T R_k^{-1} \Phi]^{-1} \Phi^T R_k^{-1} Y_k \tag{14.48}$$

If the preceding values of input and output variables are supplemented by the values corresponding to the instant $k + 1$, then the quadratic cost function (14.46) may, using eqn. (14.45), be rewritten in the form

$$J_{k+1} = \left\{ \begin{bmatrix} Y_k \\ y(k+1) \end{bmatrix} - \begin{bmatrix} Z_k \\ z(k+1) \end{bmatrix} \hat{x}(k+1|k+1) \right\}^T \left\{ \begin{bmatrix} R_k^{-1}, & 0 \\ 0, & R_{k+1}^{-1} \end{bmatrix} \right\} \cdot$$

$$\cdot \left\{ \begin{bmatrix} Y_k \\ y(k+1) \end{bmatrix} - \begin{bmatrix} Z_k \\ z(k+1) \end{bmatrix} \hat{x}(k+1|k+1) \right\}$$

$$J_{k+1} = [Y_k - Z_k \hat{x}(k+1|k+1)]^T R_k^{-1} [Y_k - Z_k \hat{x}(k+1|k+1)] +$$
$$+ [y(k+1) - z(k+1) \hat{x}(k+1|k+1)]^T \cdot$$
$$\cdot R_{k+1}^{-1} [y(k+1) - z(k+1) \hat{x}(k+1|k+1)] =$$
$$= \| Y_k - \Phi A^{-1} \hat{x}(k+1|k+1) \|_{R_k^{-1}}^2 +$$
$$+ \| y(k+1) - C^T \hat{x}(k+1|k+1) \|_{R_{k+1}^{-1}}^2 \tag{14.49}$$

Let us introduce

$$\hat{x}(k+1|k+1) = A \hat{x}(k|k) + \Delta\hat{x}(k+1|k) \tag{14.50}$$

The last expression in (14.49) assumes on substitution of eqn. (14.50) the form

$$J_{k+1} = \| Y_k - \Phi[\hat{x}(k|k) + A^{-1} \Delta\hat{x}(k+1|k)] \|_{R_k^{-1}}^2 +$$
$$+ \| y(k+1) - C^T[A \hat{x}(k|k) + \Delta\hat{x}(k+1|k)] \|_{R_{k+1}^{-1}}^2 \tag{14.51}$$

The minimum of the cost function J_{k+1} is determined according to the condition

$$\frac{\partial J_{k+1}}{\partial \Delta\hat{x}(k+1|k)} = 0 \tag{14.52}$$

At the same time

$$\frac{\partial}{\partial \Delta\hat{x}(k+1|k)} \left[\frac{\partial J_{k+1}}{\partial \Delta\hat{x}(k+1|k)} \right]^T$$

must be positive definite.

We obtain

$$A^{-T}\Phi^T R_k^{-1}[Y_k - \Phi \hat{x}(k|k) - \Phi A^{-1} \Delta\hat{x}(k+1|k)] +$$
$$+ CR_{k+1}^{-1}[y(k+1) - C^T A \hat{x}(k|k) - C^T \Delta\hat{x}(k+1|k)] = 0$$
$$A^{-T}\Phi^T R_k^{-1} Y_k - A^{-T}\Phi^T R_k^{-1} \Phi \hat{x}(k|k) - A^{-T}\Phi^T R_k^{-1} \Phi A^{-1} \Delta\hat{x}(k+1|k) +$$
$$+ CR_{k+1}^{-1} y(k+1) - CR_{k+1}^{-1} C^T A \hat{x}(k|k) - CR_{k+1}^{-1} C^T \Delta\hat{x}(k+1|k) = 0$$
$$\tag{14.53}$$

Substituting for $\hat{x}(k|k)$ in the second term of the last equation the expression from eqn. (14.48), then, evidently, the difference of the first two terms will be equal to zero and from the equation simplified in this way we calculate

$$\Delta \hat{x}(k+1 \mid k) = [A^{-T}\Phi^T R_k^{-1} \Phi A^{-1} + CR_{k+1}^{-1}C^T]^{-1} CR_{k+1}^{-1} \cdot$$
$$\cdot [y(k+1) - C^T A \,\hat{x}(k \mid k)] \quad (14.54)$$

Let us now define the matrices

$$P(k \mid k) = (\Phi^T R_k^{-1} \Phi)^{-1}$$

$$P(k+1 \mid k+1) = (A^{-T}\Phi^T R_k^{-1} \Phi A^{-1} + CR_{k+1}^{-1}C^T)^{-1} \quad (14.55)$$

so that the change in the state vector estimate may be expressed as

$$\Delta \hat{x}(k+1 \mid k) = P(k+1 \mid k+1)\, CR_{k+1}^{-1}[y(k+1) - C^T A \,\hat{x}(k \mid k)] \quad (14.56)$$

and, therefore, the estimate $\hat{x}(k+1 \mid k+1)$ is, according to eqns. (14.50) and (14.56),

$$\hat{x}(k+1 \mid k+1) =$$
$$= A\,\hat{x}(k \mid k) + P(k+1 \mid k+1)\, CR_{k+1}^{-1}[y(k+1) - C^T A \,\hat{x}(k \mid k)] \quad (14.57)$$

The calculation of the covariance matrix $P(k+1 \mid k)$ of the estimate error will be simplified if the inversion of the relevant expression (14.55) is carried out according to the matrix inversion lemma (see Appendix A)

$$P(k+1 \mid k+1) =$$
$$= A\,P(k \mid k)\,A^T - A\,P(k \mid k)\,A^T C [C^T A\,P(k \mid k)\,A^T C + R_{k+1}]^{-1} \cdot$$
$$\cdot C^T A\,P(k \mid k)\,A^T \quad (14.58)$$

where, for the single-output case, C^T is a row matrix and R_{k+1} is a scalar. Therefore, the inversion in eqn. (14.58) is a scalar value inversion.

Introducing

$$P(k+1 \mid k) = A\,P(k \mid k)\,A^T$$
$$K(k+1) = P(k+1 \mid k)\, C [C^T P(k+1 \mid k)\, C + R_{k+1}]^{-1}$$

eqn. (14.58) may be rewritten in the form

$$P(k+1 \mid k+1) = [E - K(k+1)\, C^T]\, P(k+1 \mid k)$$

where the last three expressions are identical with the expressions (14.41) through (14.43) in Theorem 14.2 except for the simplification in eqn. (14.42) resulting from the presumption that $w(k) = 0$, $k = 0, 1, 2, \ldots$.

If $w(k) \neq 0$, $k = 0, 1, \ldots$, the quadratic cost function (14.49) must be modified in the form

$$J = \|\hat{W}_k\|^2_{Q_k^{-1}} + \|Y_k - \Phi[\hat{x}(k \mid k) + A^{-1} \Delta\hat{x}(k + 1 \mid k)]\|^2_{R_k^{-1}} +$$

$$+ \|y(k + 1) - C^T[A \hat{x}(k \mid k) + \Delta\hat{x}(k + 1 \mid k)]\|^2_{R_{k+1}^{-1}} \tag{14.59}$$

Besides the condition $\partial J/\partial \Delta x(k + 1 \mid k) = 0$ it is necessary to apply also the condition $\partial J/\partial \hat{W}_k = 0$. The resultant relations of the solution remain the same except for the expression for $P(k + 1 \mid k)$ which now assumes the form (14.42) as stated in Theorem 14.2. The proof of the statement (c) in Theorem 14.2 is performed analogously to that of the statement (b) in Theorem 14.1, Q. E. D.

Note 14.1: If we compare the resultant relations given in Theorem 14.2 with those of eqns. (13.77) through (13.80) we find that the results in the present paragraph are formally identical with those of Par. 13.2.5 although the subject in Par. 13.2.5 was a parameter estimate while in this paragraph we have to do with a state estimate of a plant. However, in both cases we used an identical estimate cost function, i.e. the least sum of square errors.

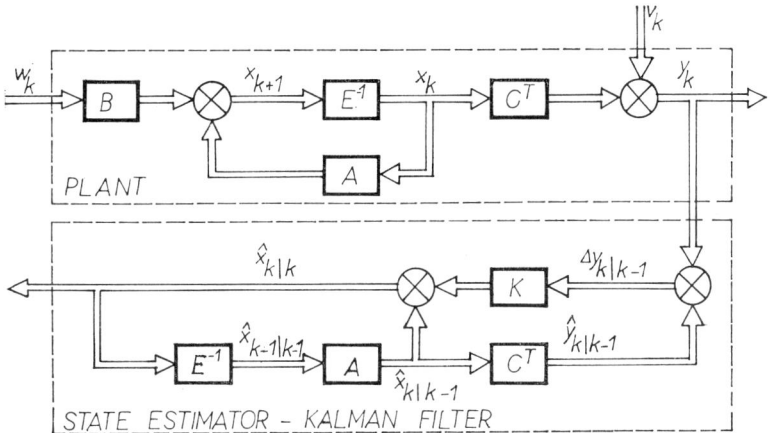

Fig. 14.2. Controlled plant with a Kalman filter for the state estimation.

Note 14.2. Expressions (14.40) through (14.43) represent a mathematical model of the Kalman filter. The block diagram of the plant with the Kalman filter designed for the estimation of the state $\hat{x}(k \mid k)$ is shown in Fig. 14.2. The block diagram corresponds to eqn. (14.40) with the time variable k shifted one sampling interval backwards. From this diagram the calculation procedure is directly apparent. In each stage it begins with the knowledge of $\hat{x}(k - 1 \mid k - 1)$ and $y(k)$. To start the calculation it is necessary to know $\hat{x}(0 \mid 0)$ and $P(0 \mid 0)$. Here $\Delta\hat{x}(0 \mid 0) = x(0) -$

$- \hat{x}(0 \mid 0) = x(0)$ so that $\hat{x}(0 \mid 0) = 0$; $P(0 \mid 0) = \mathscr{E}[\Delta \hat{x}(0 \mid 0) \Delta \hat{x}^T(0 \mid 0)] =$
$= \mathscr{E}[x(0) x^T(0)] = P(0)$.

For calculation, the matrices A, B, C, Q, R_k, R_{k+1} and $\hat{x}(k-1 \mid k-1)$ must be stored in the computer memory. In a time-variant case the complete time history of these matrices must be stored. The past measured outputs and state vectors need not be stored, however.

Note 14.3: An important property of Kalman's filtering is that the matrices Γ and W_k in eqn. (14.45) do not appear in the resultant relations. One can manage with a mere measurement of the output y and with the knowledge of the covariance matrices Q and R.

Note 14.4: $P(k+1 \mid k)$ is a predicted value of the covariance matrix $P(k+1 \mid k+1)$, where $P(k+1 \mid k+1) = \text{cov}\{x(k+1) - \hat{x}(k+1 \mid k+1)\}$, while $P(k+1 \mid k) = \text{cov}\{x(k+1) - \hat{x}(k+1 \mid k)\} = \text{cov}\{x(k+1) - A\hat{x}(k \mid k)\}$.

Note 14.5: Equation (14.58) is a discrete modification of the Riccati matrix equation, whose usefulness appears particularly in the synthesis of control systems according to quadratic cost functions.

Note 14.6: In specialized literature we may find recursive state vector estimate relations derived, in contrast to Theorem 14.1, on the assumption that $\hat{x}(j \mid j) = \mathscr{E}[x(j) \mid y(1), y(2), \ldots, y(j-1)]$. This, is therefore, not a case of filtering but a single-stage prediction. Equations (14.40) through (14.42) then assume the form

$$\hat{x}(k+1 \mid k) = A \hat{x}(k \mid k-1) + K(k+1) [y(k) - C^T \hat{x}(k \mid k-1)]$$
$$K(k+1) = A P(k \mid k-1) C[C^T P(k \mid k-1) C + R_k]^{-1}$$
$$P(k+1 \mid k) = A P(k \mid k-1) A^T + B Q_k B^T -$$
$$- A P(k \mid k-1) C[C^T P(k \mid k-1) C + R_k]^{-1} \cdot$$
$$\cdot C^T P(k \mid k-1) A^T =$$
$$= A P(k \mid k-1) A^T + B Q_k B^T -$$
$$- K(k+1) [C^T P(k \mid k-1) C + R_k]^{-1} K^T(k+1) =$$
$$= A P(k \mid k-1) A^T + B Q_k B^T - K(k+1) C^T P(k \mid k-1) A^T$$

Note 14.7: The covariance matrices of the state vector estimate errors as expressed by eqns. (14.42) and (14.43) are also designated as a priori and a posteriori error covariance matrices.

Example 14.1: Consider a controlled plant whose state equations are

$$x(k+1) = \begin{bmatrix} 2, & 1 \\ 0, & 1 \end{bmatrix} x(k) + \begin{bmatrix} 1 \\ 2 \end{bmatrix} w(k)$$

$$y(k) = [1, 0] x(k) + v(k)$$

SYSTEM STATE VECTOR ESTIMATION

The noise at the plant input and output is time-invariant with the dispersion $Q = \sigma_w^2 = 0.5$ and $R = \sigma_v^2 = 0.2$. Let the initial error covariance matrix be

$$P(0 \mid 0) = \begin{bmatrix} 5, & 0 \\ 0, & 5 \end{bmatrix}$$

Calculate $K(k+1)$ for $k = 0, 1, 2, 3$.

Solution: According to eqn. (14.42) we calculate the a priori covariance matrix of the state vector estimate errors

$$P(1 \mid 0) = \begin{bmatrix} 25, & 5 \\ 5, & 5 \end{bmatrix} + \begin{bmatrix} 0.5, & 1 \\ 1, & 2 \end{bmatrix} = \begin{bmatrix} 25.5, & 6 \\ 6, & 7 \end{bmatrix}$$

Next we determine, by (14.41), the gain matrix

$$K(1) = \begin{bmatrix} 0.99 \\ 0.23 \end{bmatrix}$$

Using the two preceding results we determine the a posteriori covariance matrix of the state vector estimate errors according to eqn. (14.43):

$$P(1 \mid 1) = \begin{bmatrix} 0.20, & 0.05 \\ 0.05, & 5.60 \end{bmatrix}$$

Using this estimate we determine, again by (14.42),

$$P(2 \mid 1) = \begin{bmatrix} 7.08, & 6.69 \\ 6.69, & 7.60 \end{bmatrix} \text{ etc.}$$

The numerical solution converges toward the matrices

$$P(3 \mid 3) = \begin{bmatrix} 0.19, & 0.15 \\ 0.15, & 1.28 \end{bmatrix}, \quad P(4 \mid 3) = \begin{bmatrix} 3.15, & 2.59 \\ 2.59, & 3.28 \end{bmatrix}$$

$$K(4) = \begin{bmatrix} 0.94 \\ 0.77 \end{bmatrix}, \quad P(4 \mid 4) = P(3 \mid 3)$$

The values of the elements $K_1(k+1)$, $K_2(k+1)$ of the gain matrix $K(k+1)$ are given in the following table:

k	0	1	2	3	4
$K_1(k+1)$	0.99	0.97	0.95	0.94	0.94
$K_2(k+1)$	0.23	0.92	0.77	0.77	0.77

14.2.2 State estimate by the maximum likelihood method

The basic properties of the maximum likelihood method were discussed in Sec. 13.3. Therefore, in this paragraph we shall only briefly indicate the application of this method to the system state estimate. Another reason for not going into details is that the resultant solution in the sense of maximum likelihood is identical with the results of minimum error variance.

Consider a system with the Gaussian white noise acting, for simplicity, only at its output:

$$x(k + 1) = A\, x(k)$$
$$y(k) = C^T x(k) + v(k) \tag{14.60}$$

where

$$\left.\begin{array}{l} \mathscr{E}[v(k)\, v^T(j)] = R\delta_{kj} \\ \mathscr{E}[v(k)\, x^T(j)] = 0 \\ \mathscr{E}[v(k)] = 0 \end{array}\right\} \quad \text{for an arbitrary } k, j = 0, 1, 2, \ldots$$

and

$$x \in X^n, \quad y \in Y^p, \quad v \in V^p$$

The problem may now be formulated as follows: Let us have the matrices A, C^T, R and the set of output vectors $y_0, y_1, \ldots$ and determine the set $x_0, x_1, \ldots$ for which the outputs $y_0, y_1, \ldots$ are realized with maximum likelihood. For the state vector estimate $\hat{x}(k)$, $k = 1, 2, \ldots$, the likelihood function has the form

$$L[x_{(1)}^{(k)}; R] = f[y_{(1)}^{(k)} \mid x(0); R] \tag{14.61}$$

where f is the conditional probability density function. The maximum of the likelihood function is found according to the condition

$$\frac{\partial L[x_{(1)}^{(k)}; R]}{\partial x_{(1)}^{(k)}} = 0 \tag{14.62}$$

which enables to determine the estimate $\hat{x}(k)$.

At the same time

$$\frac{\partial}{\partial x(k)}\left[\frac{\partial L[x_{(1)}^{(k)}; R]}{\partial x(k)}\right]^T$$

must be negative definite.

SYSTEM STATE VECTOR ESTIMATION

It holds that

$$f[y_{(1)}^{(k)} \mid x(0); R] = \frac{f[y_{(1)}^{(k)}, x(0); R]}{f[x(0)]} = \frac{f[v_{(1)}^{(k)}, x(0); R]}{f[x(0)]} =$$

$$= \frac{f[v_{(1)}^{(k)}; R] f[x(0)]}{f[x(0)]} = f[v_{(1)}^{(k)}; R] \qquad (14.63)$$

$$f[v_{(1)}^{(k)}; R] = \prod_{i=1}^{k} [(2\pi)^{-p/2} \mid R_i \mid^{-1/2} \exp\{-\tfrac{1}{2}[y(i) - C^T x(i)]^T R_i^{-1} [y(i) - C^T x(i)]\}] \qquad (14.64)$$

It is apparent that the maximum of the likelihood function (14.64) is identical with the minimum of the sum of squares

$$J = \tfrac{1}{2} \sum_{i=1}^{k} \| y(i) - C^T x(i) \|_{R_i^{-1}}^2 \qquad (14.65)$$

with the constraint (14.60), which must be satisfied with respect to $x(i)$, $i = 1$, $2, \ldots, k$.

The cost function (14.65) is equivalent to the cost function (14.46) and can be extended to the form (14.51). Therefore, the further procedure and results will be identical with those of Par. 14.2.1, where the optimum estimate $\hat{x}(k + 1)$ in the Kalman sense was obtained by a minimization of the sum of squares of estimate errors. Further procedure will, therefore, not be repeated.

14.2.3 State estimate in the Bayes sense

In this paragraph we shall be concerned with the system state estimate in the Bayes sense. In fact, there exists a direct relationship between the method of maximum likelihood and the solution in the Bayes sense. In the preceding paragraph the basic condition of the maximum likelihood criterion was expressed using the conditional probability density function which may symbolically be expressed by the relation

$$f_4(y \mid x) = \frac{f_1(x, y)}{f_2(x)} \qquad (14.66)$$

It also holds that

$$f_5(x \mid y) = \frac{f_1(x, y)}{f_3(y)} \qquad (14.67)$$

Eliminating from these relations the joint probability density $f_1(x, y)$ we obtain the known *Bayes rule*

$$f_5(x \mid y) = \frac{f_4(y \mid x) f_2(x)}{f_3(y)} \qquad (14.68)$$

where $f_2(x)$ is the a priori probability density function of the state vector x and $f_3(y)$ the probability density function of measured outputs. Therefore, the conditional probability density function $f_5(x \mid y)$ can be determined if $f_1(x, y)$ is known or can be calculated, or if $f_4(y \mid x)$ is known or can be determined by calculation. In the case considered the joint probability density function is known because

$$f_1(x, y) = f_2(x) f_6(v) = f_2(x) f_6(y - C^T x) \tag{14.69}$$

so that we may write, according to eqn. (14.67), that

$$f_5(x \mid y) = \frac{f_2(x) f_6(y - C^T x)}{f_3(y)} \tag{14.70}$$

Now we can define:

DEFINITION 14.1: *The optimum estimate in the Bayes sense is such an estimate $\hat{x}$ for which the conditional probability density function $f_5(x \mid y)$ reaches its maximum value.*

The maximum value of $f_5(x \mid y)$ is obtained if

$$\frac{\partial f_5(x \mid y)}{\partial x} = 0 \tag{14.71}$$

and if

$$\frac{\partial}{\partial x} \left[\frac{\partial f_5(x \mid y)}{\partial x} \right]^T \tag{14.72}$$

is negative definite.

Consider again a system given by the equations

$$x(k + 1) = A\, x(k) + B\, w(k)$$
$$y(k) \;\;\;\;\;\;= C^T x(k) + v(k) \tag{14.73}$$

where $v(k)$ and $w(k)$ are white Gaussian noises. It holds that

$$\mathcal{E}[w(k)] \;\;\;\;\;= \mathcal{E}[v(k)] = 0$$
$$\mathcal{E}[w(k)\, w^T(j)] = Q_k \delta_{kj}$$
$$\mathcal{E}[v(k)\, v^T(j)] \;= R_k \delta_{kj}$$
$$\mathcal{E}\{x(k) \mid Y(k)\} = \hat{x}(k \mid k)$$
$$\mathrm{cov}\,\{x(k) \mid Y(k)\} = P(k \mid k) \tag{14.74}$$

where $f[x(k) \mid Y(k)] = f[x(k) \mid y(0), y(1), \ldots, y(k)]$. The joint probability density function of the Gauss-Markov noise is

$$f[w(k), v(k+1) \mid x(k), y(k)] = f[w(k), v(k+1)] = f[w(k)] f[v(k+1)] \quad (14.75)$$

It can be proved that [64.2]:

(1) $f[x(k+1) \mid Y(k)]$ is a Gaussian conditional probability density function with the following properties:

$$\mathscr{E}[x(k+1) \mid Y(k)] = A \hat{x}(k \mid k)$$

$$\operatorname{cov}\{x(k+1) \mid Y(k)\} = A P(k \mid k) A^T + B Q_k B^T = P(k+1 \mid k)$$

(2) $f[y(k+1) \mid Y(k)]$ is a Gaussian conditional probability density function with the properties:

$$\mathscr{E}[y(k+1) \mid Y(k)] = C^T A \hat{x}(k \mid k)$$

$$\operatorname{cov}\{y(k+1) \mid Y(k)\} = C^T P(k+1 \mid k) C + R_{k+1}$$

(3) $f[y(k+1) \mid x(k+1)]$ is a Gaussian conditional probability density function with the properties:

$$\mathscr{E}[y(k+1) \mid x(k+1)] = C^T x(k+1)$$

$$\operatorname{cov}\{y(k+1) \mid x(k+1)\} = R_{k+1} \quad (14.76)$$

It should be noted that we have introduced $\hat{x}(k \mid k) = \mathscr{E}[x(k) \mid Y(k)]$ which is a conditional mean estimate. So far, however, we do not know that the estimate in question is an optimum one and, therefore, until further notice, $\hat{x}(k \mid k)$ is only a parameter.

Now we can proceed to determine the optimum filter for which the conditional probability density function $f[x(k+1) \mid Y(k+1)]$ reaches its maximum value. As in the case of eqn. (14.70) we can derive that

$$f[x(k+1) \mid Y(k+1)] = \frac{f[x(k+1), Y(k+1)]}{f[Y(k+1)]} =$$

$$= \frac{f[x(k+1), y(k+1) \mid Y(k)] f[Y(k)]}{f[y(k+1) \mid Y(k)] f[Y(k)]} =$$

$$= \frac{f[y(k+1) \mid x(k+1)] f[x(k+1) \mid Y(k)]}{f[y(k+1) \mid Y(k)]} \quad (14.77)$$

where $f[y(k+1) \mid x(k+1)] = f[y(k+1) - C^T x(k+1)] = f[v(k+1)]$.

Using relations (14.76), the probability density function (14.77) may now be expressed in the form

$$f[x(k+1) \mid Y(k+1)] = \text{const.} \exp\left(-\tfrac{1}{2}J\right) =$$

$$= \text{const.} \exp\left\{-\tfrac{1}{2}[[y(k+1) - C^T x(k+1)]^T R_{k+1}^{-1}[y(k+1) - C^T x(k+1)] + \right.$$
$$+ [x(k+1) - A\hat{x}(k \mid k)]^T P^{-1}(k+1 \mid k)[x(k+1) - A\hat{x}(k \mid k)] -$$
$$- [y(k+1) - C^T A \hat{x}(k \mid k)]^T \cdot$$
$$\left. \cdot [C^T P(k+1 \mid k) C + R_{k+1}]^{-1} [y(k+1) - C^T A \hat{x}(k \mid k)]]\right\} \qquad (14.78)$$

The function (14.78) reaches its maximum at such a value of $x(k+1) = \hat{x}(k+1 \mid k+1)$ for which the exponent in eqn. (14.78) assumes its minimum value, i.e. for

$$\frac{\partial J}{\partial x(k+1)} = 0 \qquad (14.79)$$

and for

$$\frac{\partial}{\partial x(k+1)} \left[\frac{\partial J}{\partial x(k+1)}\right]^T$$

being positive definite.

According to the condition (14.79) we obtain

$$-CR_{k+1}^{-1}[y(k+1) - C^T \hat{x}(k+1 \mid k+1)] +$$
$$+ P^{-1}(k+1 \mid k)[\hat{x}(k+1 \mid k+1) - A\hat{x}(k \mid k)] = 0 \qquad (14.80)$$

$$[P^{-1}(k+1 \mid k) + CR_{k+1}^{-1}C^T]\hat{x}(k+1 \mid k+1) =$$
$$= CR_{k+1}^{-1} y(k+1) + P^{-1}(k+1 \mid k) A \hat{x}(k \mid k) \qquad (14.81)$$

$$\hat{x}(k+1 \mid k+1) =$$
$$= [P^{-1}(k+1 \mid k) + CR_{k+1}^{-1}C^T]^{-1} [CR_{k+1}^{-1} y(k+1) + P^{-1}(k+1 \mid k) A \hat{x}(k \mid k)] \qquad (14.82)$$

The inversion of the first expression on the right-hand side of eqn. (14.82) is expressed according to the matrix inversion lemma (Appendix A) and, after rearranging, we obtain

$$\hat{x}(k+1 \mid k+1) = A\hat{x}(k \mid k) - P(k+1 \mid k) C[C^T P(k+1 \mid k) C + R_{k+1}]^{-1} \cdot$$
$$\cdot C^T A \hat{x}(k \mid k) + P(k+1 \mid k) CR_{k+1}^{-1} y(k+1) +$$
$$+ P(k+1 \mid k) C[C^T P(k+1 \mid k) C + R_{k+1}]^{-1} \cdot$$
$$\cdot C^T P(k+1 \mid k) CR_{k+1}^{-1} y(k+1) \qquad (14.83)$$

$$\hat{x}(k+1\mid k+1) = A\,\hat{x}(k\mid k) + P(k+1\mid k)\,C\,.$$
$$\cdot\left[C^T P(k+1\mid k)\,C + R_{k+1}\right]^{-1}\left[y(k+1) - C^T A\,\hat{x}(k\mid k)\right] =$$
$$= A\,\hat{x}(k\mid k) + K(k+1)\left[y(k+1) - C^T A\,\hat{x}(k\mid k)\right] \tag{14.84}$$

where

$$K(k+1) = P(k+1\mid k)\,C\left[C^T P(k+1\mid k)\,C + R_{k+1}\right]^{-1} \tag{14.85}$$

and according to eqns. (14.41) through (14.43) we have

$$P(k+1\mid k) = A\,P(k\mid k)\,A^T + B Q_k B^T \tag{14.86}$$

$$P(k+1\mid k+1) = \left[P^{-1}(k+1\mid k) + C R_{k+1}^{-1} C^T\right]^{-1} =$$
$$= P(k+1\mid k) - P(k+1\mid k)\,C\left[C^T P(k+1\mid k)\,C + R_{k+1}\right]^{-1} C^T P(k+1\mid k) =$$
$$= \left[E - K(k+1)\,C^T\right] P(k+1\mid k) \tag{14.87}$$

It is obvious that eqns. (14.84) through (14.87) are identical with eqns. (14.40) through (14.43). It is, therefore, again a case of the Kalman recursive filtering where eqns. (14.84) through (14.87) represent the Kalman filter. For the numerical solution the same remarks will apply as in the case of the least squares method.

14.2.4 Parameter estimate using the Kalman filter

The method of state estimation given in Par. 14.2.1 may be modified to apply to the estimation of parameters of the state model. In order that the recursive relations (14.40) through (14.43) might be used directly, the parameters being estimated must be elements of a vector and the state variables must be measurable. The possibility of using the Kalman filter for the estimate of state model parameters was pointed out by D. Q. Mayne [63.6].

To derive the Mayne estimator for the estimation of the elements of the matrices A, B of the equation of dynamics

$$x(k+1) = A\,x(k) + Bu(k) + w(k) \tag{14.88}$$

let us introduce the parameter vector

$$\Theta(k) = [A_1(k), A_2(k), \ldots, A_n(k)\mid B_1(k), B_2(k), \ldots, B_n(k)]^T \tag{14.89}$$

where $A_i(k)$ and $B_j(k)$, $i, j = 1, 2, \ldots, n$, are rows of the matrices A and B, respectively. The argument k in eqn. (14.89) has the meaning of time as well as of the iteration step, since iterations always take place after the measurement of new values at sampling instants. Equation (14.88) may now be rewritten in the form

$$x(k+1) = Z(k)\,\Theta(k) + w(k) \tag{14.90}$$

where

$$Z(k) = \begin{bmatrix} x^T(k), & 0, & \ldots, & 0 & u^T(k), & 0, & \ldots, & 0 \\ 0, & x^T(k), & \ldots, & 0 & 0, & u^T(k), & \ldots, & 0 \\ \multicolumn{8}{c}{\dotfill} \\ 0, & 0, & \ldots, & x^T(k) & 0, & 0, & \ldots, & u^T(k) \end{bmatrix}$$

This equation has the form of a state model output equation, where $x(k + 1) \sim y(k)$, $Z(k) \sim C^T(k)$, $\Theta(k) \sim x(k)$ and $w(k) \sim v(k)$.

Considering that the assumed plant is k-invariant, we have

$$\Theta(k + 1) = \Theta(k) \tag{14.91}$$

i.e. in the equation of dynamics (14.91) $A = E_n$ is an identity matrix of the dimension $(n; n)$ and $B = 0$.

Using eqns. (14.40) through (14.43), the parameter estimate may be written as

$$\hat{\Theta}(k + 1 \mid k + 1) = \hat{\Theta}(k \mid k) + K(k + 1)\left[x(k + 1) - Z(k)\,\hat{\Theta}(k \mid k)\right] \tag{14.92}$$

$$K(k + 1) = P(k + 1 \mid k)\,Z(k)\left[Z^T(k)\,P(k + 1 \mid k)\,Z(k) + Q^w(k + 1)\right]^{-1} \tag{14.93}$$

$$P(k + 1 \mid k) = P(k \mid k) + Q^w(k) \tag{14.94}$$

$$P(k + 1 \mid k + 1) = \left[E - K(k + 1)\,Z^T(k)\right] P(k + 1 \mid k) \tag{14.95}$$

where $\mathscr{E}\left[w(j)\,w^T(k)\right] = Q^w(k)\,\delta_{jk}$, $j, k = 0, 1, 2, \ldots$, has the dimension $(n; n)$ and where

$$\mathscr{E}\left[(\Theta(k) - \hat{\Theta}(k \mid k))\,(\Theta(k) - \hat{\Theta}(k \mid k))^T\right] = P(k)$$

To identify the parameters it is necessary to know $P(0 \mid 0)$ and $Q^w(k)$.

From the foregoing it is evident that the same procedure may be used for the calculation of the elements of the matrix C in the plant output equation.

15

Principles of Deterministic Synthesis

In automatic control the term "synthesis" means finding such a sequence of values of plant input variables that the plant output variables satisfy the given conditions. If synthesis is performed in the state space, then the plant state includes complete information about the past behaviour of the plant so that a knowledge of the state allows to determine its present as well as future behaviour. In other words, if the state $x(t_1)$ at the instant $t = t_1$ is known, it must be possible to determine an input $u(t), t \geq t_1$, that would satisfy the requirements imposed on the plant output. The knowledge and use of the states $x(t), t < t_1$, will in no case bring an improvement. In the linear cases we therefore express the plant input vector always as a linear combination of the state vector components, i.e. in a discrete case by the equation

$$u(k) = -K(k)\,x(k) + r(k) \qquad (15.1)$$

where, for time-invariant plants, $K(k) = K = \text{const.}$ and $r(k)$ is the command variable. It should be emphasized that eqn. (15.1) has the form of the output equation of a state description and, therefore, a controller defined by eqn. (15.1) has no dynamics.

If the plant state is non-measurable it must be estimated. In deterministic and stochastic cases this is accomplished by the estimator. For an estimate of the present state $x(t)$ we need, of course, the past and present values of the plant input and output variables. It is plain that the calculation of the state estimate will be carried out separately from the synthesis and from the calculation of the vector of the inputs u. That is to say: For the synthesis in state space we can make use of the principle of separability. We first determine the state estimate $\hat{x}(k)$ and calculate the plant input vector using this estimate

$$u(k) = -K\,\hat{x}(k) + r(k) \qquad (15.2)$$

if $x(k)$ cannot be measured directly.

The advantage of the state space over the description by means of input and output variables lies particularly in the fact that the results derived for single-input/single-output plants may easily be generalized for multi-input/multi-output plants. In the subsequent chapters this possibility will, therefore, be always kept in view.

In some cases it is desirable to transfer the computing approach from the state space to classical models and their solutions or vice versa. The first case is solved by expressing the whole control loop, or the controller and plant separately, in state space in such a canonical form, e.g. in the canonical form of controllability, which yields directly the coefficients of the discrete transfer function or of the appropriate difference equation. Since the state vector components are uniquely defined for this canonical form of the state model (see Sec. 6.1 and 6.2), the transcription to the difference equation or transfer function will make no particular difficulties. Namely, for the controller of the type (15.2) the transcription is very simple. For example, for a single-input/single-output system we have $K = [k_{11}, k_{12}, ..., k_{1n}]$ and

$$u(k) = k_{11} x_1(k) + k_{12} x_2(k) + ... + k_{1n} x_n(k) + r(k) \tag{15.3}$$

The second case, i.e. the transcription of the transfer function or difference equation to the state space representation, is likewise very simple if a suitable canonical form of the state model is used.

Let us assume for a while that it is desired to preserve the concept of control, accepted in the classical theory of linear feedback loops, even in the state space. In such a case, e.g. in a control loop with one command input variable and one controlled output variable, the control error is $e(k) = r(k) - y(k)$. This error is a scalar quantity and acts at the controller input. In contrast with the controller defined by eqn. (15.1) it is now, in general, necessary for the controller to have dynamical elements if the desired quality of control is to be ensured. In the state description, therefore, the controller is described not only by the output equation but also by the equation of dynamics. To determine the state description of a closed control loop we may use eqns. (8.15) and (8.17) for the feedback configuration of two systems of which one may be a dynamic controller and the other a dynamic controlled plant.

Let the controller be described by the equations

$$^1x(k + 1) = {}^1A\, {}^1x(k) + {}^1b\, e(k)$$
$$^1y(k) = {}^1c^T\, {}^1x(k) + {}^1d\, e(k) = {}^2u(k)$$

and the controlled plant by the equations

$$^2x(k + 1) = {}^2A\, {}^2x(k) + {}^2b\, {}^2u(k)$$
$$^2y(k) = {}^2c^T\, {}^2x(k)$$

Let the dimension of the controller state vector ^1x be identical with the dimension of the state vector ^2x of the controlled plant.

The control error is defined by the expression

$$e(k) = r(k) - {}^2y(k) = r(k) - {}^2c^T\, {}^2x(k)$$

PRINCIPLES OF DETERMINISTIC SYNTHESIS

The command variable is denoted by $r(k)$. According to eqns. (8.15) and (8.17), and for $^2d = 0$ and $K = 1$, the state description of the control loop is

$$\zeta(k+1) = \begin{bmatrix} {}^1A & -{}^1b\,{}^2c^T \\ \hline {}^2b\,{}^1c^T & {}^2A - {}^2b\,{}^1d\,{}^2c^T \end{bmatrix} \zeta(k) + \begin{bmatrix} {}^1b \\ \hline {}^2b\,{}^1d \end{bmatrix} r(k)$$

$$^2y(k) = [0,\,{}^2c^T]\,\zeta(k), \quad \zeta(k) = \begin{bmatrix} {}^1x(k) \\ {}^2x(k) \end{bmatrix} \tag{15.4}$$

It can be seen that $^2d = 0$ led to a considerable simplification of eqns. (8.15) and (8.17). If we now choose for the description of the controller and the controlled plant the canonical form of reachability, we obtain a further simplification following from the fact that the column matrices ^1b and ^2b have the elements $^1b_1 = {}^2b_1 = 1$ and the remaining elements are equal to zero. We obtain

$$\zeta(k+1) = \begin{bmatrix} {}^1A & \begin{bmatrix} {}^2c^T \\ 0 \end{bmatrix} \\ \hline \begin{bmatrix} {}^1c^T \\ 0 \end{bmatrix} & {}^2A - \begin{bmatrix} {}^1d\,{}^2c^T \\ 0 \end{bmatrix} \end{bmatrix} \zeta(k) + \begin{bmatrix} 1 \\ 0 \\ \vdots \\ \hline {}^1d \\ 0 \\ \vdots \end{bmatrix} r(k) \tag{15.5}$$

The output equation in (15.4) remains unchanged.

A similar simplification will be arrived at when using the canonical form of controllability. However, in spite of these simplifications, the dimension of the closed-loop state vector according to (15.4) remains equal to the sum of the state vector dimensions of the controller and controlled plant. A reduction in the closed-loop state vector dimension may be obtained by reducing the controller state vector dimension. This, however, also reduces the possibility of influencing the control-loop dynamics by the controller. In the boundary case, when using a controller without dynamics, the matrices ^1A, ^1b, $^1c^T$ in eqn. (15.5) are zero and the equation of the controller has the form

$$^1y(k) = {}^1d\,e(k) = {}^2u(k) \tag{15.6}$$

This is a simple proportional controller with one input and one output. For a closed control loop we may write the equations

$$^2x(k+1) = \left\{ {}^2A - \begin{bmatrix} {}^1d\,{}^2c^T \\ 0 \end{bmatrix} \right\} {}^2x(k) + \begin{bmatrix} {}^1d \\ 0 \end{bmatrix} r(k)$$

$$^2y(k) = {}^2c^T\,{}^2x(k) \tag{15.7}$$

Since ^1d is a one-element matrix, the matrix of dynamics of the closed control loop can be influenced by the controller only to a limited extent and, therefore, the qualitative control requirements can only be satisfied in so far as it is possible to manage with a single optional constant of the controller given by eqn. (15.6). It should be noted that the classical concept of linear feedback control does not make sufficient use of the advantages of the state description, particularly the possibility of using the knowledge of the controlled plant state vector. We shall show a possible approach to the solution by an example.

Let us again consider a controlled plant and a controller according to eqn. (15.1) in the form

$$u(k) = -k^T x(k) + r(k)$$

For a closed control loop we have

$$^2x(k+1) = (^2A - {}^2bk^T)\,^2x(k) + {}^2b\,r(k)$$
$$^2y(k) = (^2c^T - {}^2dk^T)\,^2x(k) + {}^2d\,r(k) \qquad (15.8)$$

In this case the control error is

$$e(k) = r(k) - {}^2y(k) =$$
$$= (^2dk^T - {}^2c^T)\,^2x(k) + (1 - {}^2d)\,r(k)$$

where the last expression was obtained using the output equation (15.8). It can be seen that the dimension of the closed control loop state vector is identical with that of the controlled plant state vector. The term $^2bk^T$ is a square matrix having the same dimension as the matrix ^2A. It can easily be checked that all coefficients of the characteristic polynomial or, which is the same, all its roots can be influenced by the controller. The controller (15.1) can, therefore, satisfy a wide range of control performance requirements.

Some further possibilities of introducing the control error will be shown in the subsequent sections in connection with the applied control cost function.

Sometimes it is more useful to express in state space only the controlled plant and to calculate the initial state vector $x(0)$ with the aim to formulate the given problem as a transition of the plant from the initial state $x(0)$ to the given final state by the action of the input u. The initial vector $x(0)$ is calculated from the equation

$$x(0) = A\,x(0) + b\,u(0^-) \qquad (15.9)$$

In this equation we know A, b and usually also $x_1(0) = y(0)$. We do not know $u(0^-)$ and $x_i(0)$, $i = 2, 3, ..., n$, so that the number of equations following from eqn. (15.9) just suffices for the determination of the unknowns. The procedure for multi-input/multi-output control loops is analogous.

15.1 Pole assignment problem

In certain problems of automatic control we take interest in the possibility of influencing the characteristic polynomial coefficients of the closed-loop system since this enables us to satisfy the requirements imposed on the roots of the characteristic polynomial or on the eigenvalues of the matrix of dynamics of the control system. As an example we mention the design of the control loop satisfying the requirements of the boundary of aperiodicity, of the finite number of control steps and, in general, any solution with specified roots of the characteristic polynomial. In this section we shall show how to proceed in such a case or, how to assign the predetermined roots to the characteristic polynomial of the closed-loop system.

Consider a reachable plant whose equation of dynamics is

$$x(k + 1) = A\,x(k) + B\,u(k) \qquad (15.10)$$

where $x \in X^n$ and $u \in U^r$, $r \leq n$. The rank $h(B) = r$ so that none of the input variables can be omitted. Let the controller be described by the equation

$$u(k) = -K\,x(k) + r(k) \qquad (15.11)$$

Determine the matrix K in such a way that the closed control loop

$$x(k + 1) = (A - BK)\,x(k) + B\,r(k) =$$
$$= A^*\,x(k) + B\,r(k) \qquad (15.12)$$

has the desired characteristic polynomial

$$P(z) = \det(zE - A^*) = \alpha_0 + \alpha_1 z + \ldots + \alpha_{n-1} z^{n-1} + z^n \qquad (15.13)$$

In the following paragraphs we shall show two possible ways of solution of this problem.

15.1.1 Use of the canonical form of controllability

Consider first the single-input/single-output reachable plant (15.10) and transform it by substituting

$$x = Q x_R$$

to the canonical form of controllability (10.70):

$$x_R(k+1) = \begin{bmatrix} 0, & 1, & \ldots & 0 \\ \multicolumn{4}{c}{\dotfill} \\ 0, & 0, & \ldots & 1 \\ -a_0, & -a_1, & \ldots & -a_{n-1} \end{bmatrix} x_R(k) + \begin{bmatrix} 0 \\ \vdots \\ 0 \\ 1 \end{bmatrix} u(k) \qquad (15.14)$$

Using for the calculation of the controlling variable $u(k)$, the feedback controller

$$u(k) = -k_R^T x_R(k) \tag{15.15}$$

where

$$k_R^T = [k_{R1}, k_{R2}, \ldots, k_{Rn}]$$

and whose input is the state vector $x_R(k)$, we obtain [65.1]

$$x_R(k+1) = \begin{bmatrix} 0, & 1, & \ldots & 0 \\ \vdots & & & \\ 0, & 0, & \ldots & 1 \\ (-k_{R1} - a_0), & (-k_{R2} - a_1), & \ldots & (-k_{Rn} - a_{n-1}) \end{bmatrix} x_R(k) = A_R^* x_R(k) \tag{15.16}$$

In eqn. (15.15) and the following equations it was possible to omit the command variable $r(t)$ since the characteristic polynomial of the control loop is invariant with respect to this variable. In this case the characteristic polynomial is

$$\det(zE - A_R^*) = (k_{R1} + a_0) + (k_{R2} + a_1)z + \ldots + (k_{Rn} + a_{n-1})z^{n-1} + z^n \tag{15.17}$$

If the desired polynomial is the polynomial (15.13), then

$$\alpha_i = k_{R(i+1)} + a_i, \quad i = 0, 1, \ldots, n-1 \tag{15.18}$$

and the elements of the controller matrix k_R are

$$k_{R(i+1)} = \alpha_i - a_i, \quad i = 0, 1, \ldots, n-1 \tag{15.19}$$

The relationship between the controlling variable u and the original state vector is expressed by the relation

$$u(k) = -k_R^T x_R(k) = -k_R^T T x(k) = -k^T x(k), \quad T = Q^{-1} \tag{15.20}$$

For a multi-input/multi-output reachable plant we use again the transformation to the canonical form of controllability, (11.67). The input variables are considered to be the components of the vector u^*, where $u^* = Mu$. Since M is always non-singular, $u = M^{-1} u^*$. For the feedback controller we have

$$u^*(k) = -K_R x_R(k) \tag{15.21}$$

where

$$K_R = \begin{bmatrix} k_{R11}, & \ldots & k_{R1n} \\ \vdots & & \vdots \\ k_{Rr1}, & \ldots & k_{Rrn} \end{bmatrix}$$

PRINCIPLES OF DETERMINISTIC SYNTHESIS

Then the matrix of dynamics of the feedback system is

$$A_R - B_R K_R = \begin{bmatrix} 0, & 1, & \ldots & 0 & 0, & & \ldots & 0 \\ \vdots & & & & \vdots & & & \vdots \\ 0, & 0, & \ldots & 1 & & & & \\ h_{11}, & h_{12}, & \ldots & h_{1n_1} & h_{1(n-n_r+1)}, & & \ldots & h_{1n} \\ \hline 0, & \ldots & & 0 & 0, & 1, & \ldots & 0 \\ \vdots & & & \vdots & & & & \\ & & & & 0, & 0, & \ldots & 1 \\ h_{r1}, & \ldots & & h_{rn_1} & h_{r(n-n_r+1)}, & & \ldots & h_{rn} \end{bmatrix} = A_R^*$$

(15.22)

where

$$h_{ij} = \alpha_{ij} - k_{Rij}, \quad i = 1, 2, \ldots, r \quad j = 1, 2, \ldots, n \quad (15.23)$$

It is plain that it is always possible to determine such a value of k_{Rij} that the resultant value h_{ij} of the control loop matrix of dynamics satisfies the given requirements. However, the question of choosing such elements of the matrix K_R that the characteristic polynomial may have the required coefficients, remains open. One possible way is to choose such elements of K_R that the matrix A_R^* be by blocks an upper or lower triangular matrix or a diagonal matrix. That is to say, the coefficients of the Frobenius diagonal blocks are then the coefficients of the characteristic polynomial. We choose, for instance,

$$h_{1i} = 0 \quad \text{for} \quad n_1 + 1 \leq i \leq n$$
$$h_{2i} = 0 \quad \text{for} \quad 1 \leq i \leq n_1 \text{ and } n_1 + n_2 + 1 \leq i \leq n$$
$$\vdots$$
$$h_{ri} = 0 \quad \text{for} \quad 1 \leq i \leq n - n_r \quad (15.24)$$

Then the characteristic polynomial is

$$P(z) = \prod_{j=1}^{r} \det(zE - A_{Rj}^*) = (-h_{11} - h_{12}z - \ldots - h_{1n_1}z^{n_1-1} + z^{n_1}) \ldots$$

$$\ldots (-h_{r(n-n_r+1)} - \ldots - h_{rn}z^{n_r-1} + z^{n_r}) = \prod_{j=1}^{r} P_j(z) \quad (15.25)$$

In this case each partial plant possessing the matrix of dynamics $A_{Rj}^*, j = 1, 2, \ldots, r$, is controllable by only one input variable u_j [68.1] so that, from the point of view of the inputs, the plant is decomposed into r partial plants. The order of the individual partial plants is determined in accordance with the procedure relating to eqn. (11.69), the highest order corresponding to the index of reachability, v.

The second possibility of determining the characteristic polynomial, which presents itself, is to choose such elements of K_R that the matrix A_R^* may attain the Frobenius form. This means, that all elements in the first superdiagonal of the matrix A_R^* must be ones, the elements in the last row are the desired coefficients of the characteristic polynomial and the remaining elements must be zero. The system is then controllable by the last input variable u_r^*.

According to Fig. 11.4, the vector of the input variables is given by the relations

$$u(k) = M^{-1} u^*(k) = -M^{-1} K_R x_R(k) = -M^{-1} K_R T x(k)$$
$$u(k) = K x(k) \qquad (15.26)$$

where T is the transformation matrix (11.71).

The product $K_R T$ may be calculated directly when writing eqn. (15.23) in terms of matrices:

$$H = N_R - K_R \qquad (15.27)$$

where K_R is the matrix (15.21), N_R is a matrix set up from the rows $n_i = \sum_{j=1}^{i} n_j$, $i = 1, 2, \ldots, r$, of the matrix (11.67), i.e.

$$N_R = \begin{bmatrix} \alpha_{11}, & \ldots & \alpha_{1n} \\ \vdots & & \vdots \\ \alpha_{r1}, & \ldots & \alpha_{rn} \end{bmatrix}$$

and H is a matrix set up from the rows $n_i = \sum_{j=1}^{i} n_j$, $i = 1, 2, \ldots, r$, of the matrix (15.22), i.e.

$$H = \begin{bmatrix} h_{11}, & \ldots & h_{1n} \\ \vdots & & \vdots \\ h_{r1}, & \ldots & h_{rn} \end{bmatrix}$$

Then

$$K_R T = N_R T - H T \qquad (15.28)$$

Since the product $N_R T$ may be calculated directly according to (11.72), the matrix T need not be inverted [68.1].

15.1.2 Use of the general form of the matrix of dynamics

The problem formulated in the introductory part of Chap. 15 may be solved even without the use of suitable canonical forms of state representation [60.4]. Let us first be concerned with the single-input/sinngle-output case. It holds that

$$x(k+1) = A x(k) + b u(k)$$
$$x(k+2) = A^2 x(k) + b u(k+1) + Ab u(k)$$
$$\vdots$$
$$x(k+n) = A^n x(k) + [b, Ab, \ldots, A^{n-1}b] \begin{bmatrix} u(k+n-1) \\ u(k+n-2) \\ \vdots \\ u(k) \end{bmatrix} \qquad (15.29)$$

PRINCIPLES OF DETERMINISTIC SYNTHESIS

If the plant is reachable it can be transferred from the initial state $x(k)$ to the desired final state $x(k + n)$ in n steps. This transfer is achieved by the input variable values

$$\begin{bmatrix} u(k + n - 1) \\ u(k + n - 2) \\ \vdots \\ u(k) \end{bmatrix} = Q_D^{-1}[x(k + n) - A^n x(k)] \quad (15.30)$$

where $Q_D = [b, Ab, \ldots, A^{n-1}b]$.

However, at the selected instant k we do not want to calculate all the future values of the input variable, but wish to express this calculation by the controller equation

$$u(k) = -k^T x(k) \quad (15.31)$$

For this purpose we only need the last row of eqn. (15.30) and we, therefore, introduce

$$\eta^T = [0, \ldots, 0, 1] Q_D^{-1} \quad (15.32)$$

In eqn. (15.31) and the following equations it was possible to omit the command variable $r(t)$ for the same reason as in eqn. (15.15).

From eqn. (15.30) we now calculate

$$u(k) = \eta^T [x(k + n) - A^n x(k)] \quad (15.33)$$

From the first of eqns. (15.29) and from eqn. (15.31) it follows that

$$x(k + 1) = (A - bk^T) x(k) = \mathscr{A} x(k) \quad (15.34)$$

so that eqn. (15.33) may be written in the form

$$u(k) = \eta^T [\mathscr{A}^n - A^n] x(k) \quad (15.35)$$

The matrix $\mathscr{A}^n$ is expressed by means of the Cayley-Hamilton theorem. It holds that

$$\alpha_0 E + \alpha_1 \mathscr{A} + \ldots + \alpha_{n-1} \mathscr{A}^{n-1} + \mathscr{A}^n = 0$$

and consequently

$$u(k) = -\eta^T (\alpha_0 E + \alpha_1 \mathscr{A} + \ldots + \alpha_{n-1} \mathscr{A}^{n-1} + A^n) x(k) \quad (15.36)$$

Now we shall show that $\mathscr{A}$ in eqn. (15.36) may be replaced by A. Since

$$\eta^T Q_D = [0, \ldots, 0, 1] \quad (15.37)$$

we have

$$\eta^T \mathscr{A} = \eta^T(A - bk^T) = \eta^T A$$
$$\eta^T \mathscr{A}^2 = \eta^T A(A - bk^T) = \eta^T A^2$$
$$\vdots$$
$$\eta^T \mathscr{A}^{n-1} = \eta^T A^{n-2}(A - bk^T) = \eta^T A^{n-1} \tag{15.38}$$

The equation of the controller (15.36) may now be written in its final form

$$u(k) = -\eta^T P(A) x(k) = -k^T x(k) \tag{15.39}$$

where $P(A) = \alpha_0 E + \alpha_1 A + \ldots + \alpha_{n-1} A^{n-1} + A^n$. It should be noted that $P(A)$ is not a characteristic polynomial of the control loop and, therefore, $P(A) \neq 0$.

A generalization of this result for the multi-input/multi-output case is easily achieved by introducing, if possible, only one input variable so that it holds

$$u(k) = \beta v(k) \tag{15.40}$$

where $u(k)$ is the vector of the input variables. The equation of dynamics of the multi-input/multi-output control loop is then

$$x(k + 1) = A x(k) + B\beta v(k) \tag{15.41}$$

The column matrix β may be selected arbitrarily provided the conditions of reachability of eqn. (15.41) remain satisfied. Since $b = B\beta$ is a column matrix of the type $(n; 1)$, the equation of the fictitious controller

$$v(k) = -k^T x(k) \tag{15.42}$$

may be determined in the same way as for the single-input/single-output case. The feedback matrix of the virtual controller, however, is equal to the product $K = \beta k^T$ and has, in this case, a rank one. The total number of optional elements in the product βk^T is $r + n$ so that in this case we cannot use all the rn elements of the matrix K to satisfy the requirements imposed on the control system as in the case where the rank was $h(K) = r$, $r \leq n$.

In Par. 15.1.1 we have shown that the feedback derived from the state vector may be designed to affect not only the characteristic polynomials of the individual partial systems but also, in the sense mentioned, the structure of the multi-input/multi-output control loop. For instance, from eqn. (15.28) the method of decomposition of the multi-input/multi-output plant with respect to the input variables is apparent. A conversion of the equation of dynamics to the canonical form of controllability is not necessary. On the contrary, from eqns. (15.26) and (15.28) it follows that the matrix of the controller is

$$K = M^{-1}(N_R T - HT) \tag{15.43}$$

where the matrices N_R, H and T are the same as in eqns. (15.27) and (15.28). The product $N_R T$ may be calculated directly from eqn. (11.72). For multi-input/multi-output systems decomposed with respect to the input variables, eqn. (15.43) may, therefore, be rewritten in the form

$$K = M^{-1} \begin{bmatrix} \eta_1^T A^{n_1} \\ \vdots \\ \eta_r^T A^{n_r} \end{bmatrix} -$$

$$- M^{-1} \begin{bmatrix} h_{11}, & \ldots & h_{1n_1} & 0, & \ldots & 0 & 0, & \ldots & 0 \\ 0, & \ldots & 0 & h_{2(n_1+1)}, & \ldots & & \vdots & & \vdots \\ \vdots & & \vdots & 0 & & 0 & & & \\ & & & \vdots & & & 0, & \ldots & 0 \\ 0, & \ldots & 0 & 0, & \ldots & 0 & h_{r(n-n_r+1)}, & \ldots & h_{rn} \end{bmatrix} \begin{bmatrix} \eta_1^T \\ \eta_1^T A \\ \vdots \\ \eta_1^T A^{n_1-1} \\ \hdashline \eta_2^T \\ \vdots \\ \hdashline \vdots \\ \eta_r^T A^{n_r-1} \end{bmatrix}$$

(15.44)

With the characteristic polynomials $P_j(z)$, $j = 1, 2, \ldots, r$, according to eqn. (15.25), we also have

$$K = M^{-1} \begin{bmatrix} \eta_1^T P_1(A) \\ \eta_2^T P_2(A) \\ \cdots \cdots \\ \eta_r^T P_r(A) \end{bmatrix} \qquad (15.45)$$

which is the generalized form of the controller matrix in the relation (15.39).

The whole procedure may be summarized as follows: first we write down the matrix of reachability Q_D and determine the first n linearly independent vectors. We arrange these vectors into the matrix

$$Q_{D2} = [b_1 A b_1, \ldots, A^{n_1-1} b_1 \mid b_2, \ldots \mid \ldots A^{n_r-1} b_r] \qquad (15.46)$$

where $n = n_1 + n_2 + \ldots + n_r$ is the order of the plant and max $(n_i, i = 1, 2, \ldots, r)$ is equal to the index of reachability. The last rows of the individual blocks of the matrix Q_{D2}^{-1} are η_i^T, $i = 1, 2, \ldots, r$.

Let $A_i(z)$, $i = 1, 2, \ldots, r$, be the desired characteristic polynomials whose degrees n_i, $i = 1, 2, \ldots, r$, correspond to the assumed blocks in the matrix (15.46). The matrix M needed in eqn. (15.45) is then determined as follows. If $A^{n_i-1} b_i$ is the last column in eqn. (15.46) corresponding to b_i, we make use of the fact that the columns $A^{n_i-1} b_{i+1}, A^{n_i-1} b_{i+2}, \ldots, A^{n_i-1} b_r$ are linearly dependent on all previously chosen linearly independent columns in eqn. (15.46) which we denote, for simplicity, by β_k. It holds that [69.1]

$$A^{n_i-1}b_j = m_{ij}A^{n_i-1}b_i + \sum_k c_k \beta_k, \quad j > i \qquad (15.47)$$

The weighting coefficient m_{ij} is an element of the matrix M whose triangular form is given in (11.67).

Substituting M, η_i, $P_i(A)$, $i = 1, 2, ..., r$, into eqn. (15.45) we obtain the feedback matrix of the controller whose input is the state vector $x(k)$. In conclusion it should be noted that the diagonal structure of the matrix (15.22) with the elements according to (15.24) enables us also to determine the minimum polynomial of the degree v, where v is the highest degree occurring with the characteristic polynomials $A_i(z)$, $i = 1, 2, ..., r$. By convention, v is the index of reachability. If we find the minimum polynomial $A_{i(v)}(z)$, then the characteristic polynomials of the rest of the partial systems may have only such roots which are the roots of the minimum polynomial as well. When forming the plant structure with the individual characteristic polynomials, one has to keep in view the complex roots: a pair of complex conjugate roots cannot be decomposed into two partial systems.

15.2 Finite number of control steps

15.2.1 Transition to equilibrium state

The finite number of control steps is sometimes designated as time optimum control. This is the control performance criterion which follows directly from the definition of controllability 9.2.

DEFINITION 15.1: *Consider a continuously working system. Then the finite number of control steps is such a finite number of values, N, of the sequence $u(k)$, $u(k + 1)$, ..., $u(k + N - 1)$, by which the system is transferred from an arbitrary initial state $x(k) \neq 0$ to the final state $x(k + N) = 0$.*

Unless the controlling variable is constrained, the time of transition from the initial state to the final state of the plant described by eqn. (15.10) is reduced proportionally to the reduction of the sampling interval. In addition to this fact, however, the values of the controlling variable u in the individual intervals increase. In the limit, for the sampling period $T \to 0$, we might arrive at a physically non-realizable solution, when the time of transition tends to zero and the value of the controlling variable increases beyond all limits. For this reason constraints are imposed on the controlling variable in continuously working plants and the solution for these constraints and for the given plant in the sense of Definition 15.1 is justifiably designated as time-optimal. With regard to the constraints of the controlling variable, this problem is a non-linear

PRINCIPLES OF DETERMINISTIC SYNTHESIS

one, the controlling variable assumes in the individual intervals just the boundary values, but the values of the intervals are unequal.

For plants described by eqn. (15.10), however, the sampling period is constant, $0 < T < \infty$, and therefore $\infty > |u(k + i)| > 0$, $i = 0, 1, \ldots, N - 1$. The transition time depends on the fixed structure of the given plant and on the sampling period T. Since, however, the sampling period is usually selectable over a rather wide range, it is better to speak of a finite number of control steps than of a time-optimal solution.

We shall show that a fixed feedback whose input is the state vector x

$$u(k) = -K\,x(k) \qquad (15.48)$$

is sufficient for attaining a finite number of control steps.

Equation (15.10) with this feedback gives the closed-loop equation

$$x(k + 1) = (A - BK)\,x(k) = A^*\,x(k) \qquad (15.49)$$

According to Definition 15.1 we require that

$$x(k + N) = A^{*N}\,x(k) = 0 \qquad (15.50)$$

Since the initial state $x(k) \neq 0$, the requirement (15.50) will only be satisfied if

$$A^{*N} = 0 \qquad (15.51)$$

This implies that the matrix A^* is nilpotent with the index N. According to the Cayley-Hamilton theorem the characteristic polynomial of the matrix A^* for $N \geq n$ has the form

$$\det\,[zE - A^*] = z^n$$

i.e. $\alpha_0 = \alpha_1 = \ldots = \alpha_{n-1} = 0$ in eqn. (15.13). If a single-input plant is described in the state space by equations in the canonical form of controllability, then, according to eqn. (15.19),

$$k_{R(i+1)} = -a_i, \quad i = 0, 1, \ldots, n - 1 \qquad (15.52)$$

The controller equation (15.48) is

$$u(k) = -k_R^T\,x_R(k) = [a_0, a_1, \ldots, a_{n-1}]\,x_R(k) \qquad (15.53)$$

If the plant equations are not converted to the canonical form of controllability, the controller equation may be expressed by virtue of relation (15.39). Since $\alpha_0 = \alpha_1 = \ldots = \alpha_{n-1} = 0$ we have in this case $P(A) = A^n$, $N = n$. We obtain

$$u(k) = -k^T\,x(k) =$$
$$= -[0, \ldots, 0, 1]\,Q_D^{-1}A^n\,x(k) \qquad (15.54)$$

where

$$Q_D = [b, Ab, \ldots, A^{n-1}b]$$

If for a single-input plant the matrix of reachability has the rank $h(Q_D) = n$ and, consequently, the minimum polynomial is identical with the characteristic polynomial of the plant, eqn. (15.51) will just be satisfied for $N = n$ and n is in this case, in general, the smallest number of control steps. This statement follows directly from the feedback equation (15.54). The plant with the controller (Fig. 15.1) is suggestive of the operation of a shift register whose memory cells contain at the instant k the values of the

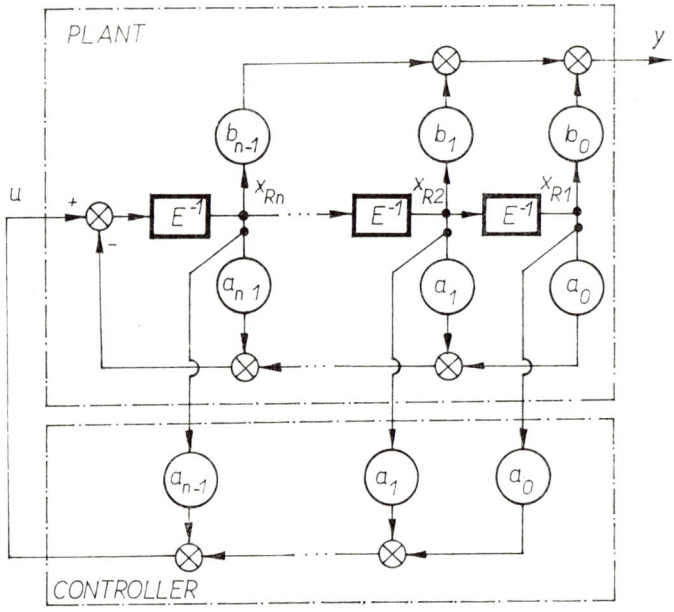

Fig. 15.1. Nullification of the initial state of a controlled plant by the action of a controller in the feedback.

components of the initial state vector $x(k)$, the contents of these cells being successively nullified in $N = n$ steps by the action of the controller.

From eqn. (15.54) it follows that

$$k^T = [0, \ldots, 0, 1] Q_D^{-1} A^n$$

This expression may be rewritten in the form

$$k^T [A^{-1}b, A^{-2}b, \ldots, A^{-n}b] = [1, 0, \ldots, 0]$$
$$k^T Q_R = [1, 0, \ldots, 0] \qquad (15.55)$$

where Q_R is the matrix of controllability. Relation (15.55) corresponds to the original solution of Kalman and Bertram [59.2], and the matrix of controllability follows

PRINCIPLES OF DETERMINISTIC SYNTHESIS

directly from the requirement to transfer the system in a finite number of steps from the initial non-zero state to a zero equilibrium state.

Example 15.1: A controlled plant with an integrating servomotor is given by the transfer function

$$S(s) = \frac{1}{s(s + 0.5)^2}$$

Find a discrete controller for transferring the plant from the initial state $x(0) = [-1, 0, 0]^T$ to the final state $x(3) = 0$. The sampling period $T = 1$.

Solution: According to eqns. (3.34) and (3.36) we have

$$x^{(1)}(t) = F x(t) + g u(t)$$
$$y(t) = h^T x(t)$$

where

$$F = \begin{bmatrix} 0, & 1, & 0 \\ 0, & 0, & 1 \\ 0, & -0.25, & -1 \end{bmatrix}; \quad g = \begin{bmatrix} 0 \\ 0 \\ 1 \end{bmatrix}; \quad h^T = [1, 0, 0]$$

and the state vector components are

$$x(t) = \begin{bmatrix} x_1(t) \\ x_2(t) \\ x_3(t) \end{bmatrix} = \begin{bmatrix} y(t) \\ y^{(1)}(t) \\ y^{(2)}(t) \end{bmatrix}$$

By virtue of (5.36) we calculate

$$A(T) = e^{FT} = \begin{bmatrix} 1, & 4 - 4e^{-0.5T} - Te^{-0.5T}, & 4 - 4e^{-0.5T} - 2Te^{-0.5T} \\ 0, & e^{-0.5T} + 0.5Te^{-0.5T}, & Te^{-0.5T} \\ 0, & -0.25Te^{-0.5T}, & e^{-0.5T} - 0.5Te^{-0.5T} \end{bmatrix}$$

$$A = \begin{bmatrix} 1, & 0.9673, & 0.3608 \\ 0, & 0.9098, & 0.6065 \\ 0, & -0.1516, & 0.3033 \end{bmatrix}$$

According to (7.54), on substituting $\vartheta = T(1 - \tau)$, we determine

$$b(T) = \int_0^T e^{F\vartheta} g \, d\vartheta = \int_0^T A(\vartheta) g \, d\vartheta = F^{-1}(e^{FT} - E) g$$

$$b = \begin{bmatrix} 0.1306 \\ 0.3608 \\ 0.6065 \end{bmatrix}$$

For the output equation it holds that $h^T = c^T$.

The matrix of reachability $Q_D = [b, Ab, A^2b]$ is

$$Q_D = \begin{bmatrix} 0.1306, & 0.6984, & 1.4184 \\ 0.3608, & 0.6961, & 0.7117 \\ 0.6065, & 0.1293, & -0.0664 \end{bmatrix}$$

Using eqn. (15.32) we find

$$\eta^T = [0, 0, 1]\, Q_D^{-1} = [1.6151, \ -1.7492, \ 0.6929]$$

and finally, by means of eqn. (15.54), we determine the equation of the controller

$$u(k) = [-1.6151, \ 1.7492, \ -0.6929]\, A^3\, x(k) =$$
$$= -[1.6151, \ 2.8460, \ 1.6083]\, x(k) = -k^T x(k)$$

For $x(0) = [-1, 0, 0]$ we calculate from the equation of the controller $u(0) = 1.6151$ and from the equation of dynamics

$$x(k + 1) = A\, x(k) + b\, u(k)$$

we obtain successively for $k = 0, 1, 2, 3, \ldots$

$$x(1) = \begin{bmatrix} -0.7891 \\ 0.5827 \\ 0.9796 \end{bmatrix}, \quad u(1) = -1.9594, \quad y(1) = -0.7891$$

$$x(2) = \begin{bmatrix} -0.1279 \\ 0.4173 \\ -0.9796 \end{bmatrix}, \quad u(2) = 0.5944, \quad y(2) = -0.1279$$

$$x(3) = 0, \quad\quad u(3) = 0, \quad\quad y(3) = 0, \ldots$$

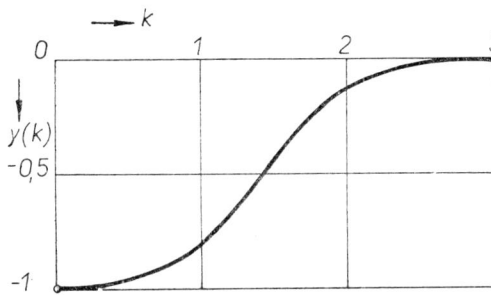

Fig. 15.2. Control process relating to Example 15.1.

The control process of $y(k)$, $k = 0, 1, 2, 3$, is shown in Fig. 15.2. The same result would be obtained when solving this problem by the procedure described in Par. 15.1.1.

PRINCIPLES OF DETERMINISTIC SYNTHESIS

We find first the characteristic polynomial of the matrix A, i.e.

$$A(z) = z^3 - 2.2130z^2 + 1.5809z - 0.3679$$

so that

$$A_R = \begin{bmatrix} 0, & 1, & 0 \\ 0, & 0, & 1 \\ 0.3679, & -1.5809, & 2.2130 \end{bmatrix}, \quad b_R = \begin{bmatrix} 0 \\ 0 \\ 1 \end{bmatrix}$$

To calculate the matrix c_R^T we must first determine, by virtue of (10.73), the transformation matrix

$$Q = \begin{bmatrix} 0.0792, & 0.4094, & 0.1306 \\ -0.2584, & -0.1024, & 0.3608 \\ 0.6064, & -1.2129, & 0.6065 \end{bmatrix}$$

and then

$$c_R^T = c^T Q = [0.0792, \; 0.4094, \; 0.1306]$$

The matrix of the controller follows from the requirement that all eigenvalues of the matrix $A_R - b_R k_R^T$ be zero. This is feasible with

$$k_R^T = [0.3679, \; -1.5809, \; 2.2130]$$

We might easily check that, according to (15.20),

$$k^T = k_R^T Q^{-1} = k_R^T T$$

or

$$k_R^T = k^T Q$$

In order that we can calculate the control process with the controller matrix k_R^T we must first determine the initial vector corresponding to the canonical form of controllability:

$$x_R(0) = Q^{-1} x(0) =$$
$$= \begin{bmatrix} 1.6151, & -1.7492, & 0.6928 \\ 1.6151, & -0.1340, & -0.2680 \\ 1.6151, & 1.4809, & 0.4201 \end{bmatrix} \begin{bmatrix} -1 \\ 0 \\ 0 \end{bmatrix} = \begin{bmatrix} -1.6151 \\ -1.6151 \\ -1.6151 \end{bmatrix}$$

The vector $x_R(0)$ may also be obtained without inverting the matrix Q. In the equilibrium state we have for $k = 0^-$ the input $u(0^-) = 0$ and the output $y(0^-) = -1$. From the equation of dynamics it follows that $x_{R1}(0^-) = x_{R2}(0^-)$ and $x_{R2}(0^-) = x_{R3}(0^-)$ and, therefore, from the output equation we may find

$$x_{R1}(0^-) = \frac{y(0^-)}{\sum_{i=1}^{3} c_{Ri}} = \frac{-1}{0.6192} \doteq -1.6151$$

$$x_{R1}(0^-) = x_{R2}(0^-) = x_{R3}(0^-) = -1.6151$$

The values of the input variable u and of the output variable y are invariant with respect to a change in base, but the state vectors are now, for $k = 1, 2, 3, \ldots,$

$$x_R(1) = \begin{bmatrix} -1.6151 \\ -1.6151 \\ 0 \end{bmatrix}, \quad x_R(2) = \begin{bmatrix} -1.6151 \\ 0 \\ 0 \end{bmatrix}, \quad x_R(3) = \begin{bmatrix} 0 \\ 0 \\ 0 \end{bmatrix} \text{ etc.}$$

Note: The example may also be regarded as the solution of a finite number of steps of a closed control loop with a given plant which was in equilibrium at the instant $k = 0^-$ and the command variable changed by a unit step at the instant $k = 0^+$. Using the classical theory, this case was solved in the literature [65.3] with the same result.

For multi-input plants the number of control steps may be $N < n$ if the matrix A^* corresponds to the minimum polynomial z^v, where v is the index of reachability. Then

$$A^{*v} = 0 \tag{15.56}$$

Since in the closed control loop all eigenvalues of the matrix A^* are equal to zero, the Frobenius form of the matrix of dynamics is identical with the Jordan form. Then the matrix of dynamics includes as many Jordan fields as many there are input variables and the index of reachability is equal to the order of the largest Jordan block.

In multi-input/multi-output systems, provided their state representation is in the canonical form of controllability, all elements h_{ij} of the matrix A^* in (15.22) may be nullified by choosing the elements of the controller matrix K_R. The matrix A^* will then have r blocks with only zero eigenvalues and the largest block determines the index of reachability v. Let us recall again that a transformation of the state representation to the canonical form of controllability need not be performed if the controller matrix is determined using the relation (15.44). Since, for a finite number of control steps, the characteristic polynomials of the individual diagonal fields are z^{n_i}, $i = 1, 2, \ldots, r$, where $\max(n_i, i = 1, 2, \ldots, r) = v$, the matrix $H = 0$ and expression (15.44) is simplified to the form [72.1]

$$K = M^{-1} \begin{bmatrix} \eta_1^T A^{n_1} \\ \vdots \\ \eta_r^T A^{n_r} \end{bmatrix} \tag{15.57}$$

The operation of the feedback controller with a multi-input/multi-output plant may again be compared to the operation of r shift registers connected in parallel, whose initial values are, successively in all partial systems, simultaneously nullified. The largest number of steps necessary for nullifying all memory cells is required for that shift register which corresponds to the partial system of the highest order. However, the output variables of the system are, as a rule, nullified all at once after completion of the last step. We have used here the term "as a rule" because, in fact, with multi-

PRINCIPLES OF DETERMINISTIC SYNTHESIS

input/multi-output systems the problem of finite number of control steps has an unlimited number of solutions differing in the form of control processes. The non-uniqueness of the solution already follows from the possibility of choosing the linearly independent columns in the matrix (15.46) since in this matrix any other column of the matrix B may now be used as the first column instead of the column b_1. This, of course, results in a change in the order of the individual partial systems. If the linearly independent columns of the matrix (15.46) have already been selected, an arbitrary linear combination of the selected columns may be used in the further evaluation without any change in the number of control steps, provided the new matrix arising in this way has the same rank as the original matrix (15.46). Hence it follows that for multi-input/multi-output systems the finite number of control steps represents a relatively weak criterion, and that of all the possible solutions we may select that one which best satisfies the suitably chosen additional conditions.

Some of the possible solutions will now be shown by a further example.

Example 15.2: Consider a double-input/double-output plant for the discrete state representation of which we know the following matrix triad:

$$A = \begin{bmatrix} 0.6, & 0.7, & 0 \\ 0, & 0.75, & 0 \\ 0, & 0, & 0.8 \end{bmatrix}, \quad B = \begin{bmatrix} 0.8, & 0.4 \\ 0, & 0.9 \\ 0.9, & 0.9 \end{bmatrix}, \quad C = \begin{bmatrix} 1, & 0, & 0 \\ 0, & 0, & 1 \end{bmatrix}$$

This is a plant already used for the determination of the canonical form of controllability in Example 11.3.

Find the controller for a finite number of control steps assuming that both input command variables of the control loop change simultaneously and have the form of a unit step.

Solution: The problem may be solved as a transfer of the plant from the initial non-zero state, where the output variables are, e.g. $y_1(0^-) = y_2(0^-) = -1$, to zero equilibrium state. With regard to the form of the matrix C we have $y_1(0^-) = x(0^-)$ and $y_2(0^-) = x_3(0^-)$. From the initial state vector it remains to determine the value of the state variable x_2 for the instant 0^-. According to eqn. (15.9) we may write

$$\begin{bmatrix} -1 \\ x_2(0^-) \\ -1 \end{bmatrix} = \begin{bmatrix} 0.6, & 0.7, & 0 \\ 0, & 0.75, & 0 \\ 0, & 0, & 0.8 \end{bmatrix} \begin{bmatrix} -1 \\ x_2(0^-) \\ -1 \end{bmatrix} + \begin{bmatrix} 0.8, & 0.4 \\ 0, & 0.9 \\ 0.9, & 0.9 \end{bmatrix} \begin{bmatrix} u_1(0^-) \\ u_2(0^-) \end{bmatrix}$$

where $x_2(0^-)$, $u_1(0^-)$ and $u_2(0^-)$ are the values of the variables corresponding to the initial state at the instant 0^-, i.e. before the unit-step change of the input variables u_1 and u_2 occurs. From the last equation we obtain $x_2(0^-) = -0.3774$.

The matrix of reachability may be used in the same form $Q_{D2} = [b_1, Ab_1, b_2]$ as in Example 11.3, where the calculation of the matrix Q_{D2}^{-1}, M and the matrix

$$\begin{bmatrix} \eta_1^T A^2 \\ \eta_2^T A \end{bmatrix}$$

has already been carried out. In this matrix η_1^T is the second row and η_2^T the third row of the matrix Q_{D2}^{-1}. By virtue of eqn. (15.57) we can, therefore, determine directly the controller matrix

$$K = \begin{bmatrix} 1, & 3.3125 \\ 0, & 1 \end{bmatrix} \begin{bmatrix} -2.25, & 7.4688, & 3.5556 \\ 0, & 0.8333, & 0 \end{bmatrix} =$$

$$= \begin{bmatrix} -2.25, & -4.7083, & 3.5556 \\ 0, & 0.8333, & 0 \end{bmatrix}$$

The matrix of the closed control loop is

$$A^* = A - BK = \begin{bmatrix} 2.4, & 4.1333, & -2.8445 \\ 0, & 0, & 0 \\ 2.025, & 3.4875, & -2.4 \end{bmatrix}$$

Since the index of reachability, v, is equal to the highest order occurring in the chosen partial systems, we have in our example, according to the form of Q_{D2}, apparently $v = n_1 = 2$. It can be checked that the matrix A^* is nilpotent, $A^{*2} = 0$. The number of control steps is, therefore, $N = 2$.

Beginning with the initial state $x(0)$ we calculate successively

$$x(0) = \begin{bmatrix} -1 \\ -0.3774 \\ -1 \end{bmatrix}, \quad x(1) = \begin{bmatrix} -1.1154 \\ 0 \\ -0.9412 \end{bmatrix}, \quad x(2) = \begin{bmatrix} 0 \\ 0 \\ 0 \end{bmatrix}$$

$$u(0) = \begin{bmatrix} 0.4713 \\ -0.3145 \end{bmatrix}, \quad u(1) = \begin{bmatrix} -0.8369 \\ 0 \end{bmatrix}, \quad u(2) = \begin{bmatrix} 0 \\ 0 \end{bmatrix}$$

In accordance with the given problem formulation, the output variables are $y_1 = x_1$ and $y_2 = x_3$.

Using for the solution the columns of the matrix of controllability instead of the columns of the matrix of reachability we have, for instance,

$$Q_{R2} = [A^{-1}b_1, A^{-1}b_2, A^{-2}b_1] = \begin{bmatrix} 1.3333, & -0.7333, & 2.2222 \\ 0, & 1.2, & 0 \\ 1.125, & 1.125, & 1.4063 \end{bmatrix}$$

Calculating Q_{R2}^{-1} and denoting the first two rows of this matrix by η_1^T, η_2^T, the controller matrix

$$K = \begin{bmatrix} \eta_1^T \\ \eta_2^T \end{bmatrix}$$

will be numerically identical with the previously given result and also the further process of solving the problem is necessarily identical. Considering that in this case none of the previous results of Example 11.3 were needed, it is clear that the procedure derived from the matrix of controllability is substantially shorter.

15.2.2 Transition to steady state

In contrast with the case considered in Par. 15.2.1 let us now be concerned with the case where the resultant state, after completion of the finite number of control steps, is not balanced but steady, corresponding to a previously given function of the command variable $r(t)$. The oldest work relating to this problem is that of R. E. Kalman and J. E. Bertram [59.2] which was followed by further publications (e.g. [60.5], [63.8] and others). A substantial generalization, especially for the so-called weak version of the finite number of control steps, discussed in Par. 15.2.3, was worked out by V. Kučera [67.3, 70.4, 72.6].

It should be noted that transition to the equilibrium state may be regarded merely as a special case of transition to the steady state where the function of the command variable $r(k) \neq $ const. From this point of view the transitions to both equilibrium and steady state may be regarded as a single case, the so-called strong version of the finite number of control steps, in the solution of which we shall express the information given in Par. 15.2.1 more precisely; this version may be defined as follows:

DEFINITION 15.2: *A strong version of the finite number of control steps is such an action exerted by the input variable on the plant, starting at the instant $k = 0$, at which the initial plant state is $x(0)$, that after completion of N sampling periods the error vector*

$$\varepsilon(k) = \hat{\omega}(k) - \hat{x}(k)$$

reaches a zero value for $k \geq N$, where $\hat{x}(k)$ is the plant state vector, $\hat{\omega}(k)$ is the state vector of the command variable $r(k)$ extended by the zero elements to the dimension of the vector $\hat{x}(k)$, and $\hat{\omega}(k)$ and $\hat{x}(k)$ are expressed in the same base.

Note 15.1: In Definition 15.2 a zero error vector is required for the strong version of the finite number of control steps. The same requirement is imposed by the equality of the derivatives of the plant command and output variables, i.e. $r^{(i)}(t) = y^{(i)}(t)$, $i = 0, 1, ..., (n - 1)$.

Since we are describing the plant in the state space, it is reasonable to express also the command variable as a signal generated by a differential equation and to determine this equation in state space representation in the adopted manner. If we confine ourselves, as in the case of plants, to such command variables whose differential equation is linear, then the command variable can be only a linear combination of the functions $t^{\varkappa}$, $\varkappa = 0, 1, 2, ...$, and $e^{\alpha t}$, $\alpha = \beta \pm j\omega$.

Let the command variable be generated by the differential equation

$$r^{(m)}(t) + A_{m-1} r^{(m-1)}(t) + ... + A_0 r(t) = 0 \qquad (15.58)$$

To express this equation in the state space we introduce the state variables

$$\omega(t) = \begin{bmatrix} \omega_1(t) \\ \omega_2(t) \\ \vdots \\ \omega_m(t) \end{bmatrix} = \begin{bmatrix} r(t) \\ r^{(1)}(t) \\ \vdots \\ r^{(m-1)}(t) \end{bmatrix} \tag{15.59}$$

Using the state vector $\omega(t)$ we may rewrite eqn. (15.58) in the form

$$\omega^{(1)}(t) = V\omega(t) \tag{15.60}$$

where

$$V = \begin{bmatrix} 0, & 1, & 0, & \ldots & 0 \\ 0, & 0, & 1, & \ldots & 0 \\ \multicolumn{5}{c}{\dotfill} \\ -A_0, & -A_1, & -A_2, & \ldots & -A_{m-1} \end{bmatrix}$$

The solution of eqn. (15.60) is

$$\omega(t) = e^{V(t-t_0)} \omega(t_0), \quad t_0 < t \tag{15.61}$$

where $\omega(t_0)$ is the vector of the initial conditions. For $t_0 = kT$, $t = (k+1)T$, where T is the sampling period, we obtain the discrete version of the state equation (15.60). Omitting, for simplicity of writing, the sampling period T in the argument, we have

$$\omega(k+1) = e^{VT}\omega(k) = W\omega(k)$$
$$W = e^{VT} \tag{15.62}$$

Equation (15.62) is the equation of dynamics of the command variable, i.e. of the control loop input variable, and the corresponding output equation has the form

$$r(k) = c_W^T \omega(k) \tag{15.63}$$

where

$$c_W^T = [1, 0, \ldots, 0]$$

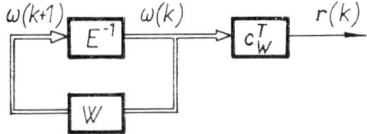

Fig. 15.3. Block diagram of the difference equation of the command variable $r(k)$.

The control loop block diagram corresponding to eqns. (15.62) and (15.63) is shown in Fig. 15.3.

We shall now give the state representation of the input variables whereby the linear combination (15.58) may be found.

PRINCIPLES OF DETERMINISTIC SYNTHESIS

Consider first the *polynomial function*

$$r(t) = p_0 + p_1 t + \ldots + p_{m-1} t^{m-1} \tag{15.64}$$

for which

$$r^{(1)}(t) = p_1 + 2p_2 t + \ldots + (m-1) p_{m-1} t^{m-2}$$
$$r^{(2)}(t) = 2p_2 + \ldots + (m-1)(m-2) p_{m-1} t^{m-3}$$
$$\vdots$$
$$r^{(m-1)}(t) = (m-1)! \, p_{m-1}$$

Expression (15.64) is the solution of the equation

$$r^{(m)}(t) = 0 \tag{15.65}$$

which may be expressed in the state space by means of the state variables

$$\omega_1(t) = r(t)$$
$$\omega_2(t) = r^{(1)}(t) = \omega_1^{(1)}(t)$$
$$\vdots$$
$$\omega_m(t) = r^{(m-1)}(t) = \omega_{m-1}^{(1)}(t)$$
$$0 = \omega_m^{(1)}(t) \tag{15.66}$$

in the form

$$\omega^{(1)}(t) = V \omega(t) \tag{15.67}$$

where

$$\omega(t) = \begin{bmatrix} \omega_1(t) \\ \vdots \\ \omega_m(t) \end{bmatrix}, \quad V = \begin{bmatrix} 0, & 1, & 0, & \ldots, & 0 \\ 0, & 0, & 1, & \ldots, & 0 \\ \multicolumn{5}{c}{\cdots\cdots\cdots\cdots} \\ 0, & 0, & 0, & \ldots, & 0 \end{bmatrix} \tag{15.68}$$

The solution of eqn. (15.67) corresponds to a whole class of the signals (15.64). The individual specific cases are determined by the vector of initial conditions

$$\omega(0) = \begin{bmatrix} p_0 \\ 1! \, p_1 \\ \vdots \\ (m-1)! \, p_{m-1} \end{bmatrix} \tag{15.69}$$

For the *sinusoidal signal* we have

$$r(t) = A \sin(\Omega t + \Phi)$$
$$r^{(1)}(t) = A\Omega \cos(\Omega t + \Phi)$$
$$r^{(2)}(t) = -A\Omega^2 \sin(\Omega t + \Phi) = -\Omega^2 r(t) \tag{15.70}$$

The differential equation has, therefore, the form

$$r^{(2)}(t) + \Omega^2 r(t) = 0 \tag{15.71}$$

or

$$\omega^{(1)}(t) = V\omega(t) \tag{15.72}$$

where

$$\omega(t) = \begin{bmatrix} r(t) \\ r^{(1)}(t) \end{bmatrix} = \begin{bmatrix} \omega_1(t) \\ \omega_2(t) \end{bmatrix}, \quad V = \begin{bmatrix} 0, & 1 \\ -\Omega^2, & 0 \end{bmatrix}, \quad \omega(0) = \begin{bmatrix} A \sin \Phi \\ \Omega A \cos \Phi \end{bmatrix}$$

Very simple is the state representation of an *exponential function*:

$$r(t) = A e^{\alpha t}$$
$$r^{(1)}(t) = \alpha A e^{\alpha t} = \alpha\, r(t) \tag{15.73}$$

The pertinent differential equation has the form

$$r^{(1)}(t) - \alpha\, r(t) = 0 \tag{15.74}$$

corresponding to the state representation as well:

$$\omega(t) = r(t), \quad V = \alpha, \quad \omega(0) = A$$
$$\omega^{(1)}(t) = V\omega(t) \tag{15.75}$$

Let us now be concerned with the question of what form the command variable may be to satisfy the strong version of the finite number of control steps in the sense of Definition 15.2.

From eqns. (15.65), (15.71) and (15.74) it is apparent that the differential equation generating the command variable has no right-hand side so that it is always possible to determine the state vector $\hat{\omega}(t)$ whose components are $r^{(i)}(t)$, $i = 0, 1, \ldots, (n-1)$. On the other hand, with $a_n = 1$ and $b_n = 0$ and making use of the equations (3.37), it is always possible to express the plant state variables in such a way that

$$\begin{bmatrix} \hat{x}_1(t) \\ \hat{x}_2(t) \\ \vdots \\ \hat{x}_n(t) \end{bmatrix} = \begin{bmatrix} y(t) \\ y^{(1)}(t) - \hat{g}_1 u(t) \\ \vdots \\ y^{(n-1)}(t) - \hat{g}_1 u^{(n-2)}(t) - \ldots - \hat{g}_{n-1} u(t) \end{bmatrix} \tag{15.76}$$

where

$$\hat{x}_i(t) = x_i(t) - a_{n-i+1} x_1(t) - a_{n-i+2} x_1^{(1)}(t) - \ldots - a_{n-1} x_1^{(i-2)}(t),$$
$$i = 2, 3, \ldots, n$$

and $\hat{g}_{n-i} = b_i$, $i = 1, 2, \ldots, (n-1)$, b_i being the parameters of the right-hand side of eqn. (3.21).

PRINCIPLES OF DETERMINISTIC SYNTHESIS

It is clear that the finite number of control steps in the sense of Definition 15.2 will be attained if, for $t \geq N$, the following equality holds:

$$\hat{\omega}(t) = \begin{bmatrix} r(t) \\ r^{(1)}(t) \\ \vdots \\ r^{(n-1)}(t) \end{bmatrix} = \begin{bmatrix} y \\ y^{(1)}(t) \\ \vdots \\ y^{(n-1)}(t) \end{bmatrix} \qquad (15.77)$$

With the exception of the trivial case, where $\hat{\omega}(t) = 0$ and where also the state vector $\hat{x}(t) = 0$, the requirement of eqn. (15.77) may be satisfied under either of the following two conditions:

(1) In eqn. (15.76) $\hat{g}_1 = \hat{g}_2 = \ldots = \hat{g}_{n-1} = 0$.

(2) Only the first m components of the command variable vector are non-zero, where $m \leq n$, the rest of them being zero.

In the latter case the necessary condition

$$\hat{g}_1 = \hat{g}_2 = \ldots = \hat{g}_{m-1} = 0 \qquad (15.78)$$

relating to eqn. (15.76) or $b_{n-1} = b_{n-2} = \ldots = b_{n-m+1} = 0$ relating to eqn. (3.21) must be satisfied so that

$$\hat{\omega}(t) = \begin{bmatrix} y(t) \\ y^{(1)}(t) \\ \vdots \\ y^{(m-1)}(t) \\ 0 \\ \vdots \\ 0 \end{bmatrix} ; \quad 0 < m \leq n \qquad (15.79)$$

It should be emphasized that the only input signal which may have a finite number of non-zero derivatives up to the order $m - 1$ is just the polynomial (15.64).

THEOREM 15.1: *If the right-hand side of the differential equation of the plant contains derivatives of up to and including the order $n - m$, $m < n$, then all input signals, which the plant is able to follow up, form a class of all polynomials (15.64) up to the degree $m - 1$. If, however, the differential equation has no input derivatives, it is of use to require the plant to follow up any command variable generated by the given linear differential equation.*

Let us designate the command signals satisfying Theorem 15.1 as *admissible command signals* for the strong version of the finite number of control steps.

If the command signal satisfies the conditions stated in Theorem 15.1 we may require that

$$\hat{\omega}(k) = Q^{-1} x(k) = \hat{x}(k), \quad k \geq N \qquad (15.80)$$

where Q^{-1} is a non-singular matrix. According to (15.80) we may also write

$$\hat{\omega}(k+1) = Q^{-1} x(k+1), \quad k \geq N$$
$$\hat{W}\hat{\omega}(k) = Q^{-1}A x(k) + Q^{-1}b u(k), \quad k \geq N$$
$$\hat{W}\hat{\omega}(k) = Q^{-1}AQ \hat{\omega}(k) + Q^{-1}b u(k), \quad k \geq N$$
$$Q^{-1}b u(k) = (\hat{W} - Q^{-1}AQ) \hat{\omega}(k), \quad k \geq N \tag{15.81}$$

Since, after completion of the finite number of control steps, $u(k) = 0$, the equality

$$\hat{W} = Q^{-1}AQ = \hat{A} \tag{15.82}$$

must be satisfied in eqn. (15.81) because in general the linear input signal cannot be permanently equal to zero after a finite time has elapsed. It can also be proved that the equality $\hat{r}(k) = \hat{x}(k)$ follows from the equality $W = Q^{-1}AQ$. It therefore holds:

THEOREM 15.2: *In order that the requirement of the strong version of the finite number of control steps in a control loop may be satisfied for admissible command signals, the matrix of dynamics of the controlled plant, A, and the matrix of dynamics of the command signal, W, must be similar. This condition is necessary and sufficient.*

From this theorem it follows that the matrices W and A must have equal eigenvalues. In addition, eqn. (15.82) applies analogously to the matrices of the continuous version, i.e.

$$\hat{V} = Q^{-1}FQ = \hat{F} \tag{15.83}$$

because in both the continuous and the discrete version there appear identically defined state vectors.

Let us apply the information gained so far to the example of a control loop with the plant

$$\hat{x}^{(1)}(t) = \begin{bmatrix} 0, & 1, & 0, & \cdots & 0 \\ 0, & 0, & 1, & \cdots & 0 \\ \multicolumn{5}{c}{\dotfill} \\ -a_0, & -a_1, & -a_2, & \cdots & -a_{n-1} \end{bmatrix} \hat{x}(t) + \begin{bmatrix} \hat{g}_1 \\ \hat{g}_2 \\ \vdots \\ \hat{g}_n \end{bmatrix} u(t) \tag{15.84}$$

and with the command input signal

$$r(t) = p_0 + p_1 t + \ldots + p_{m-1} t^{m-1}, \quad 0 < m \leq n \tag{15.85}$$

Let us further assume that the necessary condition (15.78) is satisfied so that in eqn. (15.84) $\hat{g}_1 = \hat{g}_2 = \ldots = \hat{g}_{m-1} = 0$.

It should also be noted that the order of the differential equation generating the command variable can only be $m \leq n$, where n is the plant order, so that the plant

PRINCIPLES OF DETERMINISTIC SYNTHESIS

output variable may follow up the command variable. If $m < n$, it is necessary that the m-dimensional space of the command variable be bedded in the n-dimensional plant space. This is accomplished by supplementing the m independent components of the vector $\omega(t)$ by additional dependent components to get the extended vector $\hat{\omega}(t)$ with a total number of n components and check for similarity of the extended matrix $\hat{V}$ or $\hat{W}$ to the plant matrix of dynamics. The components of the extended vector $\hat{\omega}(t)$ are

$$\begin{aligned}
\hat{\omega}_1(t) &= r(t) = \omega_1(t) \\
&\vdots \\
\hat{\omega}_m(t) &= r^{(m-1)}(t) = \omega_m(t) \\
\hat{\omega}_{m+1}(t) &= r^{(m)}(t) = 0 \\
&\vdots \\
\hat{\omega}_n(t) &= r^{(n-1)}(t) = 0
\end{aligned} \qquad (15.86)$$

so that eqn. (15.85) may be expressed by the extended set of equations

$$\begin{aligned}
\hat{\omega}_1^{(1)}(t) &= \hat{\omega}_2(t) \\
&\vdots \\
\hat{\omega}_{n-1}^{(1)}(t) &= \hat{\omega}_n(t) \\
\hat{\omega}_n^{(1)}(t) &= -A_0\,\hat{\omega}_1(t) - A_1\,\hat{\omega}_2(t) - \ldots - A_{n-1}\,\hat{\omega}_n(t)
\end{aligned} \qquad (15.87)$$

or, with the state vector $\hat{\omega}(t)$, by the equation

$$\begin{aligned}
\hat{\omega}^{(1)}(t) &= \hat{V}\,\hat{\omega}(t) \\
r(t) &= \hat{c}_V^T\,\hat{\omega}(t)
\end{aligned} \qquad (15.88)$$

The corresponding discrete form is

$$\begin{aligned}
\hat{\omega}(k+1) &= \hat{W}\,\hat{\omega}(k) \\
r(k) &= \hat{c}_W^T\,\hat{\omega}(k)
\end{aligned} \qquad (15.89)$$

where

$$\hat{W} = e^{\hat{V}T}, \quad \hat{c}_W^T = [1, 0, \ldots, 0]$$

$$\hat{V} = \begin{bmatrix} 0, & 1, & 0, & \ldots, & 0 \\ 0, & 0, & 1, & \ldots, & 0 \\ \multicolumn{5}{c}{\dotfill} \\ -A_0, & -A_1, & -A_2, & \ldots, & -A_{n-1} \end{bmatrix}; \quad \hat{\omega}(t) = \begin{bmatrix} \hat{\omega}_1(t) \\ \hat{\omega}_2(t) \\ \vdots \\ \hat{\omega}_n(t) \end{bmatrix}; \quad \hat{\omega}(0) = \begin{bmatrix} r(0) \\ \vdots \\ r^{(m-1)}(0) \\ 0 \\ \vdots \\ 0 \end{bmatrix}$$

The coefficients $A_0, \ldots, A_{m-1}$ are determined whereas the further coefficients $A_m, \ldots, A_{n-1}$ follow from the equation

$$\hat{\omega}_n^{(1)}(t) = r^{(n)}(t) = -A_0\,r(t) - A_1\,r^{(1)}(t) - \ldots - A_{n-1}\,r^{(n-1)}(t) \qquad (15.90)$$

Substituting the components of the initial vector $\hat{\omega}(0)$ into eqn. (15.90) we have

$$0 = -A_0 r(0) - \ldots - A_{m-1} r^{(m-1)}(0) - A_m 0 - \ldots - A_{n-1} 0 \qquad (15.91)$$

Equation (15.91) may be satisfied for $A_0 = A_1 = \ldots = A_{m-1} = 0$ and with arbitrary coefficients $-A_m, \ldots, -A_{n-1}$. The matrix $\hat{V}$ has, therefore, the form

$$\hat{V} = \begin{bmatrix} 0, & 1, & 0, & \ldots & 0 \\ 0, & 0, & 1, & \ldots & 0 \\ \multicolumn{5}{c}{\dotfill} \\ 0, & \ldots & 0, & -A_m, & \ldots & -A_{n-1} \end{bmatrix}$$

and condition (15.83) will be satisfied if in the matrix $\hat{F}$ in eqn. (15.84) $a_0 = a_1 = \ldots = a_{m-1} = 0$ and $a_i = A_i$, $i = m, \ldots, (n-1)$. The latter equalities may easily be satisfied since the coefficients $A_m, \ldots, A_{n-1}$ may, according to (15.91), have arbitrary values.

The requirement $a_0 = 0$ means that one eigenvalue of the plant matrix of dynamics, e.g. λ_1, must be zero. Thus, all products of eigenvalues containing λ_1 and of which the coefficients a_i, $i = 0, 1, \ldots, (n-1)$ consist, are nullified;

$$\begin{aligned}
a_0 &= \lambda_1 \lambda_2 \ldots \lambda_n \\
a_1 &= \lambda_2 \lambda_3 \ldots \lambda_n + \lambda_1 \lambda_3 \ldots \lambda_n + \ldots + \lambda_1 \lambda_2 \ldots \lambda_{n-1} \\
a_2 &= \lambda_3 \lambda_4 \ldots \lambda_n + \lambda_2 \lambda_4 \ldots \lambda_n + \lambda_2 \lambda_3 \lambda_5 \ldots \lambda_n + \ldots + \lambda_1 \lambda_2 \lambda_3 \ldots \lambda_{n-3} \\
&\vdots \\
a_{n-1} &= \lambda_1 + \lambda_2 + \ldots + \lambda_n
\end{aligned}$$

Similarly, in order that $a_0 = a_1 = 0$, there must be $\lambda_1 = \lambda_2 = 0$, etc. If all the coefficients a_i, $i = 0, 1, \ldots, (n-1)$, were to be zero, then all eigenvalues of the plant matrix of dynamics would have to be zero.

These conclusions may as well be formulated as follows:

THEOREM 15.3: *In order that a plant may, after completion of a finite number of control steps, exactly follow up a command signal having the form of a polynomial of the order $m - 1$, the characteristic equation of the plant should have m zero roots, the other roots being arbitrary, and a right-hand side of the differential equation of the order $n - m$ at most, where n is the plant order.*

Theorem 15.3 says in other words that a plant or the continuously working part of a control loop must have at least m integrators.

In conclusion it should be noted that in contrast with eqns. (15.54) and (15.55), the controller equation is now

$$u(k) = k^T \varepsilon(k) \qquad (15.92)$$

$$\varepsilon(k) = [\hat{\omega}(k) - \hat{x}(k)] \qquad (15.93)$$

where $\varepsilon(k)$ is the error vector. Equation (15.93) may be rewritten in the form

$$\varepsilon(k+1) = \hat{\omega}(k+1) - Q^{-1} x(k+1) =$$
$$= \hat{W}\hat{\omega}(k) - Q^{-1}AQ\,\hat{x}(k) - Q^{-1}b\,u(k) \qquad (15.94)$$

Using eqn. (15.82) for admissible command signals we have

$$\varepsilon(k+1) = \hat{A}\,\varepsilon(k) - \hat{b}\,u(k) \qquad (15.95)$$

where $\hat{b} = Q^{-1}b$. This result may be expressed by the following theorem.

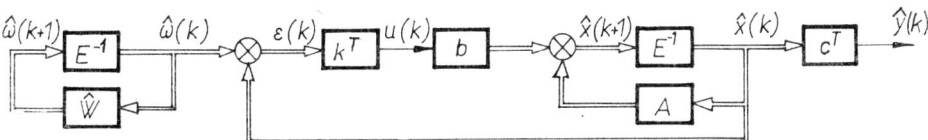

Fig. 15.4. Block diagram of a closed control loop with the controller matrix k^T.

THEOREM 15.4: *If the command signal belongs under the class of admissible inputs, then the transfer of the error vector $\varepsilon(k)$ to the zero state is given by the same difference equation, invariant with respect to the change in base, as for the transfer of the plant state $x(k)$ to zero, except for the change in sign of the input $u(k)$.*

The block diagram of the control loop is shown in Fig. 15.4.

Certain additional details may be found in the literature [67.3]. In the case of synthesis according to the strong version of the finite number of control steps the generalization for multi-input/multi-output systems may be carried out using the same procedure as in Par. 15.2.1. These problems of multi-input/multi-output system synthesis are solved in detail in [70.4].

Example 15.3: A controlled plant is given by the differential equation

$$y^{(3)}(t) + 0.5\,y^{(2)}(t) = u(t)$$

Examine whether it is possible to satisfy the condition of the strong version of the finite number of control steps for the command signal $r(t) = 0.5t$ and if so, calculate the response. Assume the sampling period for discrete state equations to be $T = 1$.

Solution: According to (3.34) we write down the state equations

$$x^{(1)}(t) = F\,x(t) + g\,u(t)$$
$$y(t) = h^T\,x(t)$$

where

$$F = \begin{bmatrix} 0, & 1, & 0 \\ 0, & 0, & 1 \\ 0, & 0, & -0.5 \end{bmatrix}, \quad g = \begin{bmatrix} 0 \\ 0 \\ 1 \end{bmatrix}, \quad h^T = [1, 0, 0]$$

According to eqn. (15.65) the command signal is generated by the differential equation

$$r^{(2)}(t) = 0$$

Using eqns. (15.66) and (15.67) we may write its state equation

$$\omega^{(1)}(t) = V\omega(t)$$

where

$$V = \begin{bmatrix} 0, & 1 \\ 0, & 0 \end{bmatrix}$$

The initial vector

$$\omega(0) = \begin{bmatrix} r(0) \\ r^{(1)}(0) \end{bmatrix} = \begin{bmatrix} 0 \\ 0.5 \end{bmatrix}$$

and for $k = 1, 2, \ldots$

$$\omega(k) = \begin{bmatrix} 0.5k \\ 0.5 \end{bmatrix}$$

The extended matrix of dynamics of the command variable has, according to (15.89), the form

$$\hat{V} = \begin{bmatrix} 0, & 1, & 0 \\ 0, & 0, & 1 \\ -A_0, & -A_1, & -A_2 \end{bmatrix}$$

so that for $A_0 = A_1 = 0$ and for $A_2 = 0.5$ we have $F = \hat{V}$. The matrices F and $\hat{V}$ are similar and have, therefore, the same characteristic polynomial $s^2(s + 0.5)$. The strong version of the finite number of control steps can, therefore, be satisfied for an arbitrary combination of command signals whose s-transform has any of the following characteristic polynomials: $s, s^2, (s + 0.5), s(s + 0.5), s^2(s + 0.5)$. In accordance with the requirement it is, therefore, possible to satisfy the strong version of the finite number of control steps for the given command signal and the number of control steps will be equal to three. The synthesis for this case will now be carried out.

First we calculate the discrete version of the plant state equations. According to (5.36) we have

$$A(T) = e^{FT} = L^{-1}\{(sE - F)^{-1}\} =$$

$$= \begin{bmatrix} 1, & T, & -2(2 - T - 2e^{-0.5T}) \\ 0, & 1, & 2(1 - e^{-0.5T}) \\ 0, & 0, & e^{-0.5T} \end{bmatrix}$$

For $T = 1$ we find

$$A = \begin{bmatrix} 1, & 1, & 0.4261 \\ 0, & 1, & 0.7869 \\ 0, & 0, & 0.6065 \end{bmatrix}$$

PRINCIPLES OF DETERMINISTIC SYNTHESIS

$$b = F^{-1}[A(T) - E]g = \left[E + \frac{F}{2!} + \frac{F^2}{3!} + \ldots\right]g = \begin{bmatrix} 0.1478 \\ 0.4262 \\ 0.7870 \end{bmatrix}$$

$$c^T = h^T$$

The controller equation is, according to eqns. (15.92) and (15.54),

$$u(k) = k^T \varepsilon(k) = [0, \ldots, 0, 1] Q_D^{-1} A^3 \varepsilon(k) =$$
$$= [1.2706, 2.4885, 1.7257] \varepsilon(k)$$

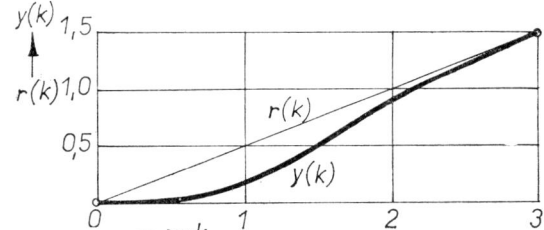

Fig. 15.5. Control process relating to Example 15.3.

The control process can be compiled by virtue of the following values:

$$\hat{\omega}(0) = \qquad \varepsilon(0) = \begin{bmatrix} 0 \\ 0.5 \\ 0 \end{bmatrix}, \qquad u(0) = 1.2443$$

$$x(1) = \begin{bmatrix} 0.1839 \\ 0.5303 \\ 0.9792 \end{bmatrix}, \qquad y(1) = 0.1839$$

$$\hat{\omega}(1) = \begin{bmatrix} 0.5 \\ 0.5 \\ 0 \end{bmatrix}, \qquad \varepsilon(1) = \begin{bmatrix} 0.3161 \\ -0.0303 \\ -0.9792 \end{bmatrix}, \qquad u(1) = -1.3636$$

$$x(2) = \begin{bmatrix} 0.9299 \\ 0.7197 \\ -0.4793 \end{bmatrix}, \qquad y(2) = 0.9299$$

$$\hat{\omega}(2) = \begin{bmatrix} 1 \\ 0.5 \\ 0 \end{bmatrix}, \qquad \varepsilon(2) = \begin{bmatrix} 0.0701 \\ -0.2197 \\ 0.4793 \end{bmatrix}, \qquad u(2) = 0.3703$$

$$x(3) = \begin{bmatrix} 1.5 \\ 0.5 \\ 0 \end{bmatrix}, \qquad y(3) = 1.5$$

$$\hat{\omega}(3) = x(3), \qquad \varepsilon(3) = 0, \qquad u(3) = 0 \quad \text{etc.}$$

The transients of the variables $r(k)$ and $y(k)$, $k = 0, 1, 2, 3$, are shown in Fig. 15.5.

If we were to determine the controller matrix by determining the characteristic polynomial of the plant matrix of dynamics expressed in the canonical form of controllability we should first find, by means of the matrix A, the characteristic polynomial $A(z)$ and then

$$A_R = \begin{bmatrix} 0, & 1, & 0 \\ 0, & 0, & 1 \\ 0.6065, & -2.2130, & 2.6065 \end{bmatrix}, \quad b_R = \begin{bmatrix} 0 \\ 0 \\ 1 \end{bmatrix}$$

The transformation matrix is determined according to eqn. (10.73)

$$Q = \begin{bmatrix} 0.1151, & 0.5241, & 0.1478 \\ -0.3608, & -0.0654, & 0.4262 \\ 0.7870, & -1.5740, & 0.7870 \end{bmatrix}, \quad Q^{-1} = \begin{bmatrix} 1.2708, & -1.3236, & 0.4781 \\ 1.2708, & -0.0528, & -0.2101 \\ 1.2708, & 1.2180, & 0.3725 \end{bmatrix}$$

and

$$c_R^T = c^T Q = [0.1151, \ 0.5241, \ 0.1478]$$

The controller matrix is directly the last row of the matrix A_R:

$$k_R^T = [0.6065, \ -2.2130, \ 2.6065] = k^T Q$$

Converting now the initial conditions according to the relation

$$\varepsilon_R(0) = Q^{-1} \varepsilon(0) = [-0.6618, \ -0.0264, \ 0.6090]^T$$

we arrive by the usual procedure at the same result as with the original procedure. The values of the state vector components will, however, be different because we changed the base.

Finally it should be noted that for the command variable $r(t) = e^{-0.5t}$, which is one of the further possible functions of the command variable, we have, according to eqn. (15.90),

$$-0.125 e^{-0.5t} = -A_0 e^{-0.5t} + 0.5 A_1 e^{-0.5t} - 0.25 A_2 e^{-0.5t}$$

$$-0.125 = -A_0 + 0.5 A_1 - 0.25 A_2 \ .$$

In the last equation any two constants may be arbitrarily selected and the third constant calculated from this equation. The further procedure is analogous to that for the originally given command signal.

15.2.3 Weak version of the finite number of control steps

A different case of the finite number of control steps is the so-called weak version of this cost function. Let us first give the corresponding definition:

PRINCIPLES OF DETERMINISTIC SYNTHESIS

DEFINITION 15.3: *The weak version of the finite number of control steps is such an action exerted by the input variable of the plant, starting at the instant $k = 0$, at which the initial plant state is $x(0)$, that after the completion of N sampling periods the error*

$$e(k, \varepsilon) = r(k, \varepsilon) - y(k, \varepsilon) = 0 \quad \text{for} \quad k \geq N, \varepsilon = 0$$

while for $0 < \varepsilon < 1$ the value of $e(k, \varepsilon)$ converges to zero for $k \to \infty$, where r is the command variable and y the output variable of the plant.

In contrast with the strong version of the finite number of control steps (Definition 15.1) we only require the value of the plant output variable at the sampling instants $k \geq N$ to be equal to the value of the command variable but do not require an equality of the corresponding derivatives of both these variables. Therefore, in general, $e(k, \varepsilon) \neq 0$, $0 < \varepsilon < 1$, i.e. between the sampling instants. The output variable is, however, equal to the command variable not only in value but also in all its derivatives for $k \to \infty$.

Although the weak version of the finite number of control steps represents a more moderate requirement than the strong version, the solution of this problem is more difficult and cannot dispense with the determination of eigenvalues of the matrix of dynamics of the closed control loop. We shall show that this matrix is identical with the inverse matrix of the plant. The solution of this problem in the state space was first worked out consistently by V. Kučera [67.3, 70.4, 72.6].

Before proceeding to the solution itself we shall define certain new terms which have not appeared in the preceding chapters.

DEFINITION 15.4: *Let a plant be described by the equations*

$$x(k + 1) = A\,x(k) + b\,u(k)$$
$$y(k) \quad\;\; = c^T x(k) + d\,u(k)$$

where $x \in X^n$, $u \in U^1$ and $y = Y^1$. Denoting

$$h_0 = d$$
$$h_i = c^T A^{i-1} b, \quad i = 1, 2, \ldots \tag{15.96}$$

then m defined by the relation

$$m = \min_{i=0,1,\ldots} \{i : h_i \neq 0\}$$

is the so-called relative order of the system [65.1].

From a comparison of the relations (15.96) with the relations (7.5) it is apparent that h_0 and h_i, $i = 1, 2, \ldots$, are ordinates of the discrete impulse response. Provided that $h_0 = 0$ and $h_i = 0$, $i = 1, 2, \ldots, m - 1$, and $h_m \neq 0$, then m is the number of periods of the plant output delay with respect to the input. It should also be noted that in the discrete transfer function of the system the relative order is given by the difference between the polynomial degree of the denominator and of the numerator.

DEFINITION 15.5: *An inverse system Σ_2 corresponding to a given system Σ_1 is a system resulting from interchanging the input and output of the system Σ_1.*

Let us apply Definition 15.5 to a plant whose relative order $m = 0$. From the equation of plant dynamics we calculate recursively the expression (9.3):

$$x(k) = A^k x(0) + Q_D U(k-1, 0) \qquad (15.97)$$

where Q_D is the matrix of reachability and $U(k-1, 0) = [u(k-1), u(k-2), \ldots, u(0)]^T$. Using eqn. (15.97) we may express the plant output variable as

$$y(k) = c^T A^k x(0) + c^T Q_D U(k-1, 0) + du(k) \qquad (15.98)$$

and hence

$$u(k) = d^{-1} y(k) - d^{-1} c^T A^k x(0) - d^{-1} c^T Q_D U(k-1, 0) =$$
$$= d^{-1} y(k) - d^{-1} c^T x(k) \qquad (15.99)$$

Substituting now eqn. (15.99) into the equation of dynamics of the plant we obtain

$$x(k+1) = (A - bd^{-1}c^T) x(k) + bd^{-1} y(k)$$
$$u(k) \quad = -d^{-1}c^T x(k) + d^{-1} y(k) \qquad (15.100)$$

These equations are the state equations of the inverse plant.

In complete analogy we may derive, for the relative order $m = 1$, the relations

$$x(k+1) = [A - b(c^T b)^{-1} c^T A] x(k) + b(c^T b)^{-1} y(k+1) =$$
$$= \hat{A}_1 x(k) + b h_1^{-1} y(k+1)$$
$$u(k) \quad = -(c^T b)^{-1} c^T A x(k) + (c^T b)^{-1} y(k+1) =$$
$$= -h_1^{-1} c^T A x(k) + h_1^{-1} y(k+1) \qquad (15.101)$$

For the relative order $m > 1$ we obtain:

$$x(k+1) = [A - b(c^T A^{m-1} b)^{-1} c^T A^m] x(k) + b(c^T A^{m-1} b)^{-1} y(k+m) =$$
$$= \hat{A}_m x(k) + b h_m^{-1} y(k+m) \qquad (15.102)$$
$$u(k) \quad = -(c^T A^{m-1} b)^{-1} c^T A^m x(k) + (c^T A^{m-1} b)^{-1} y(k+m) =$$
$$= -h_m^{-1} c^T A^m x(k) + h_m^{-1} y(k+m)$$

Starting with the instant $k + m$, at which the plant output variable is zero, we have

$$y(k + m) = c^T A^m x(k) + h_m u(k) = 0$$

Consequently, the controller equation is

$$u(k) = -h_m^{-1} c^T A^m x(k) \tag{15.103}$$

Substituting eqn. (15.103) into the equatipn of dynamics of the plant we obtain

$$x(k + 1) = (A - b h_m^{-1} c^T A^m) x(k) = \hat{A}_m x(k) \tag{15.104}$$

where $\hat{A}$ is the matrix of dynamics of the closed control loop. From a comparison with eqn. (15.102) it is apparent that the matrix of dynamics of the closed loop is identical with the matrix of dynamics of the inverse plant.

The *admissible input signals* are determined according to the requirement that

$$0 = r(k) - y(k), \qquad k \geq N$$

$$0 = c_W^T \hat{\omega}(k) - c^T x(k), \quad k \geq N \tag{15.105}$$

Equation (15.105) is also valid for $k + 1$ and, therefore,

$$c_W^T \hat{W} \hat{\omega}(k) = c^T A x(k) + c^T b u(k) \tag{15.106}$$

From this equation we may find

$$u(k) = (c^T b)^{-1} \left[c_W^T \hat{W} \hat{\omega}(k) - c^T A x(k) \right], \quad k \geq N \tag{15.107}$$

The requirement of zero error $e(k; \varepsilon)$ in the sense of Definition 15.3, i.e. asymptotic stability of the control process, may now, after the introduction of the vectors $\hat{\omega}(k)$ and $x(k)$, be formulated in such a way that

$$\lim_{k \to \infty} \hat{\omega}(k) = \lim_{k \to \infty} \hat{x}(k) = Q^{-1} \lim_{k \to \infty} x(k) \tag{15.108}$$

For the controlling variable it then holds that

$$\lim_{k \to \infty} u(k) = 0 \tag{15.109}$$

From eqn. (15.108) it follows that if asymptotic stability is to be reached, the vectors $\hat{\omega}(k)$ and $x(k)$ must be expressed, as in the case solved in Par. 15.2.2, in the same base and the conditions (15.78) must be satisfied.

Using equalities (15.108), the limit of the expression (15.107) for $k \to \infty$ may be rewritten in the form

$$0 = (c^T b)^{-1} \left[c_W^T \hat{W} - c^T A Q \right] \lim_{k \to \infty} \hat{\omega}(k) \tag{15.110}$$

If the vectors $\hat{\omega}(k)$ and $x(k)$ are expressed in the same base, then $c_W^T = c^T Q = \hat{c}^T$ and eqn. (15.110) assumes the form

$$0 = (c^T b)^{-1} c^T Q [\hat{W} - Q^{-1} A Q] \lim_{k \to \infty} \hat{\omega}(k) \tag{15.111}$$

The necessary and sufficient condition for satisfying eqn. (15.111) is that either

$$\hat{W} = Q^{-1} A Q \tag{15.112}$$

which is the same requirement as (15.82), or

$$\|W^k\| \to 0 \quad \text{for} \quad k \to \infty \tag{15.113}$$

THEOREM 15.5: *In order that the command signal may belong under the class of admissible inputs, with which it is possible to attain the weak version of the finite number of control steps, there should be either $\hat{W} = Q^{-1} A Q$ or $\|W^k\| \to 0$ for $k \to \infty$.*

The meaning of the individual symbols in Theorem 15.5 follows from the preceding text.

In the case where the command signal satisfies condition (15.112) and the matrix of dynamics of the inverse plant has altogether stable eigenvalues, the controller is obtained directly according to eqn. (15.103). If, however, $\hat{A}_m$ has at least one unstable eigenvalue we proceed in a way similar to that used in the case of the strong version of the finite number of control steps.

From the inverse plant matrix of dynamics $\hat{A}_m$ we leave in the resultant solution the components corresponding to the non-zero stable eigenvalues of this matrix. Let us denote these eigenvalues by z_i^+ and their corresponding eigenvectors by a_i^+, $i = 1, 2, \ldots, s$. We replace the other eigenvalues, provided they are non-zero, by zero eigenvalues. Thus we compensate for the unstable eigenvalues of the inverse plant in the same manner as we compensate for unstable root factors in the numerator of the plant transfer function.

It should be noted that the matrix $\hat{A}_m$ has always m zero eigenvalues corresponding to the chain of generalized eigenvectors $b, Ab, \ldots, A^{m-1} b$. The controller matrix for a single-input/single-output plant is

$$k^T = [0, \ldots, 0, 1] [A^m a_s^+, \ldots, A^m a_1^+, b, Ab, \ldots, A^{n-s-1} b]^{-1} A^{n-s} \tag{15.114}$$

where n is the plant order. Equation (15.114) may as well be given the form [72.6]

$$k^T [A^{-1} b, \ldots, A^{-m} b, \ldots, A^{-m-\mu} b, A^{-\mu} a_1^+, \ldots, A^{-\mu} a_s^+] = [1, 0, \ldots, 0] \tag{15.115}$$

where $\mu = n - s - m$ is the number of unstable eigenvalues of the matrix $\hat{A}_m$. In eqn. (15.115), $A^{-1} b, \ldots, A^{-m} b$ is the chain of generalized eigenvectors corresponding to zero eigenvalues of the inverse plant matrix $\hat{A}_m$, and the other generalized eigenvectors $A^{-m-1}b, \ldots, A^{-m-\mu}b$ of this chain correspond to the zero eigenvalues by which we have replaced the unstable eigenvalues of $\hat{A}_m$.

The eigenvectors corresponding to the stable eigenvalues of the matrix $\hat{A}_m$ are multiplied by $A^{-\mu}$. The state vector $x(k) \in X^n$ created as a linear combination of the columns of the matrix $[A^{-1}b, \ldots, A^{-\mu}a_s^+]$ in eqn. (15.115), changes, by the action of the feedback satisfying eqn. (15.115), into

$$x(k+1) = (A - bk^T) x(k)$$

where $x(k+1) \in \mathscr{X}^{n-1}$, where $\mathscr{X}^k$ is the k-dimensional subspace of the space X^n.

After μ steps, μ generalized eigenvectors of the chain corresponding to zero eigenvalues of the matrix $\hat{A}_m$ are nullified and the resultant state of the system will be $x(k+\mu) \in \mathscr{X}^{n-\mu}$. After the next m steps the state will be $x(k+\mu+m) \in \mathscr{X}^s$, $s = n - m - \mu$. The system remains in the resultant s-dimensional state space and all components influencing the system dynamics are, by definition, stable. The minimum number of control steps is apparently $n_k = m + \mu$.

The same solution will also be attained in the case where in the canonical form of controllability of the plant we compensate, according to Par. 15.1.1, for those coefficients of the characteristic polynomial not contained in the desired characteristic polynomial of the closed control loop, i.e. in the polynomial

$$P(z) = z^{n-s}(z - z_1^+) \ldots (z - z_s^+) = z^n + \beta_1 z^{n-1} + \ldots + \beta_s z^{n-s} \quad (15.116)$$

If, according to (15.14), the coefficients of the characteristic polynomial of the plant are a_i, $i = 0, 1, \ldots, n-1$, then the controller matrix

$$k^T = -[a_0, a_1, \ldots, a_{n-s-1}, a_{n-s} - \beta_s, \ldots, a_{n-1} - \beta_1] \quad (15.117)$$

However, the calculation of the controller will be different if the command variable satisfies only the condition (15.113) while the command signal dynamics differs from the plant dynamics. In this case the plant state representation must be extended by the state representation of the command signal.

Let the plant equation of dynamics and the command variables be of the form

$$x(k+1 = A x(k) + b u(k) \quad (15.118)$$
$$\omega(k+1) = W \omega(k) \quad (15.119)$$

Then the common equation of dynamics is

$$\psi(k+1) = \Phi \psi(k) + \beta u(k) \quad (15.120)$$

where

$$\psi(k) = \begin{bmatrix} x(k) \\ \omega(k) \end{bmatrix}$$

is an $(n + m)$-dimensional state vector and

$$\Phi = \begin{bmatrix} A, & 0 \\ 0, & W \end{bmatrix}, \quad \beta = \begin{bmatrix} b \\ 0 \end{bmatrix}$$

Similarly, we replace the output equations

$$y(k) = c^T x(k) \tag{15.121}$$
$$r(k) = c_W \omega(k) \tag{15.122}$$

by a single equation

$$v(k) = \gamma \psi(k) \tag{15.123}$$

where

$$v(k) = \begin{bmatrix} y(k) \\ r(k) \end{bmatrix} \quad \text{and} \quad \gamma = \begin{bmatrix} c^T, & 0 \\ 0, & c_W^T \end{bmatrix}$$

The synthesis of the composite plant obtained in this way may then proceed as in the previously described weak version of the finite number of control steps.

Example 15.4: Modify the problem solved in Example 15.1 for the weak version of the finite number of control steps.

Solution: Using the representation in the canonical form of controllability, the inverse matrix of the plant is

$$\hat{A}_1 = \begin{bmatrix} 0, & 1, & 0 \\ 0, & 0, & 1 \\ 0, & -0.6065, & -3.1348 \end{bmatrix}$$

and its eigenvalues are

$$z_1 = 0$$
$$z_2^+ = -0.2071$$
$$z_3^- = -2.9276$$

Consequently, in this example $n = 3$, $s = 1$, $m = 1$ and $\mu = 1$. According to eqn. (15.114) the controller matrix is

$$k_R^T = [0, \ldots, 0, 1] [A_R a_1^+, b_R, A_R b_R]^{-1} A_R^2$$

where a_1^+ is the eigenvector corresponding to the eigenvalue z_2^+:

$$a_1^T = \begin{bmatrix} 4.8286 \\ -1 \\ 0.2071 \end{bmatrix}$$

We find

$$k_R^T = [0, \ldots, 0, 1] \begin{bmatrix} -1, & 0, & 0 \\ 0.2071, & 0, & 1 \\ 3.8157, & 1, & 2.2130 \end{bmatrix}^{-1} \begin{bmatrix} 0, & 0, & 1 \\ 0.3679, & -1.5809, & 2.2130 \\ 0.8142, & -3.4985, & 4.8974 \end{bmatrix}$$

$$k_R^T = [0.3679, -1.5809, 2.4201]$$

PRINCIPLES OF DETERMINISTIC SYNTHESIS

The individual variables of the control process have the values:

$$x_R(0) = \begin{bmatrix} -1.6151 \\ -1.6151 \\ -1.6151 \end{bmatrix}, \quad u(0) = 1.9496, \quad y(0) = -1$$

$$x_R(1) = \begin{bmatrix} -1.6151 \\ -1.6151 \\ -0.3345 \end{bmatrix}, \quad u(1) = -2.7686, \quad y(1) = -0.7455$$

$$x_R(2) = \begin{bmatrix} -1.6151 \\ 0.3345 \\ -0.0692 \end{bmatrix}, \quad u(2) = 1.2905, \quad y(2) = 0$$

$$x_R(3) = \begin{bmatrix} 0.3345 \\ -0.0692 \\ 0.0144 \end{bmatrix}, \quad u(3) = -0.2673, \quad y(3) = 0$$

$$x_R(4) = \begin{bmatrix} -0.0692 \\ 0.0144 \\ 0.0030 \end{bmatrix}, \quad u(4) = 0.0555, \quad y(4) = 0 \quad \text{etc.}$$

The problem may as well be solved by pole assignment to the characteristic polynomial. By virtue of (15.116) we have, for our example,

$$P(z) = z^2(z + 0.2071) = z^3 + 0.2071z^2$$

so that $\beta_1 = 0.2071$. Equation (15.117) enables us to determine directly the controller matrix

$$k_R^T = [0.3679, -1.5809, 2.4201]$$

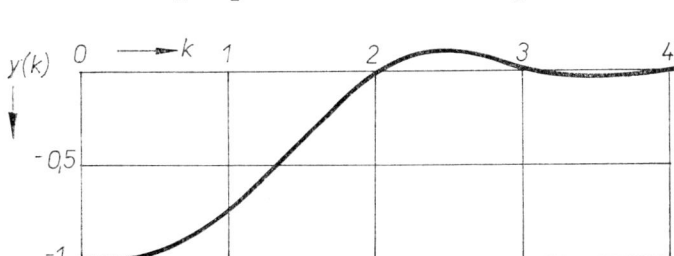

Fig. 15.6. Control process relating to Example 15.4.

As can be seen, the calculation of the controller matrix by pole assignment to the characteristic polynomial of the closed loop requires a smaller number of numerical operations.

The control process is shown in Fig. 15.6.

15.3 Determination of the characteristic polynomial of an estimator

In Par. 14.1.1 and 14.1.2 we derived the basic relations for the design of a deterministic estimator enabling us to estimate the system state vector by means of the measured input and output. Now we shall show how to determine the feedback matrix H and the characteristic polynomial of an estimator for a given state representation of the controlled plant so that the estimator may answer its purpose as well as possible. In this section we shall successively deal with the n-th order estimator, the reduced-order estimator and the estimator with zero poles of the characteristic equation.

Let us first consider the *n-th order estimator*.

Compare eqn. (14.7)

$$Q(z) = \det(zE - A + H_E C) = \sum_{i=0}^{n} \beta_i z^i, \quad \beta_n = 1 \qquad (15.124)$$

with eqns. (15.12) and (15.13)

$$\det(zE - A + BK) = \sum_{i=0}^{n} \alpha_i z^i, \quad \alpha_n = 1 \qquad (15.125)$$

Since the determinant of a matrix does not change with matrix transposition, eqn. (14.7) may be transposed, so that

$$\det(zE - A^T + C^T H_E^T) = \sum_{i=0}^{n} \beta_i z^i, \quad \beta_n = 1 \qquad (15.126)$$

For $A = A^T$, $B = C^T$ and $K = H_E^T$, eqn. (15.125) is identical with eqn. (15.126). In this case also $\alpha_i = \beta_i$, $i = 0, 1, 2, ..., n$.

For a system with r inputs and a single output, i.e. if $p = 1$ and $H_E = h_E^T$ we have, according to eqn. (15.32),

$$h_E^T = \varphi^T Q(A^T) = [0, ..., 0, 1][c, A^T c, ..., (A^T)^{n-1} c]^{-1} Q(A^T)$$

$$h_E = Q(A) \varphi =$$

$$= Q(A) \begin{bmatrix} c^T \\ c^T A \\ \vdots \\ c^T A^{n-1} \end{bmatrix}^{-1} \begin{bmatrix} 0 \\ \vdots \\ 0 \\ 1 \end{bmatrix} \qquad (15.127)$$

Equation (15.127) indicates how to calculate the column matrix h_E of the parameters of a deterministic estimator. We substitute the matrix of dynamics, A, into the characteristic polynomial Q and multiply it by the last column φ of the inverted matrix of observability.

In analogy with Par. 15.1.1 where it was possible to use the canonical form of controllability for the determination of the controller matrix we may in this case use the canonical form of reconstructability for the determination of the estimator matrix. According to eqn. (14.2) we may write

$$\hat{x}_K(k+1) = \begin{bmatrix} 0, & \ldots & 0, & -a_0 & -h_{EK1} \\ 1, & \ldots & 0, & -a_1 & -h_{EK2} \\ \vdots & & & & \\ 0, & \ldots & 1, & -a_{n-1} & -h_{EKn} \end{bmatrix} \hat{x}_K(k) + b_K u(k) + \begin{bmatrix} h_{EK1} \\ \vdots \\ h_{EKn} \end{bmatrix} y(k) \quad (15.128)$$

Comparing the coefficients of the appropriate characteristic equation

$$(a_0 + h_{EK1}) + (a_1 + h_{EK2}) z + \ldots (a_{n-1} + h_{EKn}) z^{n-1} + z^n = 0 \quad (15.129)$$

with the coefficients of eqn. (15.124) we can easily determine the estimator parameters

$$h_{EKi} = \beta_{i-1} - a_{i-1}, \quad i = 1, 2, \ldots, n \quad (15.130)$$

If the number of output variables $p > 1$ we may obtain an unlimited number of estimator solutions by choosing the matrix H.

As in the case of eqns. (15.40) and (15.41) we may, for cyclic systems, form such a linear combination of the output variables

$$v(k) = \gamma^T y(k) \quad (15.131)$$

that the complete system is observable by the scalar variable $v(k)$. Further procedure is identical with the previously described case of a single output variable.

In eqn. (15.127), however, we substitute

$$c^T = \gamma^T C \quad (15.132)$$

because, according to (15.131), it holds that

$$v(k) = \gamma^T C\, x(k) \quad (15.133)$$

In this case the rank of the estimator matrix $H_E = h\gamma^T$ is equal to one.

By virtue of the canonical form of reconstructability for multi-input/multi-output systems given by (11.74) we may, however, determine the matrix H_E of the rank p. In this case we assign to each output variable a partial system whose state is observable by the respective variable and nullify the couplings to the other partial systems. In this way it is possible to design a distinct estimator for each output variable. As in other similar cases we choose, if possible, approximately the same order of the individual partial systems. From Fig. 11.7 it can be seen that it is reasonable to introduce the vector

$$\zeta = L^{-1} y$$

instead of the vector y so that in fact we assign the individual partial systems to the components of the vector ζ.

Using relation (11.74) we have in the error equation (14.6)

$$A_K - H_{EK}C_K = A_K - H_{EK}LE_K = A_K - RE_K \qquad (15.134)$$

where

$$R = H_{EK}L \qquad (15.135)$$

Therefore, if we determine first the matrix R according to the requirements imposed on the coefficients of the characteristic polynomials of the individual partial systems we can obtain from eqn. (15.135)

$$H_{EK} = RL^{-1} \qquad (15.136)$$

since the matrix L is always non-singular.

Let

$$A_K - RE_K = \begin{bmatrix} 0, & \ldots & 0, & p_{11} & 0, & \ldots & 0, & p_{1p} \\ 1, & \ldots & 0, & p_{21} & \vdots & & \vdots & \vdots \\ \ldots & \ldots & \ldots & \ldots & & & & \\ 0, & \ldots & 1, & p_{n_1 1} & 0, & \ldots & 0, & p_{n_1 p} \\ \hline 0, & \ldots & 0, & \vdots & 0, & \ldots & 0, & \vdots \\ \vdots & & \vdots & \vdots & 1, & \ldots & 0, & \vdots \\ & & & & \ldots & \ldots & & \vdots \\ 0, & \ldots & 0, & p_{n1} & 0, & \ldots & 1, & p_{np} \end{bmatrix} \qquad (15.137)$$

Any element p_{ij}, $i = 1, 2, \ldots, n$, $j = 1, 2, \ldots, p$, of the matrix (15.137) may, in general, be expressed as the difference

$$p_{ij} = k_{ij} - r_{ij} \qquad (15.138)$$

so that by choosing the elements r_{ij} of the matrix R we may eliminate the coupling members and leave in (15.137) only the diagonal fields with the required properties. In such a case we have

$$p_{i1} = 0 \quad \text{for} \quad n_1 + 1 \leq i \leq n$$
$$p_{i2} = 0 \quad \text{for} \quad 1 \leq i \leq n_1 \quad \text{and} \quad n_1 + n_2 + 1 \leq i \leq n$$
$$\vdots$$
$$p_{ip} = 0 \quad \text{for} \quad 1 \leq i \leq n_1 + \ldots + n_{p-1} \qquad (15.139)$$

By this structural arrangement of the matrix (15.137) the disconnection of the partial system outputs is achieved. The characteristic polynomial of the entire multi-input/multi-output system is determined as the product of the characteristic polynomials of the individual blocks of the matrix (15.137)

$$\prod_{j=1}^{p} Q_j(z) = \prod_{j=1}^{p} (-p_{v,j} - p_{v+1,j} z - \ldots - p_{v+n_j,j} z^{n_j-1} + z^{n_j}) \quad (15.140)$$

where

$$v = n_1 + \ldots + n_{j-1} + 1$$

If the plant is described in the state space by the matrix triad A, B, C, we may determine multi-input/multi-output system estimators, for instance, by the following procedure. According to eqn. (14.8) the estimator equation is, in general,

$$\hat{x}(k+1) = A \hat{x}(k) + B u(k) + H_E C[x(k) - \hat{x}(k)] =$$
$$= A \hat{x}(k) + B u(k) + H_E y(k) - H_E C \hat{x}(k) \quad (15.141)$$

The estimator equation transformed to the canonical form of reconstructability is

$$\hat{x}_K(k+1) = Q^{-1} A Q \hat{x}_K(k) + Q^{-1} B u(k) + Q^{-1} H_E y(k) - Q^{-1} H_E C \hat{x}_K(k) \quad (15.142)$$

Let us denote

$$H_{EK} = Q^{-1} H_E \quad (15.143)$$

According to this relation and by (15.135) we have

$$H_E = Q H_{EK} = Q R L^{-1}$$
$$QR = H_E L \quad (15.144)$$

From this expression it is evident that for the determination of the matrix H_E it is useful to know the matrix product QR. Let us write eqn. (15.138) in matrix form

$$P = N_K - R \quad (15.145)$$

where N_K is obtained from the matrix of dynamics (11.74) by omitting the zero and unity elements. From (15.145) follows the relation

$$QR = QN_K - QP \quad (15.146)$$

Comparing this result with eqn. (15.144) we obtain

$$H_E = (QN_K - QP) L^{-1}$$

The product QN_K may be determined directly according to (11.79) and for the calculation of the product QP we determine Q according to (11.78) and P according to

(15.137) through (15.140). We obtain

$$H_E = [A^{n_1}\varphi_1, \ldots, A^{n_p}\varphi_p] L^{-1} -$$
$$- [\varphi_1, \ldots, A^{n_1-1}\varphi_1, \varphi_2, \ldots, A^{n_p-1}\varphi_p] \begin{bmatrix} p_{11}, & 0, & \ldots & 0 \\ \vdots & \vdots & & \vdots \\ p_{n_1 1}, & 0, & \ldots & 0 \\ \hline 0, & p_{(n_1+1)2} & \ldots & 0 \\ \vdots & \vdots & & \vdots \\ 0, & p_{(n_1+n_2)2} & \ldots & 0 \\ \hline 0, & \ldots & 0, & p_{(n-n_p+1)p} \\ \vdots & & \vdots & \vdots \\ 0, & \ldots & 0, & p_{np} \end{bmatrix}$$

(15.147)

With the characteristic polynomials $Q_j(z)$, $j = 1, 2, \ldots, p$, according to (15.140) we also have

$$H_E = [Q_1(A) \varphi_1, \ldots, Q_p(A) \varphi_p] L^{-1} \tag{15.148}$$

This is a relation dual to (15.45) representing a generalization of the expression (15.127) for multi-input/multi-output systems. In conclusion it may be summarized that when determining the estimator we establish first the matrix of observability for the plant represented by the matrices A, B, C and then find the first n linearly independent rows which we arrange according to the form

$$Q_{P2} = \begin{bmatrix} c_1^T \\ \vdots \\ c_1^T A^{n_1-1} \\ \hline c_2^T \\ \vdots \\ \hline \vdots \\ c_p^T A^{n_p-1} \end{bmatrix} \tag{15.149}$$

where $n = n_1 + \ldots + n_p$ is the order of the plant and $\max(n_i, i = 1, 2, \ldots, p)$ is the index of observability. The last columns of the individual blocks of the matrix Q_{P2}^{-1} are φ_i, $i = 1, 2, \ldots, p$. The characteristic polynomials Q_j are determined by virtue of (15.140) and the coupling matrix L according to (11.80).

In the case of the *reduced-order estimator* we proceed analogously because problems involving this estimator may be converted into problems including the n-th order estimator. However, the dimension of the matrix H in eqn. (14.13) and in further equations is $(n - p; p)$ and the corresponding characteristic polynomial of the multidimensional estimator is, with regard to the equation of dynamics (14.17),

$$Q_{n-p}(z) = \det(zE - P + HR) = \prod_{i=1}^{n-p}(z - z_i) \qquad (15.150)$$

The state estimate is then

$$\hat{v}^* = \begin{bmatrix} \hat{v} \\ y \end{bmatrix} \qquad (15.151)$$

or in the original basis

$$\hat{x} = T_1^{-1} T_2^{-1} \hat{v}^* = \begin{bmatrix} \hat{v} + Hy \\ -C_2^{-1}C_1(\hat{v} + Hy) + C_2^{-1}y \end{bmatrix} =$$

$$= \begin{bmatrix} E, & H \\ -C_2^{-1}C_1, & C_2^{-1}(E - C_1 H) \end{bmatrix} \begin{bmatrix} \hat{v} \\ y \end{bmatrix} \qquad (15.152)$$

where T_1 and T_2 are the transformation matrices (14.11) and (14.15), and C_1 and C_2 are the output matrices (14.9) corresponding to the vectors

$$\hat{x}_a = \hat{v} + Hy$$
$$\hat{x}_b = C_2^{-1}(y - C_1 \hat{x}_a) \qquad (15.153)$$

respectively.

It should be noted that the transformation matrices T_1 and T_2 need not be calculated. It is sufficient to determine successively the partial matrices (14.9), the matrices P, Q, R, S and B_3 in eqn. (14.12) and finally the matrix H in eqn. (15.150), depending on the requirements put on the characteristic polynomial. It may be noted that the dynamic behaviour of the estimator should always be more rapid than that of the controlled plant. This consideration may lead to the conclusion that the most convenient characteristic equation of the estimator should have altogether zero roots. With such a design, however, all inaccuracies of the plant model and of the measured variables will apply unfavourably, in other words, the solution will be very sensitive to all inaccuracies as mentioned above. It is more advantageous if the estimator components are damped oscillations whose continuous transfer function has the form

$$F(s) = \frac{1}{s^2 + 2\zeta\omega_n s + \omega_n^2}, \quad \zeta\omega_n > 0 \qquad (15.154)$$

where ζ is the relative damping constant and ω_n the frequency of natural oscillations at zero damping. Ensuring that $|F(j\omega)| \leq 1$ for all ω and that the system oscillations decay as fast as possible, no resonance effects can arise. To the transfer function (15.154) corresponds in the time domain the impulse response

$$f(t) = \frac{1}{\omega_n \sqrt{(1 - \zeta^2)}} e^{-\omega_n \zeta t} \sin \omega_n t \sqrt{(1 - \zeta^2)}$$

and its discrete version has the form

$$f(kT) = \frac{1}{\omega_n \sqrt{(1-\zeta^2)}} a^k \sin k\alpha$$

where $a = \exp[-\omega_n \zeta T]$ and $\alpha = \omega_n T \sqrt{(1-\zeta^2)}$. The poles of the corresponding Z-transfer function are

$$z_{1,2} = a e^{\pm j\alpha}$$

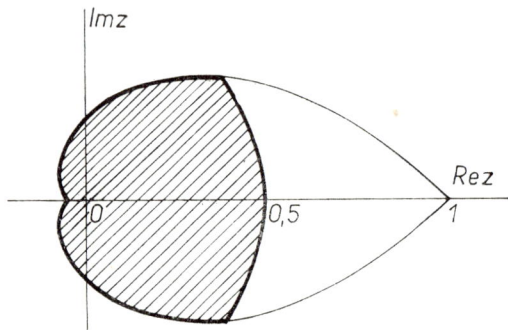

Fig. 15.7. Region of suitable estimator poles.

For instance, for $\zeta = 1/\sqrt{2}$ we have $a = \exp[-\omega_n T/\sqrt{2}]$ and $\alpha = \omega_n T/\sqrt{2}$, so that

$$z_{1,2} = e^{(-1 \pm j)\alpha}$$

For $0 < \alpha < \pi$ we obtain in the Z-plane a region satisfying the requirement mentioned earlier. Confining now the value of a, for instance, to $a = 0.5$ we obtain for the estimator poles in the Z-plane the region shown in Fig. 15.7.

If it is reasonable that the poles be selectable with regard to the results of operating tests, we may choose in the estimator equation (15.128) the column matrix h_{EK} to have an adjustable parameter p, i.e.

$$h_{EK} = p \begin{bmatrix} m_0 - a_0 \\ m_1 - a_1 \\ \dotsb \\ m_{n-1} - a_{n-1} \end{bmatrix}, \quad 0 < p < 1 \tag{15.155}$$

so that the corresponding characteristic equation of the estimator will be

$$(1-p)(a_0 + a_1 z + \dots + a_{n-1} z^{n-1} + z^n) +$$
$$+ p(m_0 + m_1 z + \dots + m_{n-1} z^{n-1} + z^n) = 0 \tag{15.156}$$

For $p = 1$, only the coefficients m_i, $i = 0, 1, \dots, n-1$, are of significance and especially for $m_i = 0$, $i = 0, 1, \dots, n-1$, we may attain a finite number of estimation steps. On the other hand, for $p = 0$ the estimator will have the same characteristic

PRINCIPLES OF DETERMINISTIC SYNTHESIS

equation as the controlled plant. It is advisable [72.1] to choose in the characteristic equation of the estimator only the double root $z_{1,2} = 0$ and the other roots within the limits $0 < z_i < 0.3$, $i = 3, 4, \ldots, n$.

Example 15.5: Calculate the equation of the reduced-order estimator for the plant in Example 15.1. The poles of the estimator are given: $q_1 = 0.1$, $q_2 = 0.2$. The individual plant matrices are:

$$A = \begin{bmatrix} 1, & 0.9673, & 0.3608 \\ 0, & 0.9098, & 0.6065 \\ 0, & -0.1516, & 0.3033 \end{bmatrix}, \quad b = \begin{bmatrix} 0.1306 \\ 0.3608 \\ 0.6065 \end{bmatrix}, \quad c^T = [1, 0, 0]$$

Solution: We perform first a regrouping of the state variables as in (14.9)

$$\begin{bmatrix} x_a(k+1) \\ x_b(k+1) \end{bmatrix} = \begin{bmatrix} 0.9098, & 0.6065, & 0 \\ -0.1516, & 0.3033, & 0 \\ \hline 0.9673, & 0.3608, & 1 \end{bmatrix} \begin{bmatrix} x_a(k) \\ x_b(k) \end{bmatrix} + \begin{bmatrix} 0.3608 \\ 0.6065 \\ \hline 0.1306 \end{bmatrix} u(k)$$

where the original state vector $x(k)$ has the components

$$\begin{bmatrix} x_a(k) \\ x_b(k) \end{bmatrix} = \begin{bmatrix} v(k) \\ y(k) \end{bmatrix} = v^*(k), \quad x_a(k) = \begin{bmatrix} x_2(k) \\ x_3(k) \end{bmatrix}$$

$$x_b(k) = x_1(k)$$

The output equation is

$$y(k) = [0, 0 \mid 1] \, v^*(k)$$

According to eqns. (14.12) we find successively

$$P = A_{11} - A_{12} C_2^{-1} C_1 = A_{11} = \begin{bmatrix} 0.9098, & 0.6065 \\ -0.1516, & 0.3033 \end{bmatrix}$$

$$Q = A_{12} C_2^{-1} = A_{12} = 0$$

$$R = C_1 A_{11} + C_2 A_{21} - (C_1 A_{12} + C_2 A_{22}) C_2^{-1} C_1 = A_{21} = [0.9673, 0.3608]$$

$$S = (C_1 A_{12} + C_2 A_{22}) C_2^{-1} = A_{22} = 1$$

$$B_3 = C_1 B_1 + C_2 B_2 = B_2 = 0.1306$$

Using eqn. (15.150) we now determine the characteristic polynomial of the estimator for the given values of the roots of its characteristic equation:

$$Q_{3-1}(z) = \det(zE - P + HR) = \prod_{i=1}^{2}(z - z_i)$$

$$\det \begin{bmatrix} z - 0.9098 + 0.9673 h_{11} & -0.6065 + 0.3608 h_{11} \\ 0.1516 + 0.9673 h_{21} & z - 0.3033 + 0.3608 h_{21} \end{bmatrix} = (z - 0.1)(z - 0.2)$$

Comparing the coefficients at equal powers of z we obtain the set of equations

$$0.9673h_{11} + 0.3608h_{21} = 0.9131$$

$$-0.3481h_{11} + 0.2584h_{21} = -0.3478$$

Solving this set of equations we have

$$H = \begin{bmatrix} h_{11} \\ h_{21} \end{bmatrix} = \begin{bmatrix} 0.9624 \\ -0.0495 \end{bmatrix}$$

The estimator equation is, according to eqn. (14.18),

$$\hat{v}(k+1) = \begin{bmatrix} -0.0211, & 0.2593 \\ -0.1037, & 0.3212 \end{bmatrix} \hat{v}(k) + \begin{bmatrix} -0.9955 \\ -0.0662 \end{bmatrix} y(k) + \begin{bmatrix} 0.2351 \\ 0.6130 \end{bmatrix} u(k)$$

The estimate of the reduced state vector is, by virtue of eqn. (15.153),

$$\hat{x}_a(k) = \hat{v}(k) + \begin{bmatrix} 0.9624 \\ -0.0495 \end{bmatrix} y(k)$$

If we now calculate for the initial vector $x(0) = [-1, 0, 0]^T$ and $u(0) = 0.25$, $u(k) = 0$ for $k > 0$ by means of the matrices A, B, C, the exact values of the components $x_2(k)$ and $x_3(k)$, and their estimates $\hat{x}_a(k) = [\hat{x}_2(k), \hat{x}_3(k)]$ according to the determined estimator equations, we may arrive at a comparison as shown in the following table:

k	$x_1(k) = y(k)$	$x_2(k)$	$\hat{x}_2(k)$	$x_3(k)$	$\hat{x}_3(k)$
0	—1.0000	0.0000		0,0000	
1	—0.9674	0.0902	0.1233	0.1516	0.2674
2	—0.8255	0.1740	0.2032	0.0323	0.0661
3	—0.6455	0.1779	0.1861	—0.0166	—0.0087
4	—0.4794	0.1518	0.1536	—0.0320	—0.0304
5	—0.3441	0.1187	0.1190	—0.0327	—0.0324
6	—0.2411	0.0882	0.0882	—0.0279	—0.0279

Further values agree exactly to four decimal places.

Note to Example 15.5: The matrix H may also be calculated by virtue of eqn. (15.127) if the matrices P, R are substituted for A, c. We obtain

$$H = Q(P) \begin{bmatrix} R \\ RP \end{bmatrix}^{-1} \begin{bmatrix} 0 \\ 1 \end{bmatrix}$$

where $Q(P) = P^2 + (q_1 + q_2) P + q_1 q_2 E$.

On substitution of the matrices P, R we find

$$H = \begin{bmatrix} 0.4829, & 0.5538 \\ -0.1384, & -0.0709 \end{bmatrix} \begin{bmatrix} 1.8536, & -0.9608 \\ -2.1979, & 2.5758 \end{bmatrix} \begin{bmatrix} 0 \\ 1 \end{bmatrix} = \begin{bmatrix} 0.9624 \\ -0.0495 \end{bmatrix}$$

which is a result identical with the original one. The procedure according to eqn. (15.127) is more convenient particularly for more complex cases and for computer-aided calculations.

Finally, let us be concerned with the *estimator with zero roots of the characteristic equation*. In this case, all eigenvalues of the matrix of dynamics in eqn. (14.18) are zero and the error $\Delta v(k+1)$ is zero after the completion of a finite number of estimation steps. In the case of a reduced-order estimator this will be after $(n-p)$ steps, where n is the order of a single-input/single-output system and p is the number of measured state variables. In this case the estimator matrix of dynamics $(P-HR)^{n-p} = 0$. This means that for single-input/single-output systems the feedback matrix H of the estimator must be determined in such a way that the eigenvalues of the matrix $(P-HR)$ are $z_i = 0$, $i = 1, 2, \ldots, n$. For multi-input/multi-output systems the matrix H may be determined in such a way that the minimum polynomial $z^{\mu-1} = 0$, where μ is the index of observability, corresponds to the matrix $(P-HR)$, and then $(P-HR)^{\mu-1} = 0$.

From eqn. (15.128) follows directly the equation of the reduced-order estimator for a single-input/single-output system:

$$\begin{bmatrix} \hat{x}_{K1}(k+1) \\ \vdots \\ \hat{x}_{K(n-1)}(k+1) \end{bmatrix} = \begin{bmatrix} 0, & \ldots & 0, & 0 \\ 1, & \ldots & 0, & 0 \\ & \cdots & & \\ 0, & \ldots & 1, & 0 \end{bmatrix} \begin{bmatrix} \hat{x}_{K1}(k) \\ \vdots \\ \hat{x}_{K(n-1)}(k) \end{bmatrix} + \begin{bmatrix} b_0 \\ \vdots \\ b_{n-2} \end{bmatrix} u(k) - \begin{bmatrix} a_0 \\ \vdots \\ a_{n-2} \end{bmatrix} y(k)$$

(15.157)

It is apparent that the elements of the feedback matrix of the estimator are

$$h_{\text{EK}i} = -a_{i-1}, \quad i = 1, 2, \ldots, n-1$$

If, for the instant $k = 0$, $\hat{x}_{K1}(0) \neq x_{K1}(0)$ we obtain successively coincidence of all state variables with their estimates in $n-1$ steps.

If the multi-input/multi-output system has the configuration shown in Fig. 11.7, then the i-th partial system of the order n_i is observed by the output variable ζ_i and $n_i - 1$ steps are needed to determine the corresponding state vector. Since for all partial systems the calculation of the corresponding state vector runs in parallel, the determination of the entire state vector terminates at the instant of determination of the state vector of the highest-order partial system. The number of sampling intervals needed is $\max(n_1, n_2, \ldots, n_p) - 1$, where p is the number of partial systems. Here again, the order of the largest partial system is equal to the index of observability μ and, consequently, the number of steps required for the determination of the entire state is $\mu - 1$.

Example 15.6: Consider a double-input/double-output plant from Example 11.3 defined by the matrices

$$A = \begin{bmatrix} 0.6, & 0.7, & 0 \\ 0, & 0.75, & 0 \\ 0, & 0, & 0.8 \end{bmatrix}, \quad B = \begin{bmatrix} 0.8, & 0.4 \\ 0, & 0.9 \\ 0.9, & 0.9 \end{bmatrix}, \quad C = \begin{bmatrix} 1, & 0, & 0 \\ 0, & 0, & 1 \end{bmatrix}$$

Find the state estimator with a finite number of estimation steps.

Solution: This example concerns the estimation of a single variable, i.e. $x_2(k)$. We arrange first the matrices A, B, C as shown in (14.9):

$$A^* = \begin{bmatrix} 0.75 & 0, & 0 \\ \hline 0.7, & 0.6, & 0 \\ 0, & 0, & 0.8 \end{bmatrix}, \quad B^* = \begin{bmatrix} 0, & 0.9 \\ \hline 0.8, & 0.4 \\ 0.9, & 0.9 \end{bmatrix}, \quad C^* = \begin{bmatrix} 0, & 1, & 0 \\ 0, & 0, & 1 \end{bmatrix}$$

We calculate the matrices (14.12):

$$P = A_{11} = 0.75$$

$$Q = 0$$

$$R = A_{21} = \begin{bmatrix} 0.7 \\ 0 \end{bmatrix}$$

$$S = A_{22} = \begin{bmatrix} 0.6, & 0 \\ 0, & 0.8 \end{bmatrix}$$

$$B_1 = [0, \ 0.9]$$

$$B_3 = B_2 = \begin{bmatrix} 0.8, & 0.4 \\ 0.9, & 0.9 \end{bmatrix}$$

The first partial system with the state variables $x_1(k)$, $x_3(k)$ is of the order $n_1 = 2$ and the second partial system with the state variable $x_2(k)$ is of the order $n_2 = 1$. The index of observability is, therefore, $\mu = 2$. The number of estimation steps will be $\mu - 1 = 1$. The matrix of dynamics of the estimator must satisfy the requirement

$$(P - HR)^{\mu-1} = P - HR = 0$$

$$0.75 - [h_{11}, h_{12}] \begin{bmatrix} 0.7 \\ 0 \end{bmatrix} = 0$$

PRINCIPLES OF DETERMINISTIC SYNTHESIS

From this condition we obtain $h_{11} = 1.0714$ and h_{12} may have an arbitrary value. Since, however, the first partial system is not observable by the variable $y_2(k)$, the matrix HS must be a row matrix of the dimension $(1; 2)$ with a zero element in the second column, and consequently $h_{12} = 0$. Therefore

$$H = [1.0714, \; 0]$$

In the estimator equation (14.18) we therefore have

$$\hat{v}(k+1) = [-0.6428, \; 0] \begin{bmatrix} y_1(k) \\ y_2(k) \end{bmatrix} + [-0.8571, \; 0.4714] \begin{bmatrix} u_1(k) \\ u_2(k) \end{bmatrix}$$

The estimated state variable $\hat{v}(k) = \hat{x}_2(k)$ exactly agrees with the actual values of $x_2(k)$ after one sampling period provided that $u(k) = $ const. For the initial vector $x^T(0) = [-1, \; -0.3774, \; -1]$, $u(0) = 1$, $u(k) = 0$, $k > 0$, the following values were calculated:

k	1	2	3	4	5	6
$x_2(k) = \hat{x}_2(k)$	0.6170	0.4627	0.3471	0.2603	0.1952	0.1465

The same results were obtained independently by means of the given matrices A, B, C, provided that all state vector components are measurable, as well as when using the calculated estimator with the estimated component $x_2(k)$.

Note to Example 15.6: The matrix H may also be obtained using eqn. (15.148) if we substitute P and R for A and C, respectively. The solution of the problem is in this case trivial because for the first partial system $P_1 = 0.75$ and $R_1 = 0.7$, while for the second partial system with the output variable $y_2(k)$ we have $P_2 = 0$ and R_2 has an arbitrary non-zero value. The matrix H is therefore

$$H = [P_1 R_1^{-1}, \; P_2 R_2^{-1}] = [1.0714, \; 0]$$

as calculated originally.

15.4 Principle of separability

If the estimator of non-measurable state vector components is used in a feedback control loop, the block diagram of such a control loop may, in general, be expressed as shown in Fig. 15.8. In this figure, z_s is a general vector of the disturbing variables acting upon a multi-input/multi-output plant, w is the vector of the command variables, S the controlled plant, ER the estimator, K the controller and V the filter of the command variables.

Let the plant be described by the equations

$$x(k + 1) = A\, x(k) + B\, u(k) + z_s(k)$$
$$y(k) = C\, x(k) \qquad (15.158)$$

the controller by the equation

$$u(k) = r(k) - K\, \hat{x}(k) \qquad (15.159)$$

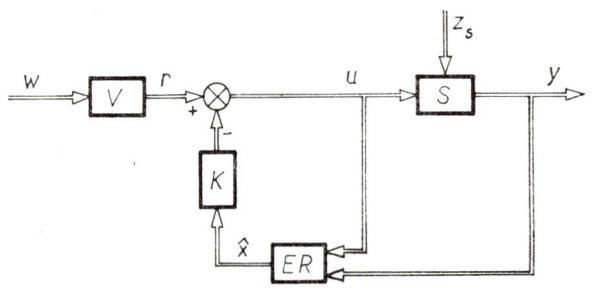

Fig. 15.8. Block diagram of a closed control loop with estimator ER and controller K.

and the estimator of the n-th order by the equation

$$\hat{x}(k + 1) = A_E\, \hat{x}(k) + B_E\, u(k) + H[y(k) - C_E\, \hat{x}(k)] \qquad (15.160)$$

The state equation for the entire system is

$$\begin{bmatrix} x(k+1) \\ \hat{x}(k+1) \end{bmatrix} = \begin{bmatrix} A, & -BK \\ HC, & A_E - HC_E - B_E K \end{bmatrix} \begin{bmatrix} x(k) \\ \hat{x}(k) \end{bmatrix} + \begin{bmatrix} B \\ B_E \end{bmatrix} r(k) + \begin{bmatrix} E \\ 0 \end{bmatrix} z_s(k)$$

$$y(k) = [C, 0] \begin{bmatrix} x(k) \\ \hat{x}(k) \end{bmatrix} \qquad (15.161)$$

If the error

$$\Delta x(k) = x(k) - \hat{x}(k)$$

is introduced into eqn. (15.161) by using the linear transformation [69.4]

$$\begin{bmatrix} x(k) \\ \Delta x(k) \end{bmatrix} = \begin{bmatrix} E, & 0 \\ E, & -E \end{bmatrix} \begin{bmatrix} x(k) \\ \hat{x}(k) \end{bmatrix} \qquad (15.162)$$

we obtain for $A = A_E$, $B = B_E$, $C = C_E$

$$\begin{bmatrix} x(k+1) \\ \Delta x(k+1) \end{bmatrix} = \begin{bmatrix} A - BK, & BK \\ 0, & A - HC \end{bmatrix} \begin{bmatrix} x(k) \\ \Delta x(k) \end{bmatrix} + \begin{bmatrix} B \\ 0 \end{bmatrix} r(k) + \begin{bmatrix} E \\ E \end{bmatrix} z_s(k)$$

$$y(k) = [C, 0] \begin{bmatrix} x(k) \\ \Delta x(k) \end{bmatrix} \qquad (15.163)$$

PRINCIPLES OF DETERMINISTIC SYNTHESIS

From the last equation it is apparent that the characteristic equation of the whole system is

$$\det(zE - A + BK)\det(zE - A + HC) = 0 \tag{15.164}$$

From eqn. (15.164) it follows that the characteristic polynomials of the control loop and of the state estimator are mutually independent and, therefore, we may assign the poles for the state estimation regardless of the pole assignment for the control loop. The process described by eqn. (15.163) consists of two mutually independent partial processes, i.e. of the estimation of the vector, $\hat{x}(k)$, and of the feedback control. This possibility of designing, in a single system, separately the operation of the estimator and of the controller is usually designated as the *principle of separability*. It is important here that the input variables r or w act on the system state at the point of the variable u. By this it is achieved, as can be seen from eqns. (15.161) and (15.163), that the command variable influences the estimator and the feedback control alone in the same manner and causes no change in the error estimate of the vector x. This advantage cannot be gained by introducing the control error $e = r - y$. On the other hand, the disturbing variable z_s influences directly the error vector Δx.

When using the reduced-order estimator, the vector x in eqn. (15.163) must be decomposed into the part x_a being estimated and the measurable part $x_b = y$. For the vector x decomposed in this way one must also change correspondingly the arrangement of the elements of the plant matrices. Let us assume that the arrangement of the plant matrices A, B, C and of the controller matrix K as well as of the vector of the disturbing variables, z_s, corresponds to the decomposition of the state vector x.

Substituting now into eqn. (15.159) for $\hat{x}$ according to eqn. (15.152), we obtain

$$u(k) = r(k) - K \begin{bmatrix} E_{n-p} \\ -C_2^{-1}C_1 \end{bmatrix} \hat{v}(k) - K \begin{bmatrix} H \\ C_2^{-1}(E_p - C_1 H) \end{bmatrix} y(k) =$$

$$= r(k) - K \begin{bmatrix} E_{n-p} & H \\ -C_2^{-1}C_1 & C_2^{-1}(E_p - C_1 H) \end{bmatrix} \begin{bmatrix} \hat{v}(k) \\ y(k) \end{bmatrix} \tag{15.165}$$

and the equation of dynamics of the closed control loop will be

$$x(k+1) = A\,x(k) + B\,r(k) - BK \begin{bmatrix} E_{n-p} \\ -C_2^{-1}C_1 \end{bmatrix} \hat{v}(k) -$$

$$- BK \begin{bmatrix} H \\ C_2^{-1}(E_p - C_1 H) \end{bmatrix} C\,x(k) + z_s(k) \tag{15.166}$$

The subscripts of the unit matrix E in eqns. (15.165) and (15.166) denote its dimensions.

Using the transformation matrices T_1 of eqn. (14.11) and T_2 of eqn. (14.15) we may calculate

$$v = x_a - HCx \tag{15.167}$$

Since
$$v = \hat{v} + \Delta v \tag{15.168}$$

relation (15.167) may be rewritten in the form

$$\hat{v}(k) = x_a(k) - HC\,x(k) - \Delta v(k) \tag{15.169}$$

Substituting now eqn. (15.169) into eqn. (15.166) we obtain

$$\begin{bmatrix} x_a(k+1) \\ x_b(k+1) \end{bmatrix} = A\,x(k) - BK \begin{bmatrix} E_{n-p} \\ -C_2^{-1}C_1 \end{bmatrix} x_a(k) + BK \begin{bmatrix} E_{n-p} \\ -C_2^{-1}C_1 \end{bmatrix} H\,y(k) +$$

$$+ BK \begin{bmatrix} E_{n-p} \\ -C_2^{-1}C_1 \end{bmatrix} \Delta v(k) - BK \begin{bmatrix} H \\ C_2^{-1}(E_p - C_1 H) \end{bmatrix} y(k) + B\,r(k) + z_s(k)$$

Adding to the right-hand side of the last equation the terms $\pm BK\,x(k)$ we have

$$\begin{bmatrix} x_a(k+1) \\ x_b(k+1) \end{bmatrix} = (A - BK)\,x(k) + BK \begin{bmatrix} E_{n-p} \\ -C_2^{-1}C_1 \end{bmatrix} \Delta v(k) + B\,r(k) + z_s(k) +$$

$$+ BK \left\{ x(k) - \begin{bmatrix} E_{n-p} & H \\ -C_2^{-1}C_1 & C_2^{-1}(E_p - C_1 H) \end{bmatrix} \begin{bmatrix} x_a(k) \\ y(k) \end{bmatrix} + \begin{bmatrix} E_{n-p} \\ -C_2^{-1}C_1 \end{bmatrix} H\,y(k) \right\}$$
(15.170)

where the last term is equal to zero. To this equation it is still necessary to add the error state equation with the input vector $z_s = [z_{sa}, z_{sb}]^T$. As in eqn. (15.167) we calculate the vector

$$\zeta_{n-p}(k) = z_{sa}(k) - HC\,z_s(k) \tag{15.171}$$

which represents the input disturbing vector of the error equation (14.19). Equation (15.163) modified for the estimator of the order $n - p$ has, therefore, the form

$$\begin{bmatrix} x_a(k+1) \\ x_b(k+1) \\ \hline \Delta v(k+1) \end{bmatrix} = \begin{bmatrix} A - BK & BK \begin{bmatrix} E_{n-p} \\ -C_2^{-1}C_1 \end{bmatrix} \\ \hline 0 & P - HR \end{bmatrix} \begin{bmatrix} x_a(k) \\ x_b(k) \\ \hline \Delta v(k) \end{bmatrix} +$$

$$+ \begin{bmatrix} B \\ \hline 0 \end{bmatrix} r(k) + \begin{bmatrix} z_s(k) \\ \hline z_{sa}(k) - HC\,z_s(k) \end{bmatrix}$$

$$y(k) = [C,\ 0] \begin{bmatrix} x(k) \\ \Delta v(k) \end{bmatrix} \tag{15.172}$$

PRINCIPLES OF DETERMINISTIC SYNTHESIS 313

The characteristic equation of the complete system is

$$\det [zE - A + BK] \det [zE - P + HR] = 0 \qquad (15.173)$$

and the conclusions applying to it are the same as those stated for eqn. (15.164).

The advantage of the above solution is the fact that it is available for both single-input/single-output and multi-input/multi-output systems as well as for time-variant systems [71.7].

15.5 Invariants

From the resultant equations (15.163) and (15.172) of Sec. 15.4 it is apparent that the feedback matrix K of the controller affects only the matrix of dynamics of the resulting system but not the corresponding matrices B and C. It could be shown that in single-input/single-output systems the matrix K does not affect the coefficients of the transfer function numerator, in other words, these coefficients do not change with a change of the matrix K. In multi-input/multi-output systems with equal number of input and output variables, the coefficients in the numerator of the determinant of the transfer function matrix of the closed control loops are also independent of the changes of the matrix K. In general, the properties of a system which do not change with the changes of the feedback controller matrix K and with the changes of the basis of the input variables are designated as invariants of the system.

More detailed information on invariants may be obtained by investigating the structural changes in system reachability, due to the changes of the feedback controller matrix K and the changes in the basis of input variables.

For $z_s(k) = 0$ the controlled plant (15.158) and the controller (15.159) give the closed control loop equation

$$x(k + 1) = (A - BK)\, x(k) + B\, r(k) = A^* x(k) + B\, r(k)$$

where r is the input variable.

Let us now compare the matrix of reachability of the plant alone with the matrix of reachability of the closed control loop. We find

$$[B, AB, \ldots, A^{i-1}B] =$$

$$= [B, A^*B, \ldots, A^{*i-1}B] \begin{bmatrix} E_r, & KB, & KAB, & \ldots, & KA^{i-2}B \\ 0, & E_r, & KB, & \ldots, & KA^{i-3}B \\ 0, & 0, & E_r, & \ldots, & KA^{i-4}B \\ \multicolumn{5}{c}{\dotfill} \\ 0, & 0, & 0, & \ldots, & E_r \end{bmatrix} \qquad (15.174)$$

The correctness of eqn. (15.174) may be checked by multiplying the matrices to the right of the equality sign. Denote the individual matrices in eqn. (15.174) from the left to the right by Q_{oi}, Q_{ci} and T_i and write this equation in the abbreviated form

$$Q_{oi} = Q_{ci} T_i, \quad i = 1, 2, \ldots \tag{15.175}$$

where the subscripts o and c denote appurtenance to an open and closed control loop, respectively. Since T_i is always non-singular, the rank of the matrices Q_{oi} and Q_{ci} satisfies the relation

$$h(Q_{oi}) = h(Q_{ci}), \quad i = 1, 2, \ldots \tag{15.176}$$

For a single-input/single-output plant and for $i = n$ we may formulate the following theorem:

THEOREM 15.6: *A feedback system of the order n is reachable if the controlled plant of the order n is reachable.*

For multi-input/multi-output systems with the index of reachability v and for $i = v$ we have the theorem:

THEOREM 15.7: *The index of reachability of multi-input/multi-output systems does not change with feedback changes derived from the state vector.*

Equation (15.176) may also be extended by writing out the next block $A^i B$ into individual columns. This modification is justified in such cases where the number of columns of the matrix Q_{oi} is $ri < n$ and $r(i + 1) > n$. Then the equality

$$h(Q_{oi}, A^i b_1, \ldots, A^i b_j) = h(Q_{ci}, A^{*i} b_1, \ldots, A^{*i} b_j) \tag{15.177}$$

where $j = 1, 2, \ldots, r$, represents a mutual equality of the ranks of the matrices defined column by column in contrast to eqn. (15.176) where the matrices are defined block by block.

THEOREM 15.8: *If the column $A^i b_j$ is linearly independent of the columns located in the matrix $Q_{o(i+1)}$ to the right of $A^i b_j$, the same applies to the columns $A^{*i} b_j$ in the matrix $Q_{c(i+1)}$.*

COROLLARY 15.1: *If we seek out in the matrix Q_o of a multi-input/multi-output system such first n linearly independent columns, forming the matrix Q_{D2} ((11.69) or (15.46)) and determining the orders of partial systems in such a way that $n_1 + n_2 + \ldots + n_r = n$, then the order of the individual partial systems will not depend on the changes of the feedback derived from the state vector.*

The effect of change in basis in the space of the input variables may be considered according to the transformation of the input vector w into the vector r by the non-singular transformation matrix V:

$$r = Vw \qquad (15.178)$$

According to Fig. 15.9 it holds that

$$Bu = B(Vw - Kx) = Br - BKx \qquad (15.179)$$

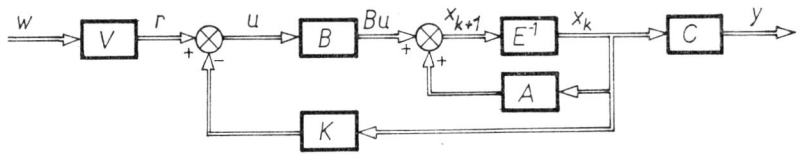

Fig. 15.9. Block diagram of a closed control loop with a linear transformation of the command variables by means of the matrix V.

where K is the feedback matrix. Equation (15.179) may be rewritten in the form

$$Bu = BV(w - V^{-1}Kx) = B^*(w - K^*x) \qquad (15.180)$$

corresponding to Fig. 15.10 that represents the changed part of Fig. 15.9. To this input and feedback arrangement eqn. (15.174) applies if B^* and K^* is substituted for B and K, respectively. Denoting now in eqn. (15.174) the individual matrices by Q^*_{oi}, Q^*_{ci} and T^*_i we have

$$Q^*_{oi} = Q^*_{ci} T^*_i \qquad (15.181)$$

and it holds, for the same reason as in eqn. (15.176), that

$$h(Q^*_{oi}) = h(Q^*_{ci}) = h(Q_{oi}), \quad i = 1, 2, \ldots \qquad (15.182)$$

Relation (15.182) applies to the block arrangement of the matrices Q_{oi} and Q^*_{ci} but an equation similar to eqn. (15.177), where the matrices Q_{oi} and Q_{ci} would be defined

Fig. 15.10. Application of the transformation matrix V in the input matrix and in the controller matrix of the system.

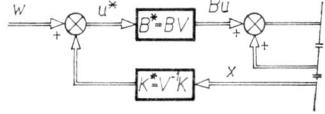

column by column, will not apply since the matrices B and B^* differ from each other. Nevertheless, for multi-input/multi-output systems the orders of the individual partial systems do not change with the changes of the feedback matrix K, but their sequence may be changed by the choice of the transformation matrix V. The sequence is usually chosen so that $n_1 \geq n_2 \geq \ldots \geq n_r$. The individual orders $n_i = v_i$, $i = 1, 2, \ldots, r$, are called *Kronecker indicators or reachability indicators* and form a complete set of invariants of the system with respect to the feedback derived from the state vector and with respect to the changes in the basis of the input variables.

By means of reachability indicators we may derive a general canonical form of multi-input/multi-output systems which involves all systems with the plant

$$x(k+1) = A\,x(k) + B\,u(k)$$

$$y(k) = C\,x(k)$$

where the rank $h(B) = r \leq n$, and with the inputs

$$u(k) = V w(k) - K\,x(k)$$

We may proceed in such a way that we determine first by means of the matrix

$$Q_{on} = [B, AB, \ldots, A^{n-1}B]$$

beginning from the left, n linearly independent columns forming the matrix (11.61), where n_i, $i = 1, 2, \ldots, r$, are the orders of the individual partial systems. We arrange the reachability indicators corresponding to the individual partial systems by their size, so that $v_1 \geq v_2 \geq \ldots \geq v_r$. The index of reachability is then $v = v_1$.

We transform the system to the canonical form of controllability (11.67) and arrange the transformation matrix (11.71) so that the i-th partial system of the order n_i corresponds to the reachability indicator v_i. We thus obtain the set of equations

$$x_R(k+1) = TAT^{-1}\,x_R(k) + TB\,u(k) = A_R\,x_R(k) + E_R M\,u(k) \quad (15.183)$$

$$y(k) = CT^{-1}\,x_R(k) = C_R\,x_R(k)$$

$$M\,u(k) = MV w(k) - MKT^{-1}\,x_R(k)$$

$$x_R(k) = T x(k)$$

where E_R and M are matrices described in (11.67).

The feedback matrix K is arbitrary but it is reasonable to partition it into two blocks:

$$K = K_1 + K_2 \quad (15.184)$$

Suppose that K_2 is the arbitrary part and that K_1 is determined so that the matrix

$$A_0 = A_R - E_R M K_1 T^{-1} \quad (15.185)$$

in eqn. (15.183) is of the form

$$A_0 = \begin{bmatrix} \begin{matrix} 0, 1, \ldots 0 \\ \ldots \ldots \ldots \\ 0, 0, \ldots 1 \\ 0, 0, \ldots 0 \end{matrix} & 0 \\ 0 & \begin{matrix} 0, 1, \ldots 0 \\ \ldots \ldots \ldots \\ 0, 0, \ldots 1 \\ 0, 0, \ldots 0 \end{matrix} \end{bmatrix} \begin{matrix} \} v_1 \\ \\ \} v_r \end{matrix}$$

The equation of the system may now be rewritten as

$$x_R(k+1) = A_0 \, x_R(k) + E_R[MVw(k) - MK_2 T^{-1} x_R(k)] \qquad (15.186)$$

Since V is an arbitrary non-singular matrix and K_2 is also an arbitrary matrix, the matrices

$$V^* = MV$$

$$K^* = MK_2 T^{-1}$$

are arbitrary as well.

In conclusion, the following theorems based on the preceding considerations may be formulated:

THEOREM 15.9: *When transforming the vector w of the input variables of a multi-input/multi-output system to the state vector x_R, the only invariants are the reachability indicators v_i, $i = 1, 2, \ldots, r$. The total number of invariants is r.*

THEOREM 15.10: *When transforming the vector w of the input variables of a multi-input/multi-output system to the output vector y, the only invariants are the reachability indicators v_i, $i = 1, 2, \ldots, r$, and the parameters of the matrix $C_R = CT^{-1}$, where T is the transformation matrix* (11.71). *The total number of invariants is $r + pn$.*

15.6 Command control

In Sec. 15.4 we have learned that when using the n-th order estimator for the determination of the estimate of the state vector $x(k)$ and, equally, when using the reduced-order estimator for the determination of the estimate of the non-measurable part of the state vector $x(k)$, the command variable $r(k)$ has no effect on the error $\Delta x(k)$ or $\Delta v(k)$. This statement follows from eqns. (15.163) and (15.172). This means that the dynamic properties of an estimator do not affect the control process, this depending only on the properties of observability and reachability of the system, i.e. on the properties following from the equations

$$x(k + 1) = (A - BK) x(k) + B r(k)$$
$$y(k) = C x(k) \qquad (15.187)$$

We shall show that for single-input systems the matrix of the feedback controller is uniquely determined only by the requirements imposed on the characteristic polynomial of the system. Let the controlled plant be determined in the state space by the matrices (A, b, c^T). Then, for zero initial conditions, we obtain according to (7.31) the plant transfer function

$$G(z) = c^T(zE - A)^{-1} b = \frac{b_0 + b_1 z + \ldots + b_{n-1} z^{n-1}}{a_0 + a_1 z + \ldots + a_{n-1} z^{n-1} + z^n} \qquad (15.188)$$

The transfer function of a closed loop with the controller matrix k^T has the form

$$K_W(z) = c^T(zE - A + bk^T)^{-1} b =$$
$$= \frac{b_0 + b_1 z + \ldots + b_{n-1} z^{n-1}}{\alpha_0 + \alpha_1 z + \ldots + \alpha_{n-1} z^{n-1} + z^n} \qquad (15.189)$$

If A, b, c^T are the matrices of the canonical form of controllability (10.70), then from eqn. (15.16) it is immediately apparent that the controller matrix k^T affects only the characteristic polynomial of the closed control loop. Therefore, if the characteristic polynomial of the closed loop is required to have a certain form we may fulfil this requirement by choosing the matrix k^T, but cannot fulfil any further requirements.

The coefficients of the transfer function numerator do not depend on the matrix k^T. If, however, the coefficients in the denominator of the transfer function (15.189) could be chosen so as to give common root factors in the numerator and denominator, then these root factors could be cancelled. With this procedure, however, the modes corresponding to the cancelled root factors will be unobservable by the output variable y.

If the input variable of the control loop is the command variable w and if we transform it by means of the matrix V to the input variable r in such a way that

$$r(k) = V w(k) \qquad (15.190)$$

or

$$R(z) = \tilde{V}(z) W(z) \qquad (15.191)$$

then the transfer function (15.189) of the closed loop assumes the form

$$K_{WR}(z) = c^T(zE - A + bk^T)^{-1} b \, \tilde{V}(z) \qquad (15.192)$$

In this case we may choose such a $\tilde{V}(z)$ that the denominator has the same root factors as the numerator of the transfer function (15.189), cancel out these factors and possibly replace the complete original numerator of (15.189) by the numerator of $\tilde{V}(z)$. Although this does not change the denominator of the transfer function (15.189), the dynamic behaviour of the system still changes because the weighting factors of the individual output components change.

In any case, as known from the classical theory of discrete control, we may cancel out only those root factors of the numerator of (15.189) whose zeros lie within the stable region, i.e. within the unit circle with the centre in the origin of the z-plane.

In the multi-input/multi-output case, eqn. (15.192) assumes the form

$$K_{WR}(z) = C^T(zE - A + BK)^{-1} B \tilde{V}(z) \qquad (15.193)$$

As already stated in Par. 15.1.1, the feedback controller matrix K is not uniquely determined by the required characteristic polynomial of the multi-input/multi-output system and we can, therefore, fulfil some further requirements by the selection of the matrix K. An alternative procedure is to fulfil first these further requirements and finally to check how many coefficients of the characteristic polynomial of the system can still be affected by the choice of the mattix K. In this connection many special requirements of control quality could be mentioned, which can be solved by the indicated procedure. Among these possible requirements belongs also the autonomous or decoupled control where each command variable w_i affects only the appropriate output variable y_j, $i = j$, and not the rest of the output variables. Even for this special case of synthesis we could mention a number of practical modifications of which one is represented by the autonomous control achieved after a finite number of control steps, while at the beginning of the control process the disconnection of the individual control loops is not ensured. This case will now be demonstrated by an example.

Example 15.7: Consider again the double-input/double-output plant for which we were solving in Example 15.2 (page 277) the finite number of control steps and in Example 15.6 (page 308) the estimator with a finite number of estimation steps. Let us now calculate the filter V of eqn. (15.190) so that for $w(k) = $ const., e.g. $w(k) = [1, 1]^T$, $k = 0, 1, 2, \ldots$, the output vector $y(k) = w(k)$ after the decay of the finite transient response. It is assumed that the sampling of the input variables of the system takes place at the instants k^+, while the state vector is considered for k^-, $k = 0, 1, 2, \ldots$.

Solution: In Example 15.6, the plant was described by the matrices

$$A^* = \begin{bmatrix} 0.75, & 0, & 0 \\ \hline 0.7, & 0.6, & 0 \\ 0, & 0, & 0.8 \end{bmatrix}, \quad B^* = \begin{bmatrix} 0, & 0.9 \\ \hline 0.8, & 0.4 \\ 0.9, & 0.9 \end{bmatrix}, \quad C^* = \begin{bmatrix} 0, & 1, & 0 \\ 0, & 0, & 1 \end{bmatrix}$$

where the state vector was decomposed into the estimated part x_a and the measurable part x_b:

$$x = \begin{bmatrix} x_a \\ x_b \end{bmatrix}, \quad x_a = x_2, \quad x_b = \begin{bmatrix} x_1 \\ x_3 \end{bmatrix}$$

In Example 15.2 a controller feedback matrix was calculated for which the characteristic polynomial of the system has the form z^3. In view of the change in sequence of the state vector components, the matrix K has now the form

$$K = \begin{bmatrix} -4.7083, & -2.25, & 3.5556 \\ 0.8333, & 0, & 0 \end{bmatrix}$$

From the block diagram in Fig. 15.8 it follows that

$$u(k) = r(k) - K\,x(k)$$

We replace the unknown vector $x(k)$ by the estimate $\hat{x}(k)$ by virtue of eqn. (15.152). Since in the example being solved $C_1 = 0$ and $C_2 = E$, we have

$$\hat{x}(k) = L\begin{bmatrix} \hat{v} \\ y \end{bmatrix} = \begin{bmatrix} E, & H \\ 0, & E \end{bmatrix}\begin{bmatrix} \hat{v} \\ y \end{bmatrix} = \begin{bmatrix} 1 & 1.0174, & 0 \\ \hline 0 & 1, & 0 \\ 0 & 0, & 1 \end{bmatrix}\begin{bmatrix} \hat{v} \\ y_1 \\ y_2 \end{bmatrix}$$

where the matrix H was taken over from Example 15.6. The estimator equation may be taken over from this example as well. In view of the fact that $P - HR = 0$ and $Q = 0$, we have

$$\hat{v}(k+1) = -HS\,y(k) + (B_1 - HB_3)\,u(k)$$
$$\hat{v}(k+1) = M\,y(k) + N\,u(k) = [-0.6428,\ 0]\,y(k) + [-0.8571,\ 0.4714]\,u(k)$$

The filter V is calculated according to the problem formulation by means of the equilibrium state in order that $y(k) = w(k)$ for $k \to \infty$. In this case we also have

$$x(k+1) = x(k)$$
$$x_1(k) \quad = y_1(k) = w_1(k)$$
$$x_3(k) \quad = y_2(k) = w_2(k)$$
$$\hat{v}(k+1) = \hat{v}(k)$$
$$(E - A^*)\,x(k) = B^*\,u(k)$$

$$\begin{bmatrix} 0.25, & 0, & 0 \\ -0.7, & 0.4, & 0 \\ 0, & 0, & 0.2 \end{bmatrix}\begin{bmatrix} x_2(k) \\ x_1(k) \\ x_3(k) \end{bmatrix} = \begin{bmatrix} 0, & 0.9 \\ 0.8, & 0.4 \\ 0.9, & 0.9 \end{bmatrix}\begin{bmatrix} u_1(k) \\ u_2(k) \end{bmatrix}$$

Solving the last equation we obtain

$$\begin{bmatrix} u_1(k) \\ u_2(k) \end{bmatrix} = \begin{bmatrix} -0.1887, & 0.3061 \\ 0.1887, & -0.0839 \end{bmatrix} \begin{bmatrix} x_1(k) \\ x_3(k) \end{bmatrix}$$

$$u(k) = Tw(k)$$

The equation

$$u = r - K\hat{x}$$

may now be modified in the form

$$r = u + K\hat{x} = Tw + KL \begin{bmatrix} \hat{v} \\ y \end{bmatrix}$$

$$r = Tw + KL \begin{bmatrix} My + NTw \\ y \end{bmatrix}$$

Since in the equilibrium state $y = w$, we may rewrite the last expression in the sought form

$$r = Vw$$

where

$$V = T + KL \begin{bmatrix} M + NT \\ E \end{bmatrix}$$

$$V = \begin{bmatrix} -5.6370, & 5.2831 \\ 0.7548, & -0.3355 \end{bmatrix}$$

Substituting $r = Vw$ in eqn. (15.172) we may use this equation for the calculation of the response to the given vectors of the command and disturbing variables. In the problem solved the vector of disturbing variables is zero and the vector of command variables $w(k) = [1, 1]^T$. Equation (15.172) has in this problem the form

$$x^*(k+1) = \begin{bmatrix} 0, & 0, & 0 & | & 0.75 \\ 4.1333, & 2.4, & -2.8445 & | & -3.4333 \\ 3.4875, & 2.025, & -2.4 & | & -3.4875 \\ \hline 0, & 0, & 0 & | & 0 \end{bmatrix} x^*(k) +$$

$$+ \begin{bmatrix} 0.6793, & -0.3020 \\ -4.2077, & 4.0923 \\ -4.3940, & 4.4528 \\ \hline 0, & 0 \end{bmatrix} w(k) = \mathscr{A} x^*(k) + \mathscr{B} w(k)$$

where

$$x^*(k) = \begin{bmatrix} x_2(k) \\ x_1(k) \\ x_3(k) \\ \hline \Delta v(k) \end{bmatrix}$$

Since the index of reachability $v = 2$, the control process terminates after two steps and, therefore, the upper left block of the matrix $\mathscr{A}^2$ is zero:

$$\mathscr{A}^2 = \begin{bmatrix} 0, & 0, & 0, & 0 \\ 0, & 0, & 0, & 4.7802 \\ 0, & 0, & 0, & 4.0332 \\ \hline 0, & 0, & 0, & 0 \end{bmatrix}$$

According to eqn. (7.1)

$$x^*(2) = \mathscr{A}^2 x^*(0^-) + \mathscr{A}\mathscr{B} w(0^+) + \mathscr{B} w(1)$$

Let us choose the initial vector $x^*(0) = 0$. Since the disturbing variables do not affect the system and since $w(0^+) = w(1) = w(k), k = 0^+, 1, 2, \ldots$, we have

$$x^*(2) = (\mathscr{A}\mathscr{B} + \mathscr{B}) w(0^+)$$

where

$$\mathscr{A}\mathscr{B} + \mathscr{B} = \begin{bmatrix} 0, & 0 \\ 1, & 0 \\ 0, & 1 \\ 0, & 0 \end{bmatrix}$$

The output equation of the system is

$$y(k) = \begin{bmatrix} 0, & 1, & 0, & 0 \\ 0, & 0, & 1, & 0 \end{bmatrix} x^*(k) = \mathscr{C} x^*(k)$$

Choosing the initial vector $x^*(0^-) = 0$ we have

$$x^*(1) = \begin{bmatrix} 0.3773 \\ -0.1154 \\ 0.0588 \\ \hline 0 \end{bmatrix}, \quad x^*(2) = \begin{bmatrix} 0 \\ 1 \\ 1 \\ \hline 0 \end{bmatrix}$$

We may conclude this example by the following summary:

(a) The vector $x^*(2)$ does not depend on the initial state $x(0^-)$ but only on the vectors $w(0^+)$ and $\Delta v(0^+)$. If $w = $ const. and $\Delta v = 0$ beginning at the instant k^+

and if the index of reachability of a multi-input/multi-output plant is v, then $y(k + v) = w(k^+)$ and $x^*(k + v)$ does not depend on $x^*(k^-)$ but only on $w(k^+)$. From the foregoing it follows that $y(\varkappa) \neq w(k^+)$ for $\varkappa = k^+, k + 1, \ldots, k + v - 1$.

(b) In this example apparently $\mathscr{A}^3 = 0$ so that even the disturbing variable $\Delta v(0^+) \neq 0$ may be compensated in three steps. In general it holds that for $w = \text{const.}$, starting with the instant k^+, $\Delta v(k^+) \neq 0$, $\Delta v(\varkappa) = 0$ for $\varkappa = k + 1$,

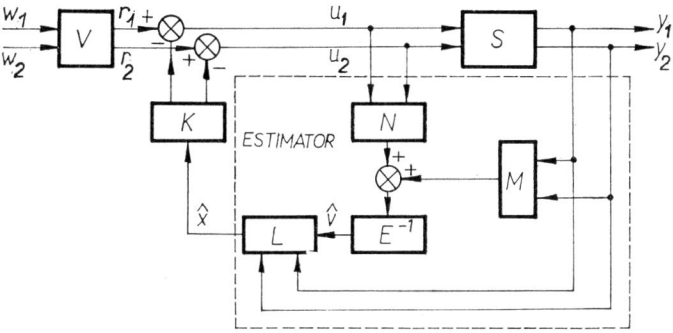

Fig. 15.11. Block diagram relating to Example 15.7.

$k + 2, \ldots,$ and for the index of reachability v of a multi-input/multi-output plant $y(k + v + 1) = w(k^+)$ and $x^*(k + v + 1)$ does not depend on $x^*(k^-)$ but only on $w(k^+)$ and $\Delta v(k^+)$. Hence it follows that $y(\varkappa) \neq w(k^+)$ for $\varkappa = k^+, k + 1, \ldots, k + v$.

(c) From the structure of the matrices $\mathscr{A}$, $\mathscr{B}$ and $\mathscr{C}$ it is apparent that an interaction between the command variables w_1 and w_2 with the output variables y_1 and y_2, respectively, occurs always only during the transient response. After its termination the input variables are autonomous. The variable w_i affects only the variable y_i, $i = 1, 2$. If the change of the command variables occurred at the instant k^+, then, after v steps, $y(k + v) = w(k^+)$, but $y(k + i) \neq w(k^+)$ for $i = 0, 1, 2, \ldots, v - 1$.

The block diagram of the control loop for this example is shown in Fig. 15.11. The detailed diagram of the controlled plant is in Fig. 11.6.

15.7 Disturbance-variable compensation

So far we have been concerned, in the preceding sections, with the action and with the compensation of disturbing variables in special cases only; in Sec. 15.6, for instance, we have considered only the disturbance $v(0^+)$ acting temporarily at the instant 0^+. In this section we shall discuss more general problems. It is useful to distinguish between the following disturbing variables which may act upon the control loop:

(1) z_w — non-measurable disturbing variables superposed on the control loop command variables w,

(2) z_y — non-measurable disturbing variables superposed on the controlled plant output variables y,

(3) z_s — non-measurable disturbing variables acting on the controlled plant,

(4) z_{sm} — measurable disturbing variables acting on the controlled plant.

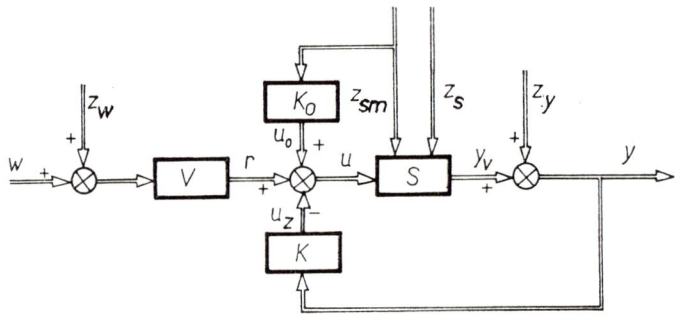

Fig. 15.12. General block diagram of a closed control loop.

Non-measurable disturbing variables may be compensated only by the action of feedback, while measurable disturbing variables may be compensated by the controller K_0 in the open loop. The block diagram of a multi-input/multi-output closed control loop with all the mentioned disturbing variables is shown in Fig. 15.12 where the input of the feedback member K are the measurable output variables y, while the output variables y_v of the plant S are non-measurable, virtual. The state equations for this general case of control loop assume the form

$$x(k+1) = A\,x(k) + B\,u(k) + z_{sm}(k) + z_s(k)$$
$$y(k) = C\,x(k) + D\,u(k) + z_y(k) \qquad (15.194)$$

where

$$u(k) = V[w(k) + z_w(k)] + K_0\,z_{sm}(k) - K\,y(k)$$

and the vector

$$u_0(k) = K_0\,z_{sm}(k) \qquad (15.195)$$

corresponds to the open loop controller and the vector

$$u_z(k) = K\,y(k) \qquad (15.196)$$

appertains to the feedback controller.

PRINCIPLES OF DETERMINISTIC SYNTHESIS 325

The individual methods of synthesis for the compensation of disturbing variables are usually confined to only a certain type of the disturbing variables mentioned. We shall, therefore, discuss in this section only some selected problems.

The compensation or elimination of the action of measurable disturbing variables z_{sm} can be solved easily. These variables may be brought to the estimator input. The equation of dynamics (15.194) is, for this purpose, modified in the form

$$x(k+1) = A\,x(k) + [B, E]\begin{bmatrix} u(k) \\ z_{sm}(k) \end{bmatrix} \quad (15.197)$$

and the design of the estimator is carried out as in the case when $z_{sm} = 0$ but with the equation of plant dynamics modified according to eqn. (15.197). In this case the state estimate error does not depend on the disturbing variables z_{sm}.

If the controlled plant is also disturbed by the non-measurable variables z_s, their action may be reduced in estimating the state vector if they can be expressed as the solution of the difference equation

$$q(k+1) = N\,q(k) + \delta_v(k)$$
$$z_s(k) \;\;\;\;\;\; = M\,q(k) + \delta_z(k) \quad (15.198)$$

where the matrices $N(s; s)$ and $M(n; s)$ are known. In eqns. (15.198) the matrices $\delta_v(s; 1)$ and $\delta_z(n; 1)$ are independent white vector noises with a zero mean. When calculating the estimator we extend the equation of the plant dynamics to the form

$$\begin{bmatrix} x(k+1) \\ q(k+1) \end{bmatrix} = \begin{bmatrix} A, & M \\ 0, & N \end{bmatrix}\begin{bmatrix} x(k) \\ q(k) \end{bmatrix} + \begin{bmatrix} B, & E \\ 0, & 0 \end{bmatrix}\begin{bmatrix} u(k) \\ z_{sm}(k) \end{bmatrix} + \begin{bmatrix} \delta_z(k) \\ \delta_v(k) \end{bmatrix} \quad (15.199)$$

where the sequences $\delta_z(k)$ and $\delta_v(k)$ may be replaced by randomly selected values of the initial state, whose action is compensated in the course of further control. The action of the subsequent values of the sequences $\delta_z(k)$ and $\delta_v(k)$ at the instants $k = 1, 2, \ldots$ is, due to the system linearity, superposed, compensated in like manner as the action of the initial state and does not influence the compensation of the earlier actions of disturbing variables.

The general form of the output equation is

$$y(k) = [C, 0]\begin{bmatrix} x(k) \\ q(k) \end{bmatrix} \quad (15.200)$$

if only such disturbing variables z_s are included in the system description, which influence the output y.

It should be noted that the difference equations of the disturbing variables represent an uncontrollable but observable part of the system, whose state can be estimated by an estimator designed for this purpose. This state estimate may then be utilized for control. In this case the plant input vector is regarded as the sum of two vectors

$$u(k) = u_R(k) + u_z(k) \tag{15.201}$$

where $u_R(k)$ is determined, for instance, by the previously described method for the plant state control and $u_z(k)$ is designed for the compensation of the action of disturbing variables. The equation of the plant then has the form

$$x(k + 1) = A\,x(k) + B\,u_R(k) + B\,u_z(k) + z_s(k) \tag{15.202}$$

The vector $u_z(k)$ must be designed so that the input vector

$$e(k) = B\,u_z(k) + z_s(k) \tag{15.203}$$

may act on the plant dynamics as little as possible. For example, the sum of squares expressed as a scalar product of input vectors

$$e^T(k)\,e(k) = [B\,u_z(k) + z_s(k)]^T\,[B\,u_z(k) + z_s(k)] \tag{15.204}$$

attains its minimum with the input

$$u_z(k) = -(B^T B)^{-1}\,B^T\,z_s(k) = -(B^T B)^{-1}\,B^T M\,q(k) \tag{15.205}$$

We replace in the control loop the vector $q(k)$ by the estimate $\hat{q}(k)$ determined by the estimator. If the disturbing variables of the vector q are acting at the plant input at the same points as the corresponding controlling variables, then $M = B$ and eqn. (15.205) is simplified to

$$u_z(k) = -q(k) \tag{15.206}$$

Example 15.8: Determine the reduced-order estimator for the plant whose equations are

$$x(k + 1) = \begin{bmatrix} 1, & 0.5 \\ 0, & 0.3 \end{bmatrix} x(k) + \begin{bmatrix} 0.4 \\ 0.6 \end{bmatrix} u(k)$$

$$y(k) = [1,\ 0]\,x(k)$$

The matrix of dynamics of this plant has one stable eigenvalue and the other one lies on the stability boundary. A disturbing variable is acting on the plant at the same point as the controlling variable. The difference equations of this disturbing variable have the form

$$q(k + 1) = q(k) + \delta_v(k)$$

$$z_s(k) = \begin{bmatrix} 0.4 \\ 0.6 \end{bmatrix} q(k)$$

$$\mathscr{E}[\delta_v(k)] = 0$$

PRINCIPLES OF DETERMINISTIC SYNTHESIS

Solution: According to eqn. (15.199)

$$\begin{bmatrix} x_1(k+1) \\ x_2(k+1) \\ q(k+1) \end{bmatrix} = \begin{bmatrix} 1, & 0.5, & 0.4 \\ 0, & 0.3, & 0.6 \\ 0, & 0, & 1 \end{bmatrix} \begin{bmatrix} x_1(k) \\ x_2(k) \\ q(k) \end{bmatrix} + \begin{bmatrix} 0.4 \\ 0.6 \\ 0 \end{bmatrix} u(k) + \begin{bmatrix} 0 \\ 0 \\ \delta_v(k) \end{bmatrix}$$

By means of the estimator we shall estimate on the one hand the state variable $x_2(k)$ and on the other hand the variable $q(k)$, since all we know about $\delta_v(k)$ is that the mean value $\mathscr{E}[\delta_v(k)] = 0$. We therefore put $\delta_v(k) = 0$ and estimate by means of the estimator the unknown state at each sampling instant, caused just by this disturbing variable. Therefore we rewrite the system equation in the form

$$\begin{bmatrix} q(k+1) \\ x_2(k+1) \\ \hline x_1(k+1) \end{bmatrix} = \begin{bmatrix} 1, & 0, & | & 0 \\ 0.6, & 0.3, & | & 0 \\ \hline 0.4, & 0.5, & | & 1 \end{bmatrix} \begin{bmatrix} q(k) \\ x_2(k) \\ \hline x_1(k) \end{bmatrix} + \begin{bmatrix} 0 \\ 0.6 \\ \hline 0.4 \end{bmatrix} u(k)$$

By virtue of eqn. (14.12) we calculate successively

$$P = A_{11} = \begin{bmatrix} 1, & 0 \\ 0.6, & 0.3 \end{bmatrix}, \quad Q = 0$$

$$R = A_{21} = [0.4, \ 0.5], \quad S = A_{22} = 1$$

$$B_1 = \begin{bmatrix} 0 \\ 0.6 \end{bmatrix}, \quad B_2 = 0.4, \quad C_1 = [0, \ 0], \quad C_2 = 1, \quad B_3 = B_2$$

For the estimator with a finite number of estimation steps we have

$$Q_{3-1}(z) = \det[zE - P + HR] = z^2$$

The matrix H is determined directly via formula (15.127), where we substitute the matrices P, R for A, c^T, respectively:

$$H = \begin{bmatrix} 1, & 0 \\ 0.6, & 0.3 \end{bmatrix}^2 \begin{bmatrix} 0.4, & 0.5 \\ 0.7, & 0.15 \end{bmatrix}^{-1} \begin{bmatrix} 0 \\ 1 \end{bmatrix} = \begin{bmatrix} 1.7241 \\ 1.2207 \end{bmatrix}$$

The sought estimator equation is, in conformity with (14.18),

$$\hat{v}(k+1) = \begin{bmatrix} 0.3104, & -0.8621 \\ 0.1117, & -0.3104 \end{bmatrix} \hat{v}(k) - \begin{bmatrix} 2.2413 \\ 1.4070 \end{bmatrix} y(k) + \begin{bmatrix} 0.6896 \\ 0.1117 \end{bmatrix} u(k)$$

where, by (14.13),

$$\begin{bmatrix} \hat{q}(k) \\ \hat{x}_2(k) \end{bmatrix} = \hat{v}(k) + \begin{bmatrix} 1.7241 \\ 1.2207 \end{bmatrix} y(k)$$

Example 15.9: Find for the plant of Example 15.8 a controller for a finite number of control steps together with a compensation of the action of the disturbing variable in the sense of the quadratic cost function.

Solution: According to eqn. (15.159), the controlling variable

$$u_R(k) = r(k) - K \hat{x}(k)$$

where the matrix K is determined using relation (15.54):

$$K = [0, 1] [B, AB]^{-1} A^2 =$$

$$= [0, 1] \begin{bmatrix} 0.4, & 0.7 \\ 0.6, & 0.18 \end{bmatrix}^{-1} \begin{bmatrix} 1, & 0.5 \\ 0, & 0.3 \end{bmatrix}^2 = [1.7241, 1.0172]$$

Changing the sequence of components in the vector $\hat{x}(k)$ we have

$$u_R(k) = r(k) - [1.0172, 1.7241] \begin{bmatrix} \hat{x}_2(k) \\ \hat{x}_1(k) \end{bmatrix}$$

Considering the disturbance compensation in the sense of the quadratic cost function, the disturbing variable $q(k)$, acting at the same point of the plant input as the controlling variable, is given by the relation (15.206). We know further that in the equilibrium state $\hat{x}_2 = 0$, $u_R = 0$ and $x_1 = y = w$. In order that these conditions may be satisfied it is necessary that $r = 1.7241w$. With these partial conclusions the last equation may be rewritten in the form

$$u_R(k) = 1.7241 \, w(k) - [1, 1.0172, 1.7241] \begin{bmatrix} \hat{q}(k) \\ \hat{x}_2(k) \\ \hat{x}_1(k) \end{bmatrix}$$

The equation of the complete closed control loop may now be expressed by (15.172) where $r(k) = 1.7241 \, w(k)$.

Using the partial results of Example 15.8 we find

$$\begin{bmatrix} q(k+1) \\ x_2(k+1) \\ x_1(k+1) \\ \Delta v(k+1) \end{bmatrix} = \begin{bmatrix} 1, & 0, & 0, & 0, & 0 \\ 0, & -0.3104, & -1.0345, & 0.6, & 0.6103 \\ 0, & 0.0931, & 0.3104, & 0.4, & 0.4069 \\ 0, & 0, & 0, & 0.3104, & -0.8621 \\ 0, & 0, & 0, & 0.1117, & -0.3104 \end{bmatrix} \begin{bmatrix} q(k) \\ x_2(k) \\ x_1(k) \\ \Delta v(k) \end{bmatrix} + \begin{bmatrix} 0 \\ 1.0345 \\ 0.6896 \\ 0 \\ 0 \end{bmatrix} w(k)$$

$$y(k) = [0, 0, 1, 0, 0] \, x^*(k)$$

where

$$x^{*T}(k) = [q(k), x_2(k), x_1(k), \Delta v(k)]$$

$$\Delta v^T(k) = [q(k) - \hat{q}(k), x_2(k) - \hat{x}_2(k)]$$

PRINCIPLES OF DETERMINISTIC SYNTHESIS

Denote in this equation the matrix of dynamics by $\mathscr{A}$ and the input matrix by $\mathscr{B}$. It holds that

$$x^*(2) = \mathscr{A}^2 x^*(0) + \mathscr{A}\mathscr{B} w(0) + \mathscr{B} w(1)$$

$$x^*(2) = \begin{bmatrix} 1, & 0, & 0, & 0, & 0 \\ 0, & 0, & 0, & -0.3456, & -1.3170 \\ 0, & 0, & 0, & 0.3496, & -0.2880 \\ 0, & 0, & 0, & 0, & 0 \\ 0, & 0, & 0, & 0, & 0 \end{bmatrix} x^*(0) + \begin{bmatrix} 0 \\ -1.0345 \\ 0.3104 \\ 0 \\ 0 \end{bmatrix} w(0) + \begin{bmatrix} 0 \\ 1.0345 \\ 0.6896 \\ 0 \\ 0 \end{bmatrix} w(1)$$

$$x^*(3) = \mathscr{A}^3 x^*(0) + \mathscr{A}^2\mathscr{B} w(0) + \mathscr{A}\mathscr{B} w(1) + \mathscr{B} w(2)$$

$$x^*(3) = \begin{bmatrix} 1, & 0, & 0, & 0, & 0 \\ 0, & 0, & 0, & -0.2544, & 0.7067 \\ 0, & 0, & 0, & 0.0763, & -0.2120 \\ 0, & 0, & 0, & 0, & 0 \\ 0, & 0, & 0, & 0, & 0 \end{bmatrix} x^*(0) + \begin{bmatrix} 0 \\ -1.0345 \\ 0.3104 \\ 0 \\ 0 \end{bmatrix} w(1) + \begin{bmatrix} 0 \\ 1.0345 \\ 0.6896 \\ 0 \\ 0 \end{bmatrix} w(2)$$

$$x^*(4) = \mathscr{A}^4 x^*(0) + \mathscr{A}^3\mathscr{B} w(0) + \mathscr{A}^2\mathscr{B} w(1) + \mathscr{A}\mathscr{B} w(2) + \mathscr{B} w(3)$$

$$x^*(4) = \begin{bmatrix} 1, & 0, & 0, & 0, & 0 \\ 0, & 0, & 0, & 0, & 0 \\ 0, & 0, & 0, & 0, & 0 \\ 0, & 0, & 0, & 0, & 0 \\ 0, & 0, & 0, & 0, & 0 \end{bmatrix} x^*(0) + \begin{bmatrix} 0 \\ -1.0345 \\ 0.3104 \\ 0 \\ 0 \end{bmatrix} w(2) + \begin{bmatrix} 0 \\ 1.0345 \\ 0.6896 \\ 0 \\ 0 \end{bmatrix} w(3)$$

since $\mathscr{A}^2\mathscr{B} = \mathscr{A}^3\mathscr{B} = 0$.

From the numerical results we may arrive at the following conclusions:

(a) For $z_s(k^+) = 0$ and $w(k^+) = $ const., $k = 0, 1, 2, \ldots$, we have $x_1(2) = w(2)$ and $x_2(2) = 0$.

(b) For $z_s(k^+) = $ const. and $w(k^+) = 0$, $k = 0, 1, 2, \ldots$, we have $\Delta v(0^+) \neq 0$, $\Delta v(1) \neq 0$, $\Delta v(2) = 0$ so that the state estimate error after two steps is zero. After the next step the state vector will be zero, $x_1(3) = x_2(3) = 0$.

(c) After two steps, the plant state vector does not depend on the initial state $x(0)$ but only on the values of $w(0^+)$ and $\Delta v(0^+)$.

(d) For $z_s(k^+) = $ const. and $w(k^+) = $ const., $k = 0, 1, 2, \ldots$, we have $\Delta v(0^+) =$ $= \Delta v(1) \neq 0$, $\Delta v(2) = 0$ and $x(0^+) = x(1) = x(2) = x(3) \neq 0$ and $x_1(4) = w(k^+) =$ $= $ const. and $x_2(4) = 0$. This means that the action of the disturbing variable is compensated in two steps and the control process corresponding to a finite number of control steps then starts with the initial vector $x^*(2)$. For instance,

$$x^*(0^+) = \begin{bmatrix} 1 \\ 1 \\ 1 \\ 1 \\ 1 \end{bmatrix}, \quad x^*(1) = \begin{bmatrix} 1 \\ 0.8999 \\ 1.9000 \\ -0.5517 \\ -0.1987 \end{bmatrix}, \quad x^*(2) = \begin{bmatrix} 1 \\ -1.6627 \\ 1.0616 \\ 0 \\ 0 \end{bmatrix}$$

$$x^*(3) = \begin{bmatrix} 1 \\ 0.4524 \\ 0.8643 \\ 0 \\ 0 \end{bmatrix}, \quad x^*(4) = \begin{bmatrix} 1 \\ 0 \\ 1 \\ 0 \\ 0 \end{bmatrix}$$

In conclusion we shall show the realization of the estimator and controller in the feedback loop. We first modify the controller equation to the form

$$u_R(k) = 1.7241\{w(k) - [0.58, 0.6, 1] \begin{bmatrix} \hat{q}(k) \\ \hat{x}_2(k) \\ \hat{x}_1(k) \end{bmatrix}\}$$

Substituting for $[\hat{q}(k), \hat{x}_2(k)]^T$ according to Example 15.8 we obtain

$$u_R(k) = 1.7241\{w(k) - [0.58, 0.6]\,\hat{v}(k) - 2.7324\,y(k)\}$$

where

$$u_v(k) = [0.58, 0.6]\,\hat{v}(k)$$

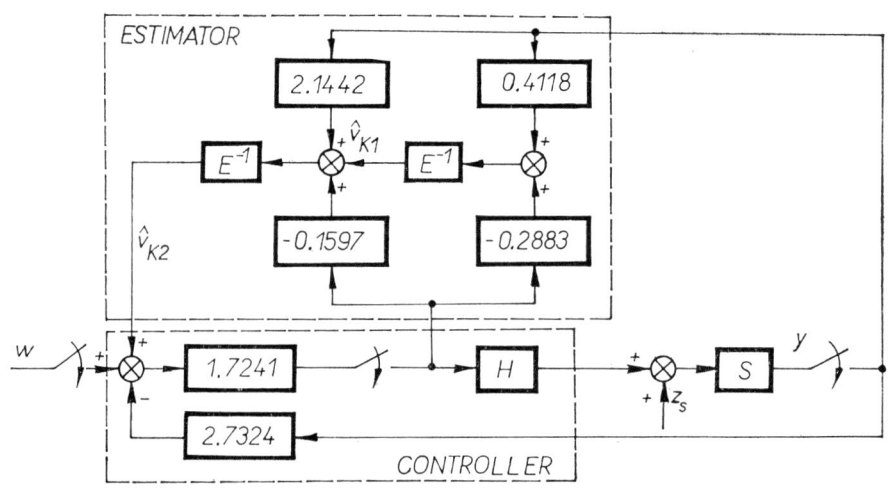

Fig. 15.13. Block diagram relating to Example 15.9.

is the estimator output equation. The estimator together with the controller might be simulated by means of the equations already determined. The number of coefficients may, however, be additionally reduced by transforming the estimator equations to the canonical form of reconstructability using the procedure outlined in Sec. 10.6. We obtain

$$\hat{v}_K(k+1) = \begin{bmatrix} 0, & 0 \\ 1, & 0 \end{bmatrix} \hat{v}_K(k) + \begin{bmatrix} 0.4118 \\ 2.1442 \end{bmatrix} y(k) + \begin{bmatrix} -0.2883 \\ -0.1597 \end{bmatrix} u(k)$$

$$u_v(k) = \begin{bmatrix} 0, & 1 \end{bmatrix} \hat{v}_K(k)$$

The block diagram of the complete control loop is shown in Fig. 15.13, where H is the holding member and S the controlled plant.

16

Deterministic Controller Design Based on Quadratic Cost Functions

There is no doubt that the quadratic cost function is one of the most useful mathematical gauges of the quality of automatic control. The procedures of controller design using the quadratic cost function, formerly elaborated for linear systems described in the s-plane by transfer functions, represent today a more or less classical approach developed for continuously or discontinuously acting single-input/single-output or multi-input/multi-output systems, for analytical or random inputs and outputs and for command control and compensation of disturbances, respectively. It is, therefore, quite natural that considerable attention was given to these cost functions in the synthesis employing the representation of the controlled system and of the complete closed control loops in the state space. It should be noted that the basic mathematical approach to the solution of linear systems is the same for control loops with time-variant or time-invariant controlled plants so that specific procedures for the time-variatnt types of control loops need not be derived. Therefore, a general approach to the solution of time-variant systems will be described in some of the later sections and the results will be simplified for time-invariant cases. However, considerable differences may appear in the details of numerical procedures for both modifications.

Consider first some general remarks:

(a) For deterministic problems the quadratic cost function may have the form

$$J = x^T(N)\,P(N)\,X(N) + \sum_{k=0}^{N-1} [x^T(k)\,Q(k)\,x(k) + u^T(k)\,R(k)\,u(k)] =$$

$$= \Theta[x(N), N] + \sum_{k=0}^{N-1} \Phi[x(k), u(k), k] \tag{16.1}$$

which, in principle, can be adapted for specific problems as required. The cost function (16.1) represents a linear combination of general squares of state variables and controlling variables, where Q and R are weighting matrices. In accordance with the second term of the expression (16.1), the squares of state variables and controlling variables are considered within the interval $0 \leq k \leq (N-1)$ while at the instant $k = N$ only the squares of state variables contribute to the cost function (16.1). The vector $u(N)$ does not influence the system considered within the assumed interval. The values of the expression (16.1) belonging to the individual time instants k represent the increments of the cost function (16.1) corresponding to the individual steps of the dynamic control process.

The cost function can be applied to either the finite control interval $0 \leq k \leq N$, $k = 0, 1, 2, ..., N$, or to the infinite interval $0 \leq k \leq \infty$, $k = 0, 1, ..., \infty$. This is in contrast to the classical design procedures requiring the control interval to be infinity.

The matrices P, Q, R may be non-symmetrical, but if they are symmetrical, the resulting relations and the numerical solution will be considerably simplified. For this reason, only the symmetrical matrices P, Q, R will be considered in the subsequent text. The matrix Q may be positive semidefinite. In some modifications of control problems the matrix R must be positive definite in order to ensure the existence of R^{-1}, but this is not always the case and, therefore, attention will be given to these particular solutions. The significance and the properties of the matric P follow from the resulting relations.

THEOREM 16.1: *If Q and R are symmetrical matrices and R is a positive definite matrix, then the function Φ in eqn. (16.1) having the more general form*

$$\Phi[x(k), u(k), k] = x^T(k) Q(k) x(k) + 2x^T(k) S(k) u(k) + u^T(k) R(k) u(k) \tag{16.2}$$

can always be transformed to the form

$$\Phi[x(k), u(k), k] = \hat{x}^T(k) \hat{Q}(k) \hat{x}(k) + \hat{u}^T(k) \hat{R}(k) \hat{u}(k) \tag{16.3}$$

where

$$\hat{Q}(k) = Q(k) - S(k) R^{-1}(k) S^T(k)$$
$$\hat{R}(k) = R(k) \tag{16.4}$$
$$\hat{x}(k) = x(k)$$
$$\hat{u}(k) = u(k) + R^{-1}(k) S^T(k) x(k) \tag{16.5}$$

Applying the transformations (16.4), the state equation

$$x(k+1) = A(k) x(k) + B(k) u(k)$$

changes into

$$\hat{x}(k+1) = \hat{A}(k) \hat{x}(k) + \hat{B}(k) \hat{u}(k) \tag{16.6}$$

where

$$\hat{A}(k) = A(k) - B(k) R^{-1}(k) S^T(k)$$
$$\hat{B}(k) = B(k) \quad (16.7)$$

and the output equation changes into

$$\hat{y}(k) = \hat{C}(k) \hat{x}(k) + \hat{D}(k) \hat{u}(k) \quad (16.8)$$

where

$$\hat{C}(k) = C(k) - R^{-1}(k) S^T(k)$$
$$\hat{D}(k) = D(k) \quad (16.9)$$

Proof: Starting with eqn. (16.2) and adding and subtracting the term $x^T(k) \cdot S(k) R^{-1}(k) S^T(k) x(k)$ and rewriting, we arrive at the resulting relations (16.3). We could also substitute for $x(k)$ and $u(k)$ in eqn. (16.2) according to eqns. (16.5).

(b) Let the control law be

$$u(k) = -K x(k) \quad (16.10)$$

(c) The control law will reach its optimum in the sense of the quadratic cost function if for any initial state $x(0) \in X^n$ of the controlled plant the following conditions hold:

(1) The functional (16.1) reaches its minimum value.
(2) The controlled loop is stable.

(d) Controlled plants with shifted output, i.e. with transportation lag, are permissible for the control synthesis.

(e) Let the matrices (A, B, C, D) and their transforms be real matrices.

16.1 Optimum control based on the second method of Lyapunov

16.1.1 General procedure

In this section we shall be concerned with the determination of such a sequence of values of the controlling variable u for which the cost function (16.1) reaches a minimum value and the system is transferred from the initial state $x(0) \neq 0$ at the instant $k = 0$ as close as possible to the desired terminal state, e.g. to the origin of the state space. First, the problem will be solved for a finite control interval with fixed times of initiation and termination, provided that at all instants of sampling the state of the controlled system is known. The solution will be based on the second method of Lyapunov and the problem will be considered to be linear.

The second method of Lyapunov attempts to give information on the stability of the equilibrium state of linear and nonlinear systems without knowing the solution

CONTROLLER DESIGN BASED ON QUADRATIC COST FUNCTIONS

of the difference equations describing the system. Roughly speaking, the procedure consists of the determination of a fictitious function called the Lyapunov function $V(x, k)$ and of testing the definiteness of this function and of its first difference.

The Lyapunov function for the given system cannot be determined in a unique way. The multiplicity of choice of such a function makes it possible to select just that kind which can be used not only for ensuring stability but also for the solution of more general tasks as, for example, the optimum control.

For discrete systems the Lyapunov function may have the form

$$V(x, k) = x^T(k) P(k) x(k) \tag{16.11}$$

where P is a positive definite matrix. Then

$$\Delta V(x, k) = \Delta[x^T(k) P(k) x(k)] =$$
$$= x^T(k+1) P(k+1) x(k+1) - x^T(k) P(k) x(k) \tag{16.12}$$

In accordance with Theorem 12.7, the difference of the Lyapunov function must be negative definite. Combining this condition with the summand of the cost function (16.1), the matrix $P(k)$ is essentially determined. We have

$$x^T(k+1) P(k+1) x(k+1) - x^T(k) P(k) x(k) =$$
$$= [x^T(k) Q(k) x(k) + u^T(k) R(k) u(k)] \tag{16.13}$$

To minimize the cost function (16.1) calculate the sum

$$\sum_{k=0}^{N-1} \Delta[x^T(k) P(k) x(k)] =$$
$$= \sum_{k=0}^{N-1} [x^T(k+1) P(k+1) x(k+1) - x^T(k) P(k) x(k)]$$

$$\sum_{k=0}^{N-1} \Delta[x^T(k) P(k) x(k)] = x^T(N) P(N) X(N) - x^T(0) P(0) x(0) \tag{16.14}$$

or, if $x(k+1)$ in eqn. (16.14) is expressed in terms of the state equation, then

$$\sum_{k=0}^{N-1} [x^T(k+1) P(k+1) x(k+1) - x^T(k) P(k) x(k)] =$$
$$= \sum_{k=0}^{N-1} [x^T(k) A^T(k) P(k+1) A(k) x(k) +$$
$$+ u^T(k) B^T(k) P(k+1) A(k) x(k) + x^T(k) A^T(k) P(k+1) B(k) u(k) +$$
$$+ u^T(k) B^T(k) P(k+1) B(k) u(k) - x^T(k) P(k) x(k)] \tag{16.15}$$

Comparing the right-hand sides of eqns. (16.14) and (16.15), the following equality evidently holds:

$$\sum_{k=0}^{N-1} [x^T(k) A^T(k) P(k+1) A(k) x(k) +$$

$$+ u^T(k) B^T(k) P(k+1) A(k) x(k) + x^T(k) A^T(k) P(k+1) B(k) u(k) +$$

$$+ u^T(k) B^T(k) P(k+1) B(k) u(k) - x^T(k) P(k) x(k)] -$$

$$- x^T(N) P(N) x(N) + x^T(0) P(0) x(0) = 0 \qquad (16.16)$$

Now, the cost function (16.1), extended by eqn. (16.16), gives

$$J = x^T(N) P(N) x(N) + \sum_{k=0}^{N-1} [x^T(k) Q(k) x(k) + u^T(k) R(k) u(k)] +$$

$$+ \sum_{k=0}^{N-1} [x^T(k) A^T(k) P(k+1) A(k) x(k) +$$

$$+ u^T(k) B^T(k) P(k+1) A(k) x(k) + x^T(k) A^T(k) P(k+1) B(k) u(k) +$$

$$+ u^T(k) B^T(k) P(k+1) B(k) u(k) - x^T(k) P(k) x(k)] -$$

$$- x^T(N) P(N) x(N) + x^T(0) P(0) x(0) \qquad (16.17)$$

The optimum vector u must satisfy the condition

$$\frac{\partial J}{\partial u} = 0 \qquad (16.18)$$

provided that the second partial derivative of the cost function with respect to the control vector u is a positive definite matrix. Condition (16.18) applied to the relation (16.17) gives

$$[R(k) + B^T(k) P(k+1) B(k)] u(k) + B^T(k) P(k+1) A(k) x(k) = 0 \qquad (16.19)$$

Provided that the inverse of the matrix $[R(k) + B^T(k) P(k+1) B(k)]$ exists, the optimum control vector u is

$$u(k) = -[B^T(k) P(k+1) B(k) + R(k)]^{-1} B^T(k) P(k+1) A(k) x(k) \qquad (16.20)$$

where

$$K(k) = [B^T(k) P(k+1) B(k) + R(k)]^{-1} B^T(k) P(k+1) A(k) \qquad (16.21)$$

represents the feedback transition matrix.

CONTROLLER DESIGN BASED ON QUADRATIC COST FUNCTIONS

In addition, the second partial derivative of the cost function J with respect to the control vector u, calculated by virtue of eqn. (16.19), yields

$$\frac{\partial^2 J}{\partial u^2} = \frac{\partial}{\partial u}\left[\frac{\partial J}{\partial u}\right]^T = R(k) + B^T(k)\, P(k+1)\, B(k) \qquad (16.22)$$

which, in accordance with the accepted assumption, is a positive definite matrix and consequently the control vector (16.20) ensures the minimum of the cost function.

The minimum value of the cost function may be calculated in the following way. Substituting for $x(k+1)$ in eqn. (16.13) from the state equation and using for $u(k)$ the relation (16.10) of the control law, eqn. (16.13) takes the form

$$x^T(k)\left[A^T(k)\, P(k+1)\, A(k) - K^T(k)\, B^T(k)\, P(k+1)\, A(k) - \right.$$
$$\left. - A^T(k)\, P(k+1)\, B(k)\, K(k) + K^T(k)\, B^T(k)\, P(k+1)\, B(k)\, K(k) - P(k)\right] x(k) =$$
$$= -x^T(k)\left[Q(k) + K^T(k)\, R(k)\, K(k)\right] x(k) \qquad (16.23)$$

Since eqn. (16.23) must be satisfied for arbitrary $x(k)$, it follows that

$$P(k) = A^T(k)\, P(k+1)\, A(k) - K^T(k)\, B^T(k)\, P(k+1)\, A(k) -$$
$$- A^T(k)\, P(k+1)\, B(k)\, K(k) +$$
$$+ K^T(k)\, B^T(k)\, P(k+1)\, B(k)\, K(k) + Q(k) + K^T(k)\, R(k)\, K(k) \qquad (16.24)$$

Now, inserting for $P(k)$ in eqn. (16.17) the relation (16.24) and for $u(k)$ the control law (16.10), the minimum value of the cost function will be

$$J = x^T(0)\, P(0)\, x(0) \qquad (16.25)$$

The difference equation of the closed control loop is

$$x(k+1) = \{A(k) - B(k)\left[B^T(k)\, P(k+1)\, B(k) + R(k)\right]^{-1} B^T(k)\, P(k+1)\, A(k)\}\, x(k) \qquad (16.26)$$

It should be noted that the matrix $R(k)$ may not necessarily be positive definite. This mathematical result corresponds to physical reality because for a discrete linear system, in contrast with the continuous linear system, the controlling variables in the vector u need not be constrained by the matrix R in order to achieve a physically realizable process. Moreover, the constraints of the components of the controlling vector u are smaller, the smaller are the values of elements of the weighting matrix R. However, for discrete linear systems the matrix R may be a positive semidefinite or even a zero matrix. Nevertheless, the resulting transient response is physically realizable since the values of the components of the controlling vector u are constrained due to the fact that all numerical output values of the computer are transformed by the holding element into a pulse of finite width and of an area proportional to the computer output. That is why the pulse height must be finite, too. In the case of amplitude modulation, which is most frequently used, the pulse has a width equal to the period of sampling and a rectangular form.

Equation (16.24) represents the matrix form of the "Riccati difference equation"

$$P(k) = A^T(k) P(k+1) A(k) -$$
$$- A^T(k) P(k+1) B(k) [B^T(k) P(k+1) B(k) + R(k)]^{-1} B^T(k) P(k+1) A(k) + Q(k) \tag{16.27}$$

Although eqn. (16.27) does not correspond exactly to the discrete matrix form of the Riccati equation, this denomination is used to stress the relation to the continuous version of this equation and to accentuate its significance. This remark applies to all other modifications of eqn. (16.27).

The solution of eqn. (16.27) determines the matrix $P(k)$ needed for the calculation of the controlling vector $u(k-1)$ by means of eqn. (16.20). Notice that $P(k+1)$ must be known if $P(k)$ is to be calculated. Hence, the matrices $P(k)$, $k = N-1$, $N-2, \ldots, 0$, can be determined successively starting with the endpoint of the process.

16.1.2 Simplified modifications

(a) For N approaching to infinity and for

$$\lim_{N \to \infty} x^T(N) P(N) x(N) = 0 \tag{16.28}$$

the cost function (16.1) changes into

$$J = \sum_{k=0}^{\infty} [x^T(k) Q(k) x(k) + u^T(k) R(k) u(k)] \tag{16.29}$$

and the problem solved according to this criterion corresponds to the minimization of the control error in the steady state and is usually called the *deterministic optimum regulator problem*.

With the simplifications carried out in eqns. (16.28) and (16.29) the final results described in a general way in Sec. 16.1.1 remain unchanged except that the matrix $P(k)$ converges for any $P(N) \geq 0$ to a sequence of semidefinite matrices as $k \to \infty$, provided that $A(k)$, $B(k)$, $Q(k)$, $R(k)$ are bounded for $k \geq 0$ and that the system is completely controllable and completely reconstructable or exponentially stable. For more details see Ref. [72.9].

(b) For N approaching to infinity, for a time-invariant controlled plant and for time-invariant weighting matrices Q and R, the relations of Par. 16.1.1 will hold if the

CONTROLLER DESIGN BASED ON QUADRATIC COST FUNCTIONS

arguments (k) and $(k + 1)$ in all the matrices are omitted. In this case the Riccati equation is simplified to the algebraic nonlinear matrix equation of the form

$$P = A^TPA - A^TPB(B^TPB + R)^{-1} B^TPA + Q \qquad (16.30)$$

(c) If we apply the simplifications formulated in a) and b) and if the weighting matrix $R = 0$, we obtain for the cost function

$$J = \sum_{k=0}^{\infty} x^T(k) \, Q \, x(k), \qquad (16.31)$$

for the controlling vector

$$u(k) = -(B^TPB)^{-1} B^TPA \, x(k), \qquad (16.32)$$

for the closed control loop

$$x(k + 1) = [A - B(B^TPB)^{-1} B^TPA] \, x(k) \qquad (16.33)$$

and the Riccati equation takes the form

$$P = A^TPA - A^TPB(B^TPB)^{-1} B^TPA + Q \qquad (16.34)$$

(d) When no input variable acts on the controlled system, its behaviour describes a homogeneous equation. Applying at the same time all other simplifications indicated in a) through c), we obtain eqn. (16.23) in the form

$$x^T(k) \, (A^TPA - P) \, x(k) = -x^T(k) \, Q \, x(k) \qquad (16.35)$$

Hence, the Riccati equation is a linear algebraic matrix equation of the form

$$A^TPA - P + Q = 0 \qquad (16.36)$$

and the minimum value of the cost function is again given by eqn. (16.25). In this particular case the solution of the homogeneous equation

$$x(k) = A^k \, x(0), \quad x(0) \ne 0 \qquad (16.37)$$

determines the control action caused by the properties of the controlled plant alone and the values of the cost function (16.25) may be denoted as the quality index of this control action.

16.2 The discrete maximum principle

16.2.1 General procedure

The problem solved in Sec. 16.1 by means of the second method of Lyapunov will now be solved by means of the calculus of variations. There are two different procedures frequently applied for the synthesis of general control problems: the Euler-Lagrange technique and the maximum principle of Pontryagin, both differing in the mathematical background. We shall draw our attention especially to the discrete version of the maximum principle and indicate some basic relations. We shall not apply a quite general formulation of problems solvable by the maximum principle, but the introductory relations will include the linear and nonlinear systems as well.

Thus, let the dynamic system be discrete and nonlinear with the state vector $x(k) \in X^n$ and the input vector $u(k) \in U^r$. Let the controlled system be described by the equations

$$x(k + 1) = f[x(k), u(k), k] \tag{16.38}$$

$$y(k) = \varphi[x(k), u(k), k] \tag{16.39}$$

and let the transient response end at the instant N. We are to determine such values of $u(k)$, $k = 0, 1, ..., N - 1$, which will minimize the cost function (16.1) subject to a constraint according to eqn. (16.38). Taking this constraint into account, the cost function (16.1) may be written in the general form

$$J^* = \Theta[x(k), k]\big|_0^N + \sum_{k=0}^{N-1} \{\Phi[x(k), u(k), k] - \lambda^T(k + 1)[x(k + 1) - f(x(k), u(k), k)]\} \tag{16.40}$$

where λ is the vector of Lagrange multipliers.

The *Hamiltonian* is defined as

$$H[x(k), u(k), \lambda(k + 1), k] = H(k) =$$
$$= \Phi[x(k), u(k), k] + \lambda^T(k + 1) f[x(k), u(k), k] \tag{16.41}$$

With relation (16.41) the cost function then becomes

$$J^* = \Theta[x(k), k]\big|_0^N + \sum_{k=0}^{N-1} [H(k) - \lambda^T(k + 1) x(k + 1)] \tag{16.42}$$

Now, to obtain the minimum of (16.42) with respect to $x(k)$ and $u(k)$, the *method of perturbations* of the calculus of variations may be used. For this purpose let us introduce for the state and input vectors the relations

$$x(k) = \hat{x}(k) + \varepsilon\, \delta(k)$$

$$u(k) = \hat{u}(k) + \varepsilon\, \eta(k) \tag{16.43}$$

where the perturbations $\delta(k)$ and $\eta(k)$ are mutually independent and their values at different time instants are independent, too.

Using the relations (16.43) the cost function (16.42) takes the form

$$J^* = \Theta[\hat{x}(N) + \varepsilon\,\delta(N), N] - \Theta[\hat{x}(0) + \varepsilon\,\delta(0), 0] +$$

$$+ \sum_{k=0}^{N-1} \{H[\hat{x}(k) + \varepsilon\,\delta(k), \hat{u}(k) + \varepsilon\,\eta(k), k] -$$

$$- \lambda^T(k+1)[\hat{x}(k+1) + \varepsilon\,\delta(k+1)]\} \tag{16.44}$$

The minimum of J^* must satisfy the conditions

$$\lim_{\varepsilon \to 0} \frac{\partial J^*}{\partial \varepsilon} = 0 \quad \text{and} \quad \lim_{\varepsilon \to 0} \frac{\partial^2 J^*}{\partial \varepsilon^2} > 0 \tag{16.45}$$

The first condition yields

$$\left[\frac{\partial \Theta(N)}{\partial x(N)}\right]^T \delta(N) + \left[\frac{\partial \Theta(0)}{\partial x(0)}\right]^T \delta(0) +$$

$$+ \sum_{k=0}^{N-1} \left\{\left[\frac{\partial H(k)}{\partial x(k)}\right]^T \delta(k) + \left[\frac{\partial H(k)}{\partial u(k)}\right]^T \eta(k) - \lambda^T(k+1)\,\delta(k+1)\right\} = 0 \tag{16.46}$$

The last term in eqn. (16.46) can be written as follows

$$\sum_{k=0}^{N-1} \lambda^T(k+1)\,\delta(k+1) = \sum_{k=1}^{N} \lambda^T(k)\,\delta(k) =$$

$$= \sum_{k=0}^{N-1} \lambda^T(k)\,\delta(k) + \lambda^T(N)\,\delta(N) - \lambda^T(0)\,\delta(0) \tag{16.47}$$

Condition (16.46) together with eqn. (16.47) may be expressed as

$$\left\{\left[\frac{\partial \Theta(N)}{\partial x(N)}\right]^T - \lambda^T(N)\right\} \delta(N) - \left\{\left[\frac{\partial \Theta(0)}{\partial x(0)}\right]^T - \lambda^T(0)\right\} \delta(0) +$$

$$+ \sum_{k=0}^{N-1} \left\{\left[\frac{\partial H(k)}{\partial x(k)}\right]^T - \lambda^T(k)\right\} \delta(k) + \sum_{k=0}^{N-1} \left[\frac{\partial H(k)}{\partial u(k)}\right]^T \eta(k) = 0 \tag{16.48}$$

Since the indicated variations are mutually independent, eqn. (16.48) may be satisfied by satisfying the individual contions

$$\frac{\partial H(k)}{\partial x(k)} = \lambda(k), \quad k = 0, 1, \ldots, N-1 \tag{16.49}$$

$$\frac{\partial H(k)}{\partial u(k)} = 0, \quad k = 0, 1, \ldots, N-1 \tag{16.50}$$

and the *transversality conditions*

$$\left\{\left[\frac{\partial \Theta(N)}{\partial x(N)}\right]^T - \lambda^T(N)\right\} \delta(N) = 0 \tag{16.51}$$

$$\left\{\left[\frac{\partial \Theta(0)}{\partial u(0)}\right]^T - \lambda^T(0)\right\} \delta(0) = 0 \tag{16.52}$$

If the value of any variable is specified, the corresponding variation vanishes and the respective conditions (16.49) through (16.52) do not apply. Particularly for given $x(0)$ and $x(N)$, the corresponding boundary conditions for $\lambda(k)$, $k = 0, N$, are satisfied by $\delta(0) = \delta(N) = 0$.

The second condition (16.45) is satisfied for all cost functions and systems of interest.

For the linear regulator problems with specified $x(0)$ and $x(N)$, eqns. (16.49) and (16.50) yield

$$\frac{\partial H(k)}{\partial x(k)} = Q(k)\,x(k) + A^T(k)\,\lambda(k+1) = \lambda(k) \tag{16.53}$$

$$\frac{\partial H(k)}{\partial u(k)} = R(k)\,u(k) + B^T(k)\,\lambda(k+1) = 0 \tag{16.54}$$

The last equations are called Euler equations for the variational problem considered. The estimated solution of these equations is

$$\lambda(k) = P(k)\,x(k) \tag{16.55}$$

Combining now eqns. (16.54), (16.55) and the equation of dynamics (16.6), we have

$$R(k)\,u(k) + B^T(k)\,P(k+1)\,[A(k)\,x(k) + B(k)\,u(k)] = 0 \tag{16.56}$$

and the controlling vector is

$$u(k) = -[B^T(k)\,P(k+1)\,B(k) + R(k)]^{-1}\,B^T(k)\,P(k+1)\,A(k)\,x(k) \tag{16.57}$$

This is exactly the same result as indicated by eqn. (16.20) and, consequently, eqn. (16.26) is valid for the maximum principle, too.

The Riccati equation can be derived by means of eqn. (16.53) if eqns. (16.55), (16.57) and (16.6) are used simultaneously. We obtain

$$P(k)\,x(k) = Q(k)\,x(k) + A^T(k)\,P(k+1)\,\{A(k)\,x(k) - B(k)\,[B^T(k)\,P(k+1)\,B(k) + \\ + R(k)]^{-1}\,B^T(k)\,P(k+1)\,A(k)\,x(k)\} \tag{16.58}$$

CONTROLLER DESIGN BASED ON QUADRATIC COST FUNCTIONS

Equation (16.58) must be valid for any vector $x(k)$. Applying this condition to eqn. (16.58) and rewriting, we obtain the same form of the Riccati equation as introduced by eqn. (16.27).

If, for example, $x(N)$ were not specified, an additional condition would have to be applied, i.e.

$$\lambda(N) = M\, x(N) \qquad (16.59)$$

with M being a positive semidefinite matrix, in order that the second variation (16.45) can be positive. Notice that for the final stage of the process the equation

$$P(N) = M \qquad (16.60)$$

must hold.

16.2.2 Simplified modifications

For N approaching to infinity, for time-invariant controlled systems and for constant weighting matrices Q and R of the cost function (16.1), relation (16.58) yields the Riccati equation in the form

$$P = A^T P A - A^T P B (B^T P B + R)^{-1} B^T P A + Q \qquad (16.61)$$

which corresponds to eqn. (16.30). According to eqn. (16.57) the controlling vector is

$$u(k) = -(B^T P B + R)^{-1} B^T P A\, x(k) \qquad (16.62)$$

From eqn. (16.47) it is evident that for a fixed $x(0)$ and $x(N)$, i.e. for $\delta(0) = \delta(N) = 0$ and for N approaching to infinity, it follows that

$$\sum_{k=0}^{N-1} \lambda^T(k+1)\,\delta(k+1) = \sum_{k=0}^{N-1} \lambda^T(k)\,\delta(k) = \sum_{k=0}^{N} \lambda^T(k)\,\delta(k) \qquad (16.63)$$

Hence, for this particular problem we may replace the index $k+1$ in eqn. (16.44) by k and modify all subsequent relations accordingly. It should be noted that the cost function assumes the form

$$J^* = \sum_{k=0}^{N-1} \left[H(k) - \lambda^T(k)\, x(k+1) \right] \qquad (16.64)$$

and the Euler equation the form

$$\frac{\partial H(k)}{\partial x(k)} = Q\, x(k) + A^T \lambda(k) = \lambda(k-1) \qquad (16.65)$$

$$\frac{\partial (Hk)}{\partial u(k)} = R\, u(k) + B^T \lambda(k) = 0 \qquad (16.66)$$

Using the same procedure for the derivation of the controlling vector as in the preceding paragraph, we have

$$u(k) = -R^{-1}B^T P\, x(k) \qquad (16.67)$$

It is evident that the last relation is simpler than the result (16.57), however, the weighting matrix R must be positive definite in order that the inverse R^{-1} can exist. However, this mathematical result is inconsistent with physical reality because, as stated in Sec. 16.1, for discrete linear systems no constraint of the controlling variables by the matrix R is necessary to ensure the physical realizability of the solution. Nevertheless, the relation (16.67) is correct if the matrix R is positive definite. It can be proved (see Ref. [72.16]) that the solution resulting from eqns. (16.53) and (16.54) corresponds to that obtained by means of eqns. (16.65) and (16.66).

To determine the Riccati equation, $\lambda(k)$ is calculated from eqn. (16.65). Substituting $k + 1$ for k and assuming the matrix A to be regular, we have

$$\lambda(k+1) = A^{-T}\lambda(k) - A^{-T}Q\, x(k+1) \qquad (16.68)$$

where

$$A^{-T} = (A^{-1})^T$$

Combining now eqns. (16.67), (16.68), (16.55) and (16.6), the Riccati equation corresponding to eqns. (16.65) and (16.66) will have the form

$$PA - (A^{-T} + A^{-T}QBR^{-1}B^T)\, P - PBR^{-1}B^T P + A^{-T}QA = 0 \qquad (16.69)$$

A comparison of eqns. (16.69) and (16.61) does not give a clear answer to whether both equations have the same solution. Moreover, the controlling vector (16.67) is not equal to the controlling vector (16.57) if the matrix $P(k)$ is the same in both relations. The question now arises, whether both vectors (16.62) and (16.67) would yield the same solution for a positive definite R.

To examine this we write the so-called Euler matrix equation

$$p(k+1) = E_M\, p(k) \qquad (16.70)$$

where $p^T(k) = [x(k), \lambda(k)]$ and $\lambda(k)$ is the vector of Lagrange multipliers introduced in eqn. (16.40).

Derive first the matrix E_M. Calculating $u(k)$ from eqn. (16.54) and substituting it into the equation of dynamics of the controlled system, we have

$$x(k+1) = A\, x(k) - BR^{-1}B^T \lambda(k+1) \qquad (16.71)$$

Equation (16.53) yields

$$\lambda(k+1) = -A^{-T}Q\, x(k) + A^{-T}\lambda(k) \qquad (16.72)$$

CONTROLLER DESIGN BASED ON QUADRATIC COST FUNCTIONS

Substituting eqn. (16.72) into eqn. (16.71), we obtain

$$x(k+1) = (A + BR^{-1}B^T A^{-T} Q) x(k) - BR^{-1}B^T A^{-T} \lambda(k) \quad (16.73)$$

Equations (16.73) and (16.72) define a system for which the Euler matrix has the form

$$E_M = \begin{bmatrix} A + BR^{-1}B^T A^{-T} Q & -BR^{-1}B^T A^{-T} \\ \hline -A^{-T} Q & A^{-T} \end{bmatrix} \quad (16.74)$$

The respective characteristic matrix is $[E_M - zE]$, where E is the identity matrix. The elements of the characteristic matrix are, in general, polynomials. For such types of matrices the eigenvalues remain unchanged if the respective matrix is rearranged using only admissible operations. The rearrangements should be selected so as to yield the simplest form of the characteristic matrix of E_M. All these rearrangements may be expressed by non-singular transformation matrices defined for the considered problem by the equation

$$[E_M - zE] =$$

$$= \begin{bmatrix} E & BR^{-1}B^T \\ \hline 0 & A \end{bmatrix} \begin{bmatrix} A + BR^{-1}B^T A^{-T} Q - zE & -BR^{-1}B^T A^{-T} \\ \hline -A^{-T} Q & A^{-T} - zE \end{bmatrix} \begin{bmatrix} E & 0 \\ \hline 0 & -z^{-1}E \end{bmatrix}$$
$$(16.75)$$

$$\det (E_M - zE) = \left| \begin{bmatrix} A - zE & BR^{-1}B^T \\ \hline -Q & A^T - z^{-1}E \end{bmatrix} \right| \quad (16.76)$$

Using the same procedure we now obtain for the Euler equations (16.65) and (16.66) the following expressions:

$$x(k+1) = A x(k) - BR^{-1}B^T \lambda(k) \quad (16.77)$$

$$\lambda(k+1) = -A^{-T} Q x(k+1) + A^{-T} \lambda(k) \quad (16.78)$$

Substituting for $x(k+1)$ in eqn. (16.78) the expression given by eqn. (16.77) we obtain

$$\lambda(k+1) = -A^{-T} QA\, x(k) + (A^{-T} QBR^{-1}B^T + A^{-T}) \lambda(k) \quad (16.79)$$

Equations (16.77) and (16.79) define the Euler matrix corresponding to eqns. (16.65) and (16.66):

$$E_M = \begin{bmatrix} A & -BR^{-1}B^T \\ \hline -A^{-T}QA & A^{-T}QBR^{-1}B^T + A^{-T} \end{bmatrix} \quad (16.80)$$

For the characteristic matrix follow the expressions

$$[E_M - zE] = \begin{bmatrix} E, & 0 \\ Q, & A^T \end{bmatrix} \begin{bmatrix} A - zE & -BR^{-1}B^T \\ -A^{-T}QA & A^{-T}QBR^{-1}B^T + A^T - zE \end{bmatrix} \begin{bmatrix} E, & 0 \\ 0, & -z^{-1}E \end{bmatrix} \tag{16.81}$$

$$\det(E_M - zE) = \left| \begin{bmatrix} A - zE & BR^{-1}B^T \\ -Q & A^T - z^{-1}E \end{bmatrix} \right| \tag{16.82}$$

Comparing the determinants (16.76) and (16.82) it is evident, that for either of the cases considered the characteristic polynomials, i.e. the determinants of the Euler matrices, have the same form and consequently the same eigenvalues.

In conclusion we can summarize:

THEOREM 16.2: *The optimum control of the time-invariant system* (16.6) *in the sense of the cost function* (16.1), *for N approaching to infinity and for a positive definite weighting matrix R, i.e.*

(1) $$u(k) = -(B^T PB + R)^{-1} B^T PA\, x(k)$$

where P satisfies eqn. (16.61), *i.e.*

$$P = A^T PA - A^T PB(B^T PB + R)^{-1} B^T PA + Q$$

is identical with the optimum control

(2) $$u(k) = -R^{-1} B^T P\, x(k)$$

where P satisfies eqn. (16.69), *i.e.*

$$PA - (A^{-T} + A^{-T}QBR^{-1}B^T) P - PBR^{-1}B^T P + A^{-T}QA = 0$$

Note 16.1: Denote the matrix satisfying eqn. (16.61) by P_1 and the matrix satisfying eqn. (16.69) by P_2. From Theorem 16.2 it follows that $P_2 \neq P_1$, whereas for the respective controlling vectors we have $u_1(k) = u_2(k)$, $k = 0, 1, 2, \ldots$, provided that all other conditions of optimum control are identical. Derive the relation between the two matrices P_1 and P_2 which would enable us to transform the Riccati equation (16.69) into the form (16.61).

Consider the matrix K given by eqn. (16.62) for the time-invariant system:

$$K = (B^T P_1 B + R)^{-1} B^T P_1 A =$$
$$= R^{-1}(B^T P_1 B + R - B^T P_1 B)(B^T P_1 B + R)^{-1} B^T P_1 A \tag{16.83}$$

where $R^{-1}(B^T P_1 B + R - B^T P_1 B)$ is an identity matrix. Hence

$$K = R^{-1}[E - B^T P_1 B(B^T P_1 B + R)^{-1}] B^T P_1 A =$$
$$= R^{-1} B^T [P_1 - P_1 B(B^T P_1 B + R)^{-1} B^T P_1] A \tag{16.84}$$

The term in square brackets corresponds, according to the matrix inversion lemma, to $(P_1^{-1} + BR^{-1}B^T)^{-1}$ so that

$$(B^T P_1 B + R)^{-1} B^T P_1 A = R^{-1} B^T (P_1^{-1} + BR^{-1}B^T)^{-1} A \tag{16.85}$$

Comparing the last expression with eqn. (16.67), it is evident, that

$$P_2 = (P_1^{-1} + BR^{-1}B^T)^{-1} A \tag{16.86}$$

This is the sought relation between the matrices P_1 and P_2. Inserting now the relation (16.86) for P in the Riccati equation (16.89), the original form (16.61) must be obtained.

To simplify this procedure it is useful to modify eqns. (16.61) and (16.69) into

$$P_1 A - A^{-T} P_1 - P_1 B(B^T P_1 B + R)^{-1} B^T P_1 A + A^{-T} Q = 0 \tag{16.87}$$

and

$$P_2 A - A^{-T} P_2 - A^{-T} Q B R^{-1} B^T P_2 - P_2 B R^{-1} B^T P_2 + A^{-T} Q A = 0 \tag{16.88}$$

respectively.

From eqn. (16.87) it follows that

$$P_1 B(B^T P_1 B + R)^{-1} B^T P_1 A = P_1 A - A^{-T} P_1 + A^{-T} Q \tag{16.89}$$

Using eqns. (16.84), (16.85) and (16.89) we obtain

$$R^{-1} B^T (P_1 A - P_1 A + A^{-T} P_1 - A^{-T} Q) = R^{-1} B^T A^{-T} (P_1 - Q) \tag{16.90}$$

Comparing the final result of eqn. (16.90) with the right-hand side of eqn. (16.67), it is obvious that

$$P_2 = A^{-T}(P_1 - Q) \tag{16.91}$$

This is a simple relation that is more suitable for substitution into the Riccati equation (16.88) than the previous equation (16.86). Using the form (16.91), we obtain

$$A^{-T}(P_1 - Q) A - A^{-2T}(P_1 - Q) - A^{-T} Q B R^{-1} B^T A^{-T}(P_1 - Q) -$$
$$- A^{-T}(P_1 - Q) B R^{-1} B^T A^{-T}(P_1 - Q) + A^{-T} Q A = 0$$
$$P_1 A - A^{-T} P_1 + A^{-T} Q - P_1 B R^{-1} B^T A^{-T}(P_1 - Q) = 0 \tag{16.92}$$

Since

$$R^{-1} B^T A^{-T}(P_1 - Q) = (B^T P_1 B + R)^{-1} B^T P_1 A \tag{16.93}$$

eqn. (16.92) can easily be modified into the Riccati equation of the form (16.87), which was to be proved. In this connection it is worth mentioning that the solution (16.67) relating to eqn. (16.69) is substantially simpler in comparison with the solution (16.32) corresponding to the Riccati equation (16.34). However, the restriction relating to the matrix R to be positive definite represents a very strong limitation. In Par. 16.5.2 attention will be given to the possibility to generalize the results of this paragraph also for the cases where matrices R are positive semidefinite.

16.3 Dynamic programming

16.3.1 General procedure

Considering again the same problem as in Sec. 16.1 and 16.2 we may designate optimum digital control as an N-stage decision process. If the optimum solution is determined by means of Bellman's dynamic programming, the results arrived at will necessarily be identically equivalent to those obtained in the preceding paragraphs. The dynamic programming method is generally known and is frequently described in specialized literature. We shall, therefore, indicate only the main solution steps which enable us to derive the desired results.

The minimum value of the quadratic cost function may be written as

$$f_{0,N} = \min_{u(0),\ldots,u(N-1)} \left\{ x^T(N) P(N) x(N) + \sum_{i=0}^{N-1} \left[x^T(i) Q(i) x(i) + u^T(i) R(i) u(i) \right] \right\}$$
(16.94)

This condition for the optimum control corresponds to the cost function (16.1). A more general form of the cost function is given by the expression

$$f_{k,N} = \min_{u(k),\ldots,u(N-1)} \left\{ x^T(N) P(N) x(N) + \sum_{i=k}^{N-1} \left[x^T(i) Q(i) x(i) + u^T(i) R(i) u(i) \right] \right\}$$
(16.95)

for $k = 0, 1, 2, \ldots, N - 1$. For $k = 0$, relation (16.95) reduces to eqn. (16.94) and the first vector $x(0)$ is determined by the initial conditions.

On the assumption that the value of the cost function corresponding to the first $k - 1$ stages is an optimum, the increment of this value in the remaining $N - k$ stages is equal to the optimum increment corresponding to the k-th stage and to the optimum increments in the subsequent $N - (k + 1)$ stages. This means that the optimum value of the cost function in the $(N - k)$-th stage is

$$f_{k,N} = \min_{u(k)} \left[x^T(k) Q(k) x(k) + u^T(k) R(k) u(k) + f_{k+1,N} \right] \quad (16.96)$$

CONTROLLER DESIGN BASED ON QUADRATIC COST FUNCTIONS

Since the cost function is a quadratic function in x, it may be estimated that

$$f_{k,N} = x^T(k) P(k) x(k) \tag{16.97}$$

for $k = 0, 1, 2, \ldots, N - 1$. With this relation eqn. (16.96) takes the form

$$f_{k,N} = \min_{u(k)} \left[x^T(k) Q(k) x(k) + u^T(k) R(k) u(k) + x^T(k + 1) P(k + 1) x(k + 1) \right] =$$
$$= \min_{u(k)} J_{k,N} \tag{16.98}$$

Substituting now for $x(k + 1)$ according to the state equation we obtain

$$J_{k,N} = \{ x^T(k) Q(k) x(k) + u^T(k) R(k) u(k) +$$
$$+ [x^T(k) A^T(k) + u^T(k) B^T(k)] P(k + 1) [B(k) u(k) + A(k) x(k)] \} \tag{16.99}$$

The derivative of this expression with respect to $u(k)$ is

$$\frac{\partial J_{k,N}}{\partial u(k)} = 2[R(k) + B^T(k) P(k + 1) B(k)] u(k) + 2B^T(k) P(k + 1) A(k) x(k)$$
$$\tag{16.100}$$

Setting the derivative (16.100) equal to zero we obtain the vector of controlling variables

$$u(k) = -[B^T(k) P(k + 1) B(k) + R(k)]^{-1} B^T(k) P(k + 1) A(k) x(k) =$$
$$= -K(k) x(k) \tag{16.101}$$

The corresponding Riccati equation may be derived by means of the relation (16.98) if we substitute (16.97) for $f_{k,N}$, (16.99) for $J_{k,N}$ and (16.101) for $u(k)$. We obtain

$$x^T(k) P(k) x(k) =$$
$$= x^T(k) Q(k) x(k) + x^T(k) A^T(k) P(k + 1) B(k) [B^T(k) P(k + 1) B(k) + R(k)]^{-1} \cdot$$
$$\cdot R(k) [B^T(k) P(k + 1) B(k) + R(k)]^{-1} B^T(k) P(k + 1) A(k) x(k) +$$
$$+ \{ x^T(k) A^T(k) - x^T(k) A^T(k) P(k + 1) B(k) [B^T(k) P(k + 1) B(k) + R(k)]^{-1} \cdot$$
$$\cdot B^T(k) \} P(k + 1) \{ -B(k) [B^T(k) P(k + 1) B(k) + R(k)]^{-1} \cdot$$
$$\cdot B^T(k) P(k + 1) A(k) x(k) + A(k) x(k) \} \tag{16.102}$$

Equation (16.102) must apply to any state vector $x(k)$ and we may, therefore, write that

$$P(k) = Q(k) + A^T(k) P(k + 1) B(k) [B^T(k) P(k + 1) B(k) + R(k)]^{-1} \cdot$$
$$\cdot \{ R(k) [B^T(k) P(k + 1) B(k) + R(k)]^{-1} - 2E + B^T(k) P(k + 1) B(k) \} \cdot$$
$$\cdot [B^T(k) P(k + 1) B(k) + R(k)]^{-1} B^T(k) P(k + 1) A(k) \} + A^T(k) P(k + 1) A(k)$$
$$\tag{16.103}$$

By several further rearrangements we obtain the already known form of the Riccati equation

$$P(k) = Q(k) - A^T(k) P(k+1) B(k) [B^T(k) P(k+1) B(k) + R(k)]^{-1} .$$
$$. B^T(k) P(k+1) A(k) + A^T(k) P(k+1) A(k) \qquad (16.104)$$

which we have derived in Par. 16.1.1 and 16.2.1 in a different manner.

16.3.2 Infinite number of control steps

Let us mention only one special case of control, i.e. the process with an infinite number of control steps, where $N \to \infty$. In this case eqn. (16.96) assumes the form

$$f_{k,\infty} = \min_{u(k)} \left[x^T(k) Q(k) x(k) + u^T(k) R(k) u(k) + f_{k+1,\infty} \right] \qquad (16.105)$$

and the expression (16.97) becomes

$$f_{k,\infty} = x^T(k) P x(k) \qquad (16.106)$$

where P is a constant matrix. Following the same procedure as used in Par. 16.3.1 we may arrive at the modified results with $P(k+1) = P(k) = P$. Since the Riccati equation (16.104) must be satisfied for an arbitrary k, and P is now a constant matrix, the weighting matrices $Q(k)$ and $R(k)$ cannot be chosen arbitrarily but, in view of the plant non-stationarity, in such a way that they satisfy the corresponding form of the Riccati equation, the fulfilment of which being a condition for achieving optimum control. For instance, if P is already known, we may select $R(k)$ for all values of k and calculate a $Q(k)$ satisfying the Riccati equation.

All results relating to time-invariant controlled plants and cost functions with constant weighting matrices follow directly from the general results in Par. 16.3.1.

16.4 Numerical solution

In the preceding Sections 16.1 to 16.3 we have derived analytic relations suitable for the determination of optimum control in the sense of the defined quadratic cost function but the numerical solution was not described. In the majority of practical cases the matrix P is calculated by means of a convenient numerical iterative method. In the case of time-variant plants with optimum control according to the cost function (16.1) it may be stated that at the end of the control process, i.e. at the stage N, the increment of the cost function is $x^T(N) Q(N) x(N)$ because the term corresponding to the vector of the controlling variables is inactive. According to (16.97) this increment is also $x^T(N) P(N) x(N)$. For $k = N$, therefore,

$$P(N) = Q(N) \qquad (16.107)$$

CONTROLLER DESIGN BASED ON QUADRATIC COST FUNCTIONS 351

If we know $P(N)$ we can, using eqn. (16.101), determine the feedback controller matrix $K(N-1)$, then we calculate the matrix $P(N-1)$ by means of the Riccati equation (16.104), then the matrix $K(N-2)$ using again eqn. (16.101), then we calculate $P(N-2)$, etc.

In the case of time-invariant plants, with constant weighting matrices of the cost function and with N approaching to infinity, we calculate the matrix P recursively from the Riccati equation by substituting an arbitrary non-singular matrix $P(k+1)$, e.g. $P_0 = E$, into eqn. (16.104) and obtain $P_1 = P(k)$. We repeat the calculation with the matrix $P(k+1) = P_1$, obtain $P(k) = P_2$, etc. As soon as $P(k+1) = P(k)$ with the desired accuracy, we regard the matrix $P(k)$ as the sought solution of P. Note that the Riccati equations (16.27) as well as (16.34) are symmetrical with respect to the matrix P so that for a symmetrical $Q(k)$ and $R(k)$ the matrix $P(k)$ is also symmetrical. On the other hand, the Riccati equation (16.69) as well as its time-variant version is not symmetrical with respect to the matrix P and, consequently, even for a symmetrical $Q(k)$ and $R(k)$, the matrix $P(k)$ will, in general, be unsymmetrical. For this reason the solution of eqn. (16.34) is considered to be less difficult than the solution of eqn. (16.69). The matrix P need not be positive definite, it may as well be positive semidefinite. This means, as we shall show in Par. 16.5.2, that, in a special case, P may even be equal to zero.

The usual solution procedure will now be shown by numerical examples.

Example 16.1: Calculate the feedback matrix k^T of a controller by the action of which the plant

$$x(k+1) = \begin{bmatrix} 1.5, & 1 \\ -0.5, & 0 \end{bmatrix} x(k) + \begin{bmatrix} 1 \\ 2 \end{bmatrix} u(k)$$

will be transferred from the initial state $x^T(0) = [x_1(0), x_2(0)]$, $x_1(0) = -1$, to the final state $x(N) = 0$. Assuming the knowledge of the complete state vector at all sampling instants, the controller has to satisfy the minimum of the cost function (16.1) with the weighting matrices

$$Q = \begin{bmatrix} 1, & 0 \\ 0, & 0 \end{bmatrix}, \quad R = 0 \quad \text{and} \quad R = 1$$

and with $N \to \infty$. The output variable is

$$y(k) = [1, 0] \, x(k)$$

Solution: We determine first, according to (15.9), the component $x_2(0)$ of the initial state:

$$\begin{bmatrix} -1 \\ x_2(0) \end{bmatrix} + \begin{bmatrix} 1.5, & 1 \\ -0.5, & 0 \end{bmatrix} \begin{bmatrix} -1 \\ x_2(0) \end{bmatrix} + \begin{bmatrix} 1 \\ 2 \end{bmatrix} u(0^-)$$

Solving this equation we find $x_2(0) = 0.5$ and $u(0^-) = 0$, so that the initial state vector

$$x(0^-) = \begin{bmatrix} -1 \\ 0.5 \end{bmatrix}$$

(a) *Case when* $R = 0$

Starting with the chosen matrix $P_0 = E$ we calculate, according to (16.34),

$$P_1 = \begin{bmatrix} 3.45, & 1.4 \\ 1.4, & 0.8 \end{bmatrix}$$

and after the 9-th iteration we already obtain the positive definite matrix P_9 identical with the matrix P_8 to five significant figures:

$$P_{8,9} = \begin{bmatrix} 1.6530, & 0.3733 \\ 0.3733, & 0.2133 \end{bmatrix}$$

By virtue of (16.32) we calculate the controller matrix

$$k^T = [0.8, \ 0.6]$$

The equation of dynamics of the control loop is

$$x(k+1) = (A - bk^T)x(k) = \begin{bmatrix} 0.7, & 0.4 \\ -2.1, & 1.2 \end{bmatrix} x(k)$$

Starting with $x(0)$ we calculate successively

$$x(0) = \begin{bmatrix} -1 \\ 0.5 \end{bmatrix}, \quad x(1) = \begin{bmatrix} -0.5 \\ 1.5 \end{bmatrix}, \quad x(2) = \begin{bmatrix} 0.25 \\ -0.75 \end{bmatrix}, \quad x(3) = \begin{bmatrix} -0.125 \\ 0.375 \end{bmatrix}$$

$$x(4) = \begin{bmatrix} 0.0625 \\ -0.1875 \end{bmatrix} \text{ etc.}$$

The value of the cost function is determined according to eqn. (16.25). We calculate

$$J_a = 1.3330$$

(b) *Case when* $R = 1$

By the same procedure we calculate in this case after five iteration cycles the positive definite matrix

$$P_{4,5} = \begin{bmatrix} 2.2402. & 0.7900 \\ 0.7900, & 0.5159 \end{bmatrix}$$

The controller matrix is now, according to (16.32),

$$k^T = [0.5695, \ 0.4514]$$

CONTROLLER DESIGN BASED ON QUADRATIC COST FUNCTIONS

and the matrix of dynamics of the closed control loop is

$$A^* = A - bk^T = \begin{bmatrix} 0.9305, & 0.5486 \\ -1.6390, & -0.9027 \end{bmatrix}$$

Using the same initial vector $x(0) = [-1, 0.5]^T$ we calculate

$$x(1) = \begin{bmatrix} -0.6562 \\ 1.1876 \end{bmatrix}, \quad x(2) = \begin{bmatrix} 0.0409 \\ 0.0034 \end{bmatrix}, \quad x(3) = \begin{bmatrix} 0.0399 \\ -0.0701 \end{bmatrix},$$

$$x(4) = \begin{bmatrix} -0.0013 \\ -0.0021 \end{bmatrix} \text{ etc.}$$

According to eqn. (16.25) we calculate the value of the cost function

$$J_b = 1.5791$$

Apparently, $J_b > J_a$ as might have been expected, since with $R = 1$ a constraint of the values of controlling variables has been introduced.

The solution of the Riccati equation by the indicated procedure leads in most practical cases to the desired aim already after a few iterations. Experience shows, however, that in certain cases, particularly in control loops with non-minimum phase plants and in ill-conditioned problems, the solution may converge slowly or not at all, although a solution does exist. This deficiency does not arise with the numerical procedure using the factorization into Cholesky's triangular matrices, which was already discussed in Sec. 13.3. For this purpose let us write the equation of the controlled plant in the form

$$x(k + 1) = [A(k), B(k)] \begin{bmatrix} x(k) \\ u(k) \end{bmatrix} = \Psi(k)\, \zeta(k) \quad (16.108)$$

where Ψ has the dimension $(n; n + r)$.

According to (16.96) and (16.97) we have

$$f_{k,N} = \min_{u_k} [x^T(k)\, Q(k)\, x(k) + 2x^T(k)\, S(k)\, u(k) + u^T(k)\, R(k)\, u(k) +$$

$$+ x^T(k + 1)\, P(k + 1)\, x(k + 1)] \quad (16.109)$$

where $k = 0, 1, \ldots, N - 1$. In eqn. (16.109) we have also made use of the mixed term with the matrix $S(k)$ as this term affects in no way the further procedure of the numerical solution.

We rewrite eqn. (16.109) in the form

$$f_{k,N} = \min_{u_k} [\zeta^T(k)\, \Phi(k)\, \zeta(k) + \zeta^T(k)\, \Psi^T(k)\, P(k + 1)\, \Psi(k)\, \zeta(k)] =$$

$$= \min_{u_k} \zeta^T(k) [\Phi(k) + \Psi^T(k)\, P(k + 1)\, \Psi(k)]\, \zeta(k) \quad (16.110)$$

where

$$\Phi(k) = \begin{bmatrix} Q(k), & S(k) \\ S(k), & R(k) \end{bmatrix} \tag{16.111}$$

is a matrix of the dimension $(n + r; n + r)$. The matrices $\Phi(k)$ and $P(k)$ may be expressed by means of Cholesky's triangular matrices. Omitting the argument at the individual matrices we have

$$\Phi + \Psi^T P \Psi = \Phi^{1/2} \Phi^{T/2} + \Psi^T P^{1/2} P^{T/2} \Psi =$$

$$= \begin{bmatrix} \Phi^{1/2}, & \Psi^T P^{1/2} \end{bmatrix} \begin{bmatrix} \Phi^{T/2} \\ P^{T/2} \Psi \end{bmatrix} = \Omega^{1/2} \Omega^{T/2} \tag{16.112}$$

so that

$$f_{k,N} = \min_{u_k} \| \Omega^{T/2}(k) \zeta(k) \|^2 \tag{16.113}$$

Here the matrix $\Omega^{T/2}$ of the quadratic form $\| \Omega^{T/2}(k) \zeta(k) \|^2$ is a lower triangular matrix whose individual blocks we denote according to the following pattern:

$$\Omega^{T/2}(k) \zeta(k) = \begin{bmatrix} \Omega_{QQ}^{T/2}(k) & 0 \\ \Omega_{RQ}^{T/2}(k) & \Omega_{RR}^{T/2}(k) \end{bmatrix} \begin{bmatrix} x(k) \\ u(k) \end{bmatrix}$$

where the dimensions of the individual matrices are $\Omega_{QQ}^{T/2}(n; n)$, $\Omega_{RQ}^{T/2}(r; n)$, $\Omega_{RR}^{2/T}(r; r)$. The triangulization of the matrix $[\Phi^{T/2}, P^{T/2} \Psi]^T$ is carried out by means of orthogonal transformations described in Appendix D. It is apparent that the minimum of the quadratic form in (16.113) may only be obtained by choosing such a value of $u(k)$ that the expression

$$\Omega_{RQ}^{T/2}(k) x(k) + \Omega_{RR}^{T/2}(k) u(k) = 0 \tag{16.114}$$

Hence

$$u(k) = -\Omega_{RR}^{-T/2}(k) \Omega_{RQ}^{T/2}(k) x(k) = -K(k) x(k) \tag{16.115}$$

Comparing eqn. (16.115) with eqn. (16.101), the equivalence of the two expressions is directly apparent. The value of $f_{k,N}$ in the $(N - k)$-th stage is then

$$f_{k,N} = \zeta^T(k) \Omega_{QQ}^{1/2}(k) \Omega_{QQ}^{T/2}(k) \zeta(k) \tag{16.116}$$

where

$$\Omega_{QQ}^{1/2}(k) \Omega_{QQ}^{T/2}(k) = P(k) \tag{16.117}$$

For a finite N and for a control loop with a time-variant plant we must know, for the numerical solution, $\Phi(k)$, $\Psi(k)$ for $k = 0, 1, \ldots, N - 1$, $\Phi(N) = Q(N)$ and the initial vector $x(0)$. According to the given procedure we start the solution with $k = N$ for which we use $P(N) = Q(N)$, $\Phi(N - 1)$ and $\Psi(N - 1)$ and determine $P(N - 1)$ and $K(N - 1)$. In the next stage with $P(N - 1)$, $\Phi(N - 2)$ and $\Psi(N - 2)$ we determine $P(N - 2)$ and $K(N - 2)$, etc., and finally we calculate $P(0)$ and $K(0)$.

Then, with the initial vector $x(0)$ and by means of the state equation, we determine successively $u(k)$, $x(k + 1)$ and $y(k)$ for $k = 0, 1, \ldots, N - 1$. For $N \to \infty$ and for control loops with time-invariant plants, $P(k) = P = \text{const}$. The matrices $\Phi(k) = \Phi$ and $\Psi(k) = \Psi$ are constant matrices as well. The matrices P and K are calculated iteratively. We start with the estimate $P_0(N) = \text{E}$, find $P_1^{T/2}$ by triangulization of the matrix $[\Phi^{T/2}, P^{T/2}\Psi]^T$, substitute $P_1^{T/2}$ for $P_0^{T/2} = \text{E}$ and repeat the calculation until $P_{k-1}^{T/2} = P_k^{T/2}$. Then the number of iteration steps is k and the sought solution is $P = P_k$ and $K = K_k$.

This procedure does not preclude the resultant matrix P to be positive semidefinite.

16.5 Analytical solution

16.5.1 Positive definite matrix R

Although in practice we always prefer the numerical methods described in the preceding section we shall still show, for the sake of completeness and elucidation of certain additional relationships, also the analytical procedure of the synthesis according to quadratic cost functions.

Let us first point out certain properties of Euler matrices:

THEOREM 16.3: *The eigenvalues of the Euler matrix* (16.74) *are symmetrically distributed in the complex z-plane with respect to the unit circle, with the centre in the origin, in that sense, that from the total number of 2n there are n eigenvalues stable, i.e.* $|z_i| < 1$, $i = 1, 2, \ldots, n$, *and n are unstable, i.e.* $|\lambda_i| > 1$, $i = 1, 2, \ldots, n$, *where* $\lambda_i = z_i^{-1}$, *provided that no eigenvalues* $z_i = 0$.

Proof: The determinant of the characteristic matrix will not change if the matrix is transposed. According to (16.76) we may write

$$\begin{vmatrix} A - z\text{E} & BR^{-1}B^T \\ -Q & A^T - z^{-1}\text{E} \end{vmatrix} = \begin{vmatrix} A^T - z\text{E} & -Q \\ BR^{-1}B^T & Q - z^{-1}\text{E} \end{vmatrix} \qquad (16.118)$$

Interchanging an even number of rows and columns, the sign of the determinant will not change so that

$$\begin{vmatrix} A^T - z\text{E} & -Q \\ BR^{-1}B^T & A - z^{-1}\text{E} \end{vmatrix} = \begin{vmatrix} A - z^{-1}\text{E} & BR^{-1}B^T \\ -Q & A^T - z\text{E} \end{vmatrix} \qquad (16.119)$$

Comparing now the first determinant in eqn. (16.118) with the last determinant in eqn. (16.119), it is apparent that the determinant and the corresponding characteristic matrix is invariant with respect to the change from z to z^{-1}. Thus, Theorem 16.3 is proved.

The homogeneous equation (16.70) has the general solution

$$p(k) = E_M^k\, p(0) = T J_{EM}^k T^{-1}\, p(0) \tag{16.120}$$

where J_{EM} is the Jordan matrix corresponding to the matrix E_M and T is a transformation matrix whose columns form a complete set of eigenvectors of the matrix E_M. According to Theorem 16.3 it is advantageous to arrange the eigenvalues in the matrix J_{EM} and the eigenvectors of the matrix E_M in such a manner that n stable eigenvalues and also the corresponding eigenvectors be located at the n first places. According to this recommendation we may introduce the notation

$$J_{EM} = \begin{bmatrix} J_{11}, & 0 \\ 0, & J_{22} \end{bmatrix} \tag{16.121}$$

$$T = \begin{bmatrix} T_{11}, & T_{12} \\ T_{21}, & T_{22} \end{bmatrix} \tag{16.122}$$

$$T^{-1} = T^* = \begin{bmatrix} T_{11}^*, & T_{12}^* \\ T_{21}^*, & T_{22}^* \end{bmatrix} \tag{16.123}$$

where all matrices with numerical subscripts have the dimension $(n; n)$ and the matrices E_M, J_{EM}, T, T^* have the dimension $(2n; 2n)$. Equation (16.120) may now be rewritten in the form

$$\begin{bmatrix} x(k) \\ \lambda(k) \end{bmatrix} = \begin{bmatrix} T_{11}, & T_{12} \\ T_{21}, & T_{22} \end{bmatrix} \begin{bmatrix} J_{11}^k, & 0 \\ 0, & J_{22}^k \end{bmatrix} \begin{bmatrix} T_{11}^*, & T_{12}^* \\ T_{21}^*, & T_{22}^* \end{bmatrix} \begin{bmatrix} x(0) \\ \lambda(0) \end{bmatrix} \tag{16.124}$$

from which follow the following two equations:

$$x(k) = (T_{11} J_{11}^k T_{11}^* + T_{12} J_{22}^k T_{21}^*)\, x(0) + (T_{11} J_{11}^k T_{12}^* + T_{12} J_{22}^k T_{22}^*)\, \lambda(0) \tag{16.125}$$

$$\lambda(k) = (T_{21} J_{11}^k T_{11}^* + T_{22} J_{22}^k T_{21}^*)\, x(0) + (T_{21} J_{11}^k T_{12}^* + T_{22} J_{22}^k T_{22}^*)\, \lambda(0) \tag{16.126}$$

Since a stable solution is required, it is necessary to eliminate in eqn. (16.125) all terms multiplied by the unstable block J_{22}^k. Therefore

$$T_{21}^*\, x(0) + T_{22}^*\, \lambda(0) = 0 \tag{16.127}$$

For the given $x(0)$ we have from eqn. (16.127)

$$\lambda(0) = -(T_{22}^*)^{-1} T_{21}^*\, x(0) \tag{16.128}$$

CONTROLLER DESIGN BASED ON QUADRATIC COST FUNCTIONS

Comparing eqn. (16.55) with eqn. (16.128), it is evidently

$$P = -(T_{22}^*)^{-1} T_{21}^* \tag{16.129}$$

Now the vector of the state variables is calculated from eqn. (16.125)

$$x(k) = T_{11} J_{11}^k [T_{11}^* x(0) + T_{12}^* \lambda(0)] \tag{16.130}$$

With $\lambda(0)$ according to (16.128) we obtain finally

$$x(k) = T_{11} J_{11}^k [T_{11}^* - T_{12}^*(T_{22}^*)^{-1} T_{21}^*] x(0) \tag{16.131}$$

By means of eqns. (16.129) and (16.131) we may, on substitution into eqn. (16.32) or into eqn. (16.67), depending on the matrix E_M used, express explicitly the vector of the controlling variables $u(k)$.

Expressing $\lambda(k)$ from eqn. (16.126)

$$\lambda(k) = T_{21} J_{11}^k [T_{11}^* x(0) + T_{12}^* \lambda(0)] \tag{16.132}$$

and substituting into this expression $\lambda(0)$ from eqn. (16.128) and $x(0)$ from eqn. (16.131) we also obtain

$$P = T_{21} T_{11}^{-1} \tag{16.133}$$

Using the matrices (16.122) and (16.123), it may easily be proved that

$$T_{21} T_{11}^{-1} = -(T_{22}^*)^{-1} T_{21}^* \tag{16.134}$$

The above procedure is a modification of the spectral factorization and elimination of unstable components of the classical Wiener-Kolmogorov filtration [65.3].

Note 16.2: In Par. 16.2.2 it was proved that the eigenvalues of the Euler matrices (16.74) and (16.80) are identical. On the other hand the two expressions (16.129) and (16.133) must give for each of these Euler matrices, in general, different results satisfying eqn. (16.86) or eqn. (16.91). For this reason the eigenvectors of the Euler matrices (16.74) and (16.80) must be different although the eigenvalues of both these matrices are identical.

THEOREM 16.4: *If R in eqn. (16.1) is a positive definite matrix, the eigenvalues of the matrix of dynamics of a closed control loop are also eigenvalues of the Euler matrix.*

Proof: Substituting the vector of controlling variables (16.67) into the equation of plant dynamics we obtain the closed control loop matrix of dynamics

$$A^* = A - BR^{-1} B^T P \tag{16.135}$$

by virtue of which the Riccati equation (16.88) takes the form

$$PA^* = \left(A^{-T} + A^{-T}QBR^{-1}B^T\right)P - A^TQA \qquad (16.136)$$

Let the matrix V transform the matrix A^* into the Jordan form so that

$$A^* = VJ_{A^*}V^{-1} \qquad (16.137)$$

and let, according to eqn. (16.133),

$$P = UV^{-1} \qquad (16.138)$$

Substituting now eqns. (16.137) and (16.138) into eqns. (16.135) and (16.136), respectively, we obtain

$$VJ_{A^*} = AV - BR^{-1}B^TU \qquad (16.139)$$

$$UJ_{A^*} = \left(A^{-T} + A^{-T}QBR^{-1}B^T\right)U - A^{-T}QAV \qquad (16.140)$$

or

$$\begin{bmatrix} V \\ U \end{bmatrix} J_{A^*} = E_M \begin{bmatrix} V \\ U \end{bmatrix}; \quad \begin{bmatrix} V \\ U \end{bmatrix} = N \qquad (16.141)$$

Let $a_1, a_2, \ldots, a_n$ be the columns of the matrix $N(2n; n)$ and let z_i, $i = 1, 2, \ldots, n$, be the eigenvalues of the matrix A^*. For distinct eigenvalues we have, according to eqn. (16.141),

$$a_i z_i = E_M a_i, \quad i = 1, 2, \ldots, n \qquad (16.142)$$

In accordance with (16.142) z_i are evidently also the eigenvalues of the Euler matrix E_M and a_i are its appropriate eigenvectors, Q.E.D. Since, by convention, the eigenvalues z_i for a closed control loop are stable, the matrices U and V in eqn. (16.141) correspond to T_{21} and T_{11} in eqn. (16.133), respectively.

Similarly, for multiple eigenvalues z_i it holds that

$$a_i z_i = E_M a_i$$

$$a_i + a_{i+1} z_i = E_M a_{i+1}$$

$$\vdots$$

$$a_{i+p_i-2} + a_{i+p_i-1} z_i = E_M a_{i+p_i-1} \qquad (16.143)$$

where p_i denotes the multiplicity of z_i and the dimension of the relevant Jordan block. Since V, being a transformation matrix, must be non-singular, $a_i \neq 0$, z_i is also an eigenvalue of the Euler matrix E_M and $a_i, a_{i+1}, \ldots, a_{i+p_i-1}$ is the corresponding chain of length p_i of the generalized eigenvectors. The matrix P may again be expressed by eqn. (16.133) or eqn. (16.138).

In Secs. 16.1 and 16.2 we supposed that the solution P of the Riccati equation is positive definite in order that we could show the relation of the two modifications (16.61) and (16.69) of the Riccati equation. This assumption, however, is not necessary if we use for the controller design a procedure not requiring such an assumption. An example of this kind of controller design is the solution resulting from the Riccati equation (16.61), whose analytical derivation given in Sec. 16.2 and 16.3, as well as the numerical solution described in Sec. 16.4, can be carried out without this assumption.

At this point it is useful to note that the knowledge gained so far is supplemented and generalized by the theorem formulated and proved as a sufficient condition by W. M. Wonham [68.10] for the continuous model of a plant in state space representation. However, the necessary and sufficient condition for the continuous and discrete version was published first by V. Kučera [72.5, 72.7].

THEOREM 16.5: *Let $BR^{-1}B^T = B^*B^{*T}$ and $Q = C^*C^{*T}$ where the matrices B^* and C^* have a full rank so that $h(B^*) = h(B)$ and $h(C^*) = h(Q)$. Here Q and R are symmetrical matrices, R is a positive definite matrix and Q a positive semi-definite matrix. Then the stabilizability of the pair (A, B^*) and the detectability of the pair (C^{*T}, A) are necessary and sufficient conditions for the unique solution of P of the Riccati equation (16.61) ensuring a stable closed control loop, where P is not a negative definite matrix.*

The following lemmata may be added to Theorem 16.5:

LEMMA 16.1: *It holds that*

$$E_M \begin{bmatrix} 0 \\ w_i \end{bmatrix} = z_i^{-1} \begin{bmatrix} 0 \\ w_i \end{bmatrix}$$

if and only if z_i is a non-zero eigenvalue of the pair (A, B^*) to which corresponds a non-controllable component.

It should be noted that the eigenvalue z_i mentioned in Lemma 16.1 is briefly denominated as the uncontrollable eigenvalue.

Proof: Substituting for E_M from eqn. (16.74) and introducing the substitutions according to Theorem 16.5 we have

$$\begin{bmatrix} -B^*B^{*T}A^{-T}w_i \\ A^{-T}w_i \end{bmatrix} = z_i^{-1} \begin{bmatrix} 0 \\ w_i \end{bmatrix}$$

which is an expression equivalent to the relations

$$w_i^T A = z_i^{-1} w_i^T$$
$$w_i^T B^* = 0$$

This is the proof of the necessary condition and the proof of the sufficient condition is trivial, Q. E. D.

LEMMA 16.2: *It holds that*

$$E_M \begin{bmatrix} q_i \\ 0 \end{bmatrix} = z_i \begin{bmatrix} q_i \\ 0 \end{bmatrix}$$

if and only if z_i is a non-zero eigenvalue of the pair (C^{*T}, A) to which corresponds an unobservable component.

It should again be noted that the eigenvalue z_i of Lemma 16.2 is briefly denominated as the unobservable eigenvalue.

Proof: Using the same procedure as in the preceding lemma we calculate

$$\begin{bmatrix} Aq_i + B^*B^{*T}A^{-T}C^*C^{*T}q_i \\ -A^{-T}C^*C^{*T}q_i \end{bmatrix} = z_i \begin{bmatrix} q_i \\ 0 \end{bmatrix}$$

which is an expression equivalent to the relations

$$Aq_i = z_i q_i$$
$$C^{*T}q_i = 0$$

This again is the proof of the necessary condition and the proof of the sufficient condition is again trivial, Q. E. D.

The result, which is already classical to-day, expressed by Theorem 16.5, was generalized and proved by V. Kučera [72.5, 72.7, 72.8] even for such cases where the pair (C^{*T}, A) has ϱ non-detectable, i.e. unstable and unobservable eigenvalues being elements of the set $\mathscr{S} = \{z_1, z_2, ..., z_\varrho\}$, $\varrho \geq 0$ whose subset is $\mathscr{S}_\alpha$, $\alpha = 1, 2, ..., \varrho$. The following theorem holds:

THEOREM 16.6: *Assume that a solution of eqn. (16.61) stabilizing the pair (A, B) exists. Then the solution P_α generated by the subsets $\mathscr{S}_\alpha$ of the set $\mathscr{S}$ forms a set of all semidefinite solutions of eqn. (16.61).*

COROLLARY: *Assume that there exists a solution of eqn. (16.61) stabilizing the pair (A, B) and that there are ϱ non-detectable eigenvalues $z_1, z_2, ..., z_\varrho$ corresponding to the pair (C^{*T}, A), $\varrho \geq 0$. Then*

(a) the number of semidefinite solutions of eqn. (16.61) is just 2^ϱ if the eigenvalues z_i, $i = 1, 2, ..., \varrho$, are cyclic and

(b) the number of semidefinite solutions of eqn. (16.61) is uncountable when any of the eigenvalues z_i is non-cyclic.

It should be noted that a cyclic eigenvalue is the eigenvalue of a cyclic matrix, which means that it lies in only one block of the Jordan form of this matrix.

Proof: (a) This statement follows directly from Theorem 16.6 since there exist just 2^ϱ subsets of the set of ϱ elements.

CONTROLLER DESIGN BASED ON QUADRATIC COST FUNCTIONS

(b) If any of the eigenvalues is non-cyclic, there will be an infinite number of possibilities of choosing independent eigenvectors and case (b) will apply to each of these possibilities, Q. E. D.

It should be emphasized that only one solution of all the positive semidefinite solutions mentioned in Theorem 16.6 and in Corollary 16.1 is stabilizing. All other solutions are unstable. In Theorem 16.6 and in Corollary 16.1 it is assumed that a stabilizing solution does exist. In other words, it is also admitted that a solution of the Riccati equation may exist but may not ensure a stable control loop, or that a solution of the Riccati equation may not exist at all. A case may also occur, where the optimum solution ensuring a minimum cost function is not stabilizing and vice versa. These cases have been discussed in the literature [73.13]. In practical cases, however, the optimum and stabilizing solutions are usually identical.

Example 16.2: Let us find the solution of the Riccati equation by means of the Euler matrix and its eigenvectors for the plant of Example 16.1 and for the weighting matrices

$$Q = \begin{bmatrix} 1, & 0 \\ 0, & 0 \end{bmatrix}, \quad R = 1$$

Solution: The Euler matrix (16.74) is

$$E_M = \begin{bmatrix} -2.5, & 1 & 4, & -7 \\ -8.5, & 0 & 8, & -14 \\ \hline 0, & 0 & 0, & 1 \\ 2, & 0 & -2, & 3 \end{bmatrix}$$

The eigenvalues of this matrix are

$$z_{1,2} = 0.0139 \pm j0.2427$$

$$z_{3,4} = 0.2361 \pm j4.1075$$

The first two eigenvalues $z_{1,2}$ are within the stable region and for these values we determine the eigenvectors

$$[a_1, a_2] = \begin{bmatrix} -0.5593 - j0.1481, & -0.5593 + j0.1481 \\ 1, & 1 \\ \hline -0.4630 - j0.3317, & -0.4630 + j0.3317 \\ 0.0740 - j0.1170, & 0.0740 + j0.1170 \end{bmatrix} = \begin{bmatrix} V \\ U \end{bmatrix}$$

$$V^{-1} = \begin{bmatrix} j3.3769, & 0.5 + j1.8888 \\ -j3.3769, & -0.5 + j1.8888 \end{bmatrix}$$

and finally, according to eqn. (16.138), we obtain

$$P = \begin{bmatrix} 2.2402, & 0.7900 \\ 0.7900, & 0.5159 \end{bmatrix}$$

which is the same result as that obtained in Example 16.1. As regards the numerical solution, however, the procedure using the Euler matrix and its eigenvectors is much less convenient, particularly because of the need to determine the Euler matrix eigenvectors.

16.5.2 Plant output optimization

In certain problems of automatic control it is required to minimize the sum of squares of the discrete output variable of the system, i.e. the cost function

$$J = \sum_{k=0}^{\infty} y^2(k) \tag{16.144}$$

In contrast with the quadratic cost function forms shown so far the, weighting matrix is now evidently $R = 0$ so that the previous results cannot be directly applied. The solution of this problem in the state space was first published by V. Kučera [72.6].

If the relative order of the controlled plant is m, the plant input variable can act at the output only after m sampling intervals. Therefore the cost function (16.144) reaches its minimum if

$$J_m = \sum_{k=0}^{\infty} y^2(k + m) \tag{16.145}$$

has the minimum value.

Using the second of eqns. (15.102) we may determine

$$y(k + m) = c^T A^m x(k) + h_m u(k) \tag{16.146}$$

and on substitution of this expression into eqn. (16.145) we obtain

$$J_m = \sum_{k=0}^{\infty} [x^T(k) Q x(k) + 2 x^T(k) S u(k) + R u^2(k)] \tag{16.147}$$

where

$$Q = (A^T)^m cc^T A^m$$
$$S = (A^T)^m ch_m$$
$$R = h_m^2 > 0 \tag{16.148}$$

Using the substitution

$$u(k) = \tilde{u}(k) - R^{-1} S^T x(k) \tag{16.149}$$

CONTROLLER DESIGN BASED ON QUADRATIC COST FUNCTIONS

we eliminate in eqn. (16.147) the terms in square brackets with the matrices S and Q. The cost function (16.147) changes into

$$J_m = \sum_{k=0}^{\infty} h_m^2 \tilde{u}^2(k) \tag{16.150}$$

The matrix of dynamics of the closed control loop with the controlling variable (16.149) follows from the equation of dynamics of the control loop:

$$x(k+1) = (A - bR^{-1}S^T) x(k) + b\tilde{u}(k) \tag{16.151}$$

According to (16.148) we have

$$bR^{-1}S^T = bh_m^{-1}c^T A^m \tag{16.152}$$

so that the matrix of dynamics of the closed control loop is

$$\hat{A}_m = A - bh_m^{-1}c^T A^m \tag{16.153}$$

and is identical with the matrix of dynamics of the inverse system (15.102) and with the matrix in (15.104), respectively. The Riccati equation now assumes the form

$$P = \hat{A}_m^T P \hat{A}_m - \hat{A}_m^T P b (h_m^2 + b^T P b)^{-1} b^T P \hat{A}_m \tag{16.154}$$

For $m > 0$ no inversion of $\hat{A}_m$ exists so that the Euler matrix E_M cannot be written down. Nevertheless, as known from the clasical theory of control, a solution does exist. Let us, however, introduce the generalized inversion $\hat{A}_m^\#$ of the matrix $\hat{A}_m$ for which it holds that

$$\hat{A}_m \hat{A}_m^\# q = q \tag{16.155}$$

$$\hat{A}_m^\# w = 0 \tag{16.156}$$

where q is an arbitrary vector from the space spanned by the eigenvectors corresponding to the non-zero stable and unstable eigenvalues of the matrix $\hat{A}_m$. On the other hand, w is an arbitrary vector from the space spanned by the eigenvectors corresponding to the zero eigenvalues. The Euler matrix (16.74) assumes now the form

$$E_M = \begin{bmatrix} \hat{A}_m, & -bh_m^{-2}b^T \hat{A}_m^{\#T} \\ 0, & \hat{A}_m^{\#T} \end{bmatrix} \tag{16.157}$$

Introducing, in addition to the column eigenvectors q_i, the row eigenvectors r_i^T corresponding to the non-multiple eigenvalues z_i it holds, for $z_i \neq 0$, that

$$\hat{A}_m q_i = z_i q_i$$
$$r_i^T \hat{A}_m = z_i r_i^T$$
$$\hat{A}_m^\# q_i = z_i^{-1} q_i$$
$$r_i^T \hat{A}_m^\# = z_i^{-1} r_i^T \tag{16.158}$$

It may easily be verified that with the Euler matrix (16.157) and with $z_i \neq 0$ it also holds that

$$E_M \begin{bmatrix} q_i \\ 0 \end{bmatrix} = z_i \begin{bmatrix} q_i \\ 0 \end{bmatrix} \qquad (16.159)$$

It should be noted that, according to Lemma 16.2, an unobservable mode corresponds to the eigenvalues z_i. Also

$$E_M \begin{bmatrix} R_i r_i \\ r_i \end{bmatrix} = z_i^{-1} \begin{bmatrix} R_i r_i \\ r_i \end{bmatrix} \qquad (16.160)$$

where

$$R_i = (z_i \hat{A}_m - E)^{-1} b h_m b^T \qquad (16.161)$$

From the equality

$$(z_i \hat{A}_m - E)^{-1} = -z_i^{-1}(z_i^{-1}E - \hat{A}_m)^{-1}$$

it also follows that the matrix $(z_i \hat{A}_m - E)^{-1}$ is non-singular if z_i^{-1} is not an eigenvalue. Here the matrix $\hat{A}_m$ always has m zero eigenvalues z_i corresponding to the chain of generalized eigenvectors

$$A^{-1}b, A^{-2}b, \ldots, A^{-m}b$$

for which we have

$$E_M \begin{bmatrix} A^{-1}b \\ 0 \end{bmatrix} = \begin{bmatrix} \hat{A}_m A^{-i}b \\ 0 \end{bmatrix} = \begin{bmatrix} AA^{-i}b - bh_m^{-1}(c^T A^{m-i}b) \\ 0 \end{bmatrix} = \begin{bmatrix} A^{-(i-1)}b \\ 0 \end{bmatrix}$$

$$i = 2, 3, \ldots, m$$

since, by convention, $(c^T A^{m-i}b)$ is equal to zero. But for $i = 1$, $c^T A^{m-1}b = h_m$ and consequently

$$E_M \begin{bmatrix} A^{-i}b \\ 0 \end{bmatrix} = \begin{bmatrix} 0 \\ 0 \end{bmatrix}$$

The modes corresponding to the chain of generalized eigenvectors appertaining to zero eigenvalues are, according to Lemma 16.2, also unobservable. Similarly, it also may be proved that

$$E_M \begin{bmatrix} 0 \\ w_i \end{bmatrix} = \begin{bmatrix} 0 \\ 0 \end{bmatrix}, \quad i = 1, 2, \ldots, m$$

for $\hat{A}_m^{\#T} w_i = 0$. It also may be noted that the eigenvectors of the form

$$\begin{bmatrix} 0 \\ w_i \end{bmatrix}$$

cannot generate a solution since in this case V^{-1} in eqn. (16.138) does not exist.

Let us now consider the solution of the Riccati equation. First of all it is clear from eqn. (16.154) that the solution $P = 0$, by virtue of which the cost function (16.25) assumes the value $J_m = 0$, always exists. From eqn. (16.150) it then follows that $\tilde{u} = 0$ and, finally, from eqn. (16.149) the controlling variable is

$$u(k) = -R^{-1}S^T x(k) \qquad (16.162)$$

Using eqn. (16.153) we also obtain

$$u(k) = -h_m^{-1} c^T A^m x(k) \qquad (16.163)$$

Since the matrix of dynamics of the closed control loop is equal to the matrix of dynamics of the inverse plant, as it follows from a comparison of eqns. (15.102) and (15.104), the solution with $P = 0$ is stable only if the inverse plant is stable or, in other words, if the plant is a so-called minimum-phase plant. In this case the optimum solution in the sense of the cost function (16.144) is identical with the optimum solution in the sense of the weak version of finite number of control steps.

In this connection it is in place to raise the question whether the solution $P = 0$ is the only solution of the Riccati equation (16.154) that satisfies the given problem. Assume that the plant (A, B, C, D) is stabilizable. Then the pair (A, B^*) is stabilizable as well. However, since $Q = 0$, the pair (C^{*T}, A) can never be observable. We may distinguish between the following two cases:

(a) If $\hat{A}_m$ is stable and the pair (C^{*T}, A) is detectable, then $P = 0$ is the only positive semidefinite solution of the given problem.

(b) If $\hat{A}_m$ is unstable, that is if (A, B, C, D) is not a minimum-phase plant, and if the pair (C^{*T}, A) is non-detectable, i.e. unstable and unobservable, then the following two semidefinite solutions of the Riccati equation (16.154) will exist:

(ba) $P = 0$ which, although it is optimum, does not ensure a stable loop,

(bb) $P \neq 0$, P is positive semidefinite, which is optimum and ensures a stable loop.

To find these solutions we must select for the Euler matrix only stabilizing eigenvectors corresponding to stable eigenvalues of the Euler matrix in order to be able to determine the matrix N defined in (16.141) and use it to obtain the sought solution (16.138). Denoting the column eigenvectors corresponding to stable non-zero eigenvalues by q_i^+, and the row eigenvectors corresponding to inverse unstable eigenvalues of the matrix $\hat{A}_m$ by $(r_i^-)^T$, we obtain, in accordance with Lemma 16.2, the relation

$$N = \begin{bmatrix} V \\ U \end{bmatrix} = \begin{bmatrix} A^{-1}b, & \ldots, & A^{-m}b, & q_1^+, & \ldots, & q_s^+, & R_1 r_1^-, & \ldots, & R_{n-m-s} r_{n-m-s}^- \\ 0, & \ldots, & 0, & 0, & \ldots, & 0, & r_1^-, & \ldots, & r_{n-m-s}^- \end{bmatrix}$$

$$(16.164)$$

where m is the number of zero eigenvalues, s the number of non-zero stable eigenvalues and $(n - m - s)$ the number of unstable eigenvalues of the matrix $\hat{A}_m$ of the dimension $(n; n)$, respectively.

From the procedure indicated it is evident that a pseudoinversion satisfying eqns. (16.155) and (16.156) need not be determined and that it is sufficient to find the eigenvalues of $\hat{A}_m$ and then the corresponding eigenvectors by means of eqn. (16.164).

With the matrix P determined by eqn. (16.138) we may now calculate a component of the controlling variable $u(k)$, e.g. according to eqn. (16.62) rewritten in the form

$$\tilde{u}(k) = -(b^T P b + h_m^2)^{-1} b^T P \hat{A}_m x(k) = -\tilde{k} \, x(k) \tag{16.165}$$

The controlling variable $u(k)$ is finally determined according to eqn. (16.149):

$$u(k) = -[(b^T P b + h_m^2)^{-1} b^T P \hat{A}_m + h_m^{-1} c^T A^m] x(k)$$
$$= -k^T x(k) \tag{16.166}$$

In this case the matrix of dynamics of the closed control loop has the form

$$\hat{A}_m^* = \hat{A}_m - b(b^T P b + h_m^2)^{-1} b^T P \hat{A}_m \tag{16.167}$$

Although a generalization of the above procedure for multi-input/multi-output systems is basically possible, it requires an inversion of the multi-input/multi-output plant. This fact leads to further complexities. Since the solution by means of the Riccati equation is for many reasons simpler and more useful than that using the Euler matrix, as it follows from a comparison of the procedures described in Sec. 16.4 and Sec. 16.5 and as it also was intended to show, the procedure for a single-input/single-output plant given in Par. 16.5.2 will not be generalized here for multi-input/multi-output plants. We only mention that the inversion of a multi-input/multi-output plant is discussed in the literature [65.1, 69.8, 69.10].

Example 16.3: Let us solve Example 16.1 with the weighting matrix $R = 0$ using the procedure given in Par. 16.5.2:

$$A = \begin{bmatrix} 1.5, & 1 \\ -0.5, & 0 \end{bmatrix}, \quad b = \begin{bmatrix} 1 \\ 2 \end{bmatrix}, \quad c^T = [1, \, 0], \quad d = 0, \quad Q = \begin{bmatrix} 1, & 0 \\ 0, & 0 \end{bmatrix}, \quad R = 0$$

Solution: We determine first the relative order of the plant according to eqn. (15.96):

$$h_0 = d = 0$$

$$h_1 = c^T A^0 b = [1, \, 0] \begin{bmatrix} 1 \\ 2 \end{bmatrix} = 1$$

CONTROLLER DESIGN BASED ON QUADRATIC COST FUNCTIONS

Since $h_1 \neq 0$, the relative order $m = 1$. According to eqn. (16.148) we determine

$$Q^* = A^T cc^T A = \begin{bmatrix} 2.25, & 1.5 \\ 1.5, & 1 \end{bmatrix}$$

$$S^* = A^T c h_1 = \begin{bmatrix} 1.5 \\ 1 \end{bmatrix}$$

$$R^* = h_1^2 = 1$$

Now we eliminate in the modified cost function (16.147) the term including the matrix S^* by means of the relations (16.4) through (16.9):

$$\hat{Q} = Q^* - S^* R^{*-1} S^{*T} = 0$$

$$\hat{R} = R^*$$

$$\hat{A}_1 = A - b R^{*-1} S^{*T} = \begin{bmatrix} 0, & 0 \\ -3.5, & -2 \end{bmatrix}$$

$$\hat{b} = b$$

$$\hat{c}^T = c^T - R^{*-1} S^{*T} = [-0.5, \; -1]$$

The eigenvalues of the matrix $\hat{A}_1$ are $z_1 = 0$ and $z_2 = -2$. The eigenvector corresponding to the eigenvalue $z_1 = 0$ is

$$A_1^{-1} b = \begin{bmatrix} 1.5, & 1 \\ -0.5, & 0 \end{bmatrix}^{-1} \begin{bmatrix} 1 \\ 2 \end{bmatrix} = \begin{bmatrix} 0, & -2 \\ 1, & 3 \end{bmatrix} \begin{bmatrix} 1 \\ 2 \end{bmatrix} = \begin{bmatrix} -4 \\ 7 \end{bmatrix}$$

The row eigenvector corresponding to the eigenvalue $z_2 = -2$ is determined according to eqns. (16.158). From the condition

$$[r_{21}, r_{22}] \begin{bmatrix} 0, & 0 \\ -3.5, & -2 \end{bmatrix} = -2 [r_{21}, r_{22}]$$

it follows that

$$r_2^T = [1.75, \; 1]$$

Using eqns. (16.161) and (16.160) we determine successively

$$R_1 = \begin{bmatrix} -1, & -2 \\ 3, & 6 \end{bmatrix}$$

$$R_1 r_2 = \begin{bmatrix} -3.75 \\ 11.25 \end{bmatrix}$$

Now we can set up the matrix (16.164)

$$N = \begin{bmatrix} V \\ U \end{bmatrix} = \begin{bmatrix} -4, & -3.75 \\ 7, & 11.25 \\ 0, & 1.75 \\ 0, & 1 \end{bmatrix}$$

and the solution of the Riccati equation (16.154) is, according to eqn. (16.138),

$$P = UV^{-1} = \begin{bmatrix} 0, & 1.75 \\ 0, & 1 \end{bmatrix} \begin{bmatrix} -4, & -3.75 \\ 7, & 11.25 \end{bmatrix}^{-1} = \begin{bmatrix} 0.6533, & 0.3733 \\ 0.3733, & 0.2133 \end{bmatrix}$$

Substituting into eqn. (16.162) we obtain the controller matrix

$$k^T = [0.8, \ 0.6]$$

and finally, using eqn. (16.167), we may calculate the matrix of dynamics of the closed control loop

$$\hat{A}_1^* = \begin{bmatrix} 0.7, & 0.4 \\ -2.1, & -1.2 \end{bmatrix}$$

The last two results agree with the solution of Example 16.1.

In conclusion, it should be noted that relations (16.158) are satisfied by the matrix

$$\hat{A}_1^* = \begin{bmatrix} 0, & 0 \\ -\tfrac{7}{8}, & -\tfrac{1}{2} \end{bmatrix}$$

and, for instance, by the eigenvectors $q^T = [0, 1]$ and $r_2^T = [1.75, 1]$.

The matrix $\hat{A}_1^{\#}$ has the eigenvalues $z_3 = 0$ and $z_4 = -0.5$. On comparison with the eigenvalues of $\hat{A}_1$ it is apparent that the matrix $\hat{A}_1^{\#}$ has also one zero eigenvalue and that the non-zero eigenvalue is $z_4 = z_2^{-1}$. Although the matrix $\hat{A}_1^{\#}$ may be determined by means of eqns. (16.155) and (16.156), this matrix is not needed for carrying out the synthesis.

16.6 Minimum sum of squares of the control error

In the foregoing sections we have been seeking the optimum control according to a linear combination of the sum of squares of the state variables and of the controlling variables in the sense of the cost function (16.1). In this section we shall determine the optimum control according to the sum of squares of the control errors. The discussion will be confined to single-input/single-output systems. The generalization of results for more complex systems makes no special difficulties. Let us define the control error

$$\varepsilon(k) = r(k) - y(k) \tag{16.168}$$

and the cost function

$$J = \sum_{k=0}^{\infty} [\sum_{i=0}^{\varkappa} \gamma_i \, \varepsilon_i^2(k) + R \, u^2(k)], \quad \varkappa \leq n - 1 \qquad (16.169)$$

where γ_i are weighting coefficients and $\varepsilon_i(k)$ is the value of the i-th derivative of the control error at the instant $k = 0, 1, 2, \ldots$. In eqn. (16.168) $r(k)$ denotes the command variable expressed by eqn. (15.63) and $y(k)$ is the output variable of a single-input/single-output plant. We know that $r(k)$ is generated by a set of homogeneous first-order differential equations having in the state space the form

$$\omega^{(1)}(t) = V \omega(t)$$

$$r(t) \;\;\;\; = c_V^T \, \omega(t) \qquad (16.170)$$

where $\omega(t) \in \Omega^m$ has as components the derivatives $r^{(i)}(t)$, $i = 0, 1, \ldots, m - 1$, and where $c_V^T = [1, 0, \ldots, 0]$. The components of the discrete form of the vector $\omega(t)$, i.e. of the vector $\omega(k)$, represent therefore the values of the derivatives $\omega^{(i)}(t)$, $i = 0, 1, \ldots, m - 1$, at the individual instants $k = 0, 1, \ldots$. The state space representation of certain command variables used most frequently for deterministic control was given in Par. 15.2.2.

Expressing the controlled plant by state space representation in such a way that the components of the state vector $x(t)$ are derivatives of the output variable $y(t)$, and if $x \in X^n$, $n \geq m$, then the components of the discrete version of the vector x, i.e. the vector $x(k)$, are again the values of the derivatives $x^{(i)}(t)$, $i = 0, 1, \ldots, (n - 1)$, at the individual instants $k = 0, 1, \ldots$. This requirement is satisfied by the state representation (3.34) and (3.35) with zero parameters $b_1, b_2, \ldots, b_n$. If in eqn. (16.169) $\gamma_i = 0$, $i = 1, 2, \ldots, (n - 1)$, but $\gamma_0 \neq 0$, the plant may as well be described by eqns. (3.40) and (3.41) or by the canonical form of observability. In both these cases all parameters of the plant may be non-zero.

Specifying for eqns. (3.34) and (3.35) the corresponding discrete version we may determine the vector error

$$\varepsilon(k) = \begin{bmatrix} \varepsilon_0(k) \\ \varepsilon_1(k) \\ \vdots \\ \varepsilon_\varkappa(k) \end{bmatrix} \qquad (16.171)$$

and rewrite the cost function (16.169) in the form

$$J = \sum_{k=0}^{\infty} [\varepsilon^T(k) \, \Gamma \, e(k) + R \, u^2(k)] \qquad (16.172)$$

where Γ is the diagonal matrix of the weighting coefficients γ_i.

Denoting the extended state vector by

$$\psi(k) = \begin{bmatrix} x(k) \\ \omega(k) \end{bmatrix} \tag{16.173}$$

we may write the extended state equations of the plant in the form

$$\psi(k+1) = \begin{bmatrix} A, & 0 \\ 0, & W \end{bmatrix} \psi(k) + \begin{bmatrix} b \\ 0 \end{bmatrix} u(k) = A^* \psi(k) + b^* u(k)$$

$$y(k) = [c^T, 0]\, \psi(k) = c^{*T}\, \psi(k) \tag{16.174}$$

and the cost function (16.172) in the form

$$J = \sum_{k=0}^{\infty} [\psi^T(k)\, Q\, \psi(k) + R\, u^2(k)] \tag{16.175}$$

where the weighting matrix Q is now

$$Q = \begin{bmatrix}
\gamma_0 & 0 & \cdots & 0 & -\gamma_0 & 0 & \cdots & 0 \\
0 & \gamma_1 & \cdots & 0 & 0 & -\gamma_1 & \cdots & 0 \\
\vdots & & & & & & & \vdots \\
0 & 0 & \cdots & \gamma_{n-1} & 0 & 0 & \cdots & -\gamma_{n-1} \\
\hline
-\gamma_0 & 0 & \cdots & 0 & \gamma_0 & 0 & \cdots & 0 \\
0 & -\gamma_1 & \cdots & 0 & 0 & \gamma_1 & \cdots & 0 \\
\vdots & & & & & & & \vdots \\
0 & 0 & \cdots & -\gamma_{n-1} & 0 & 0 & \cdots & \gamma_{n-1}
\end{bmatrix} \tag{16.176}$$

The elements not indicated are zero. For $\varkappa < (n-1)$ the weighting coefficients $\gamma_{\varkappa+1}, \ldots, \gamma_{n-1}$ are zero as well. With this formulation of the problem the further procedure of solving optimum control is identical with that given in the preceding sections.

It should nevertheless be stated that the cost function (16.169) or (16.172) considering the error vector $\varepsilon(k)$ may be applied only in special cases, where the derivatives of the output variable y can be simply expressed by means of the plant state variables or in cases where we confine ourselves to the evaluation of the error (16.168), i.e. without considering its derivatives. However, if we select a control cost function evaluating the differences of the error $\varepsilon(k)$ instead of the derivatives, the restrictions mentioned previously do not apply because differences of an arbitrary order can be expressed in terms of past values of the variables at sampling instants. The procedure for the solution will be shown by a simple example.

To solve the problem it is useful to transform the state representation of the plant into the canonical form of reachability with the output matrix $c^T = [1, 0, \ldots, 0]$,

CONTROLLER DESIGN BASED ON QUADRATIC COST FUNCTIONS

while all parameters of the plant may be non-zero. For example, the cost function

$$J = \sum_{k=0}^{\infty} \{\varepsilon^2(k) + \gamma_1[\Delta\varepsilon(k)]^2\} \qquad (16.177)$$

may be rewritten in the form

$$J = \sum_{k=0}^{\infty} [\psi^T(k) Q \psi(k) - 2\psi^T(k) S u(k) + R u^2(k)] \qquad (16.178)$$

where

$$Q = Q_1 + \gamma_1 Q_2 \qquad (16.179)$$

According to definition (16.168) we have

$$\varepsilon(k) = c_W^T \omega(k) - c^T x(k) = [-c^T, c_W^T] \psi(k) =$$
$$= [\underbrace{-1, 0, \ldots, 0}_{n}, \underbrace{1, 0, \ldots, 0}_{n}] \psi(k) \qquad (16.180)$$

so that

$$\varepsilon^2(k) = \psi^T(k) Q_1 \psi(k) \qquad (16.181)$$

where

$$Q_1 = \begin{bmatrix} -1 \\ 0 \\ \vdots \\ 0 \\ 1 \\ 0 \\ \vdots \\ 0 \end{bmatrix} [-1, 0, \ldots 0, 1, 0, \ldots 0] = \begin{bmatrix} 1, 0, \ldots & -1, 0, \ldots \\ 0, 0, \ldots & 0, 0, \ldots \\ \cdots\cdots\cdots\cdots\cdots\cdots \\ -1, 0, \ldots & 1, 0, \ldots \\ 0, 0, \ldots & 0, 0, \ldots \\ \cdots\cdots\cdots\cdots\cdots\cdots \end{bmatrix}$$

$$(16.182)$$

For the first difference of the error $\varepsilon(k)$ we have

$$\Delta\varepsilon(k) = \varepsilon(k+1) - \varepsilon(k) =$$
$$= r(k+1) - y(k+1) - r(k) + y(k) =$$
$$= c_W^T(W - E)\omega(k) - c^T(A - E)x(k) - c^T b u(k) =$$
$$= [-c^T(A - E), c_W^T(W - E)] \psi(k) - c^T b u(k) \qquad (16.183)$$

so that

$$[\Delta\varepsilon(k)]^2 = \psi^T(k) Q_2 \psi(k) - 2\psi^T(k) S u(k) + R u^2(k) \qquad (16.184)$$

where

$$Q_2 = [-c^T(A - E), c_W^T(W - E)]^T [-c^T(A - E), c_W^T(W - E)]$$
$$S = [-c^T(A - E), c_W^T(W - E)]^T c^T b$$
$$R = b^T c c^T b$$

Example 16.4: For the plant given in Example 16.1 calculate the optimum control satisfying the cost function (16.169) if the command variable has the form of a unit step, and if $\varkappa = 0$. The initial vector $\Psi^T(0) = [0, 0, 1]$. The weighting matrix $R = 0$ and the sampling period $T = 1$.

It should be noted that Examples 16.1 and 16.4 differ only in the formulation of the problem. While Example 16.1 involves only a transition of the plant from the non-zero initial state to the desired zero state, the problem in Example 16.4 is formulated as the control of a control loop with a given command variable $r(t)$.

Solution: For the unit-step command variable we have, according to (15.68), $V = 0$, and, according to (15.62), $W = 1$. Therefore, the extended matrix of dynamics in eqn. (16.174) is

$$A^* = \begin{bmatrix} A, & 0 \\ 0, & W \end{bmatrix} = \begin{bmatrix} 1.5, & 1, & 0 \\ -0.5, & 0, & 0 \\ 0, & 0, & 1 \end{bmatrix}$$

and the extended input matrix is

$$b^* = \begin{bmatrix} 1 \\ 2 \\ 0 \end{bmatrix}$$

The extended output matrix

$$c^{*T} = [1, 0, 0]$$

For our example the weighting matrix Q of the cost function (16.175) is, according to (16.176),

$$Q = \begin{bmatrix} 1, & 0, & -1 \\ 0, & 0, & 0 \\ -1, & 0, & 1 \end{bmatrix}$$

The solution of the Riccati equation corresponding to the matrices A^*, b^*, Q, R where, in compliance with the problem formulation, $R = 0$, is

$$P_{15,16} = \begin{bmatrix} 1.6533, & 0.3733, & -1.4667 \\ 0.3733, & 0.2133, & -0.2667 \\ -1.4667, & -0.2667, & 2.3333 \end{bmatrix}$$

Using relation (16.32), the controller matrix is

$$k^T = [0.8, 0.6, -0.5]$$

Comparing the matrices P and k^T with the corresponding results of Example 16.1, it is apparent that the elements $P_{11}, P_{12}, P_{21}, P_{22}$ of the matrix P and the elements k_{11}, k_{12} of the matrix k^T are identical with the previously given results.

CONTROLLER DESIGN BASED ON QUADRATIC COST FUNCTIONS

Using the extended state equation we calculate

k	0	1	2	3	4
$x_1(k)$	0	0.5	1.25	0.875	1.0625
$x_2(k)$	0	1	−1.25	−0.125	−0.6875
$r(k)$	1	1	1	1	1
$u(k)$	0.5	−0.5	0.25	−0.25	0.125

where $y(k) = x_1(k)$.

Figure 16.1 shows the behaviour of the controlled variable $y(k) = x_1(k)$, $k = 0, 1, 2, \ldots$, of the closed control loop when the command variable $r(0) = 0$ and $r(k)$, $k > 0$, alternately assumes the values ± 1 at the random instants kT. For illustration, the individual points of the solution, corresponding to the instants kT, $k = 0, 1, 2, \ldots$, are connected by lines.

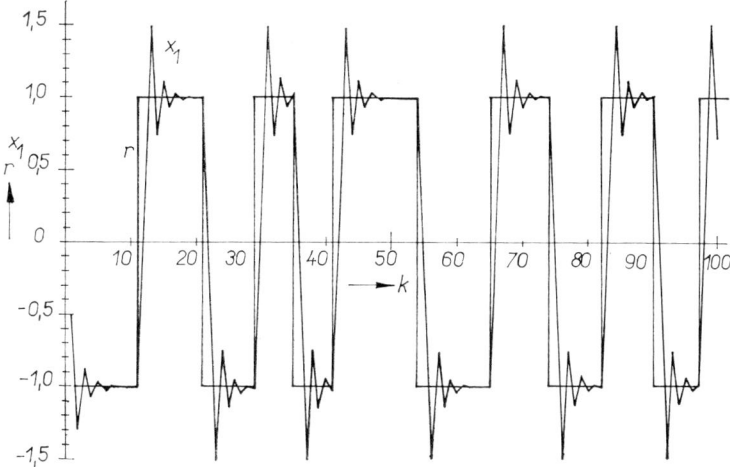

Fig. 16.1. Transient response of the controlled variable $y(k) = x_1(k)$, relating to Example 16.4.

Example 16.5: Solve Example 16.4 for a unit velocity step command variable, i.e. $r(t) = t$. The initial vector is $\Psi^T(0) = [0, 0, 0, 1]$.

Solution: As in the preceding example, we determine, using eqns. (15.68) and (15.62), the matrices of dynamics of the command variable:

$$V = \begin{bmatrix} 0, & 1 \\ 0, & 0 \end{bmatrix}, \quad W = \begin{bmatrix} 1, & 1 \\ 0, & 1 \end{bmatrix}$$

The matrices of the extended state representation are

$$A^* = \begin{bmatrix} 1.5, & 1, & 0, & 0 \\ -0.5, & 0, & 0, & 0 \\ 0, & 0, & 1, & 1 \\ 0, & 0, & 0, & 1 \end{bmatrix}, \quad b^* = \begin{bmatrix} 1 \\ 2 \\ 0 \\ 0 \end{bmatrix}, \quad c^{*T} = [1, 0, 0, 0]$$

The weighting matrix Q is now

$$Q = \begin{bmatrix} 1, & 0, & -1, & 0 \\ 0, & 0, & 0, & 0 \\ -1, & 0, & 1, & 0 \\ 0, & 0, & 0, & 0 \end{bmatrix}$$

and the weighting matrix $R = 0$ as required.

The solution of the Riccati equation is

$$P_{19,20} = \begin{bmatrix} 1.6533, & 0.3733, & -1.4667, & -0.3111 \\ 0.3733, & 0.2133, & -0.2667, & -0.1778 \\ -1.4667, & -0.2667, & 2.3333, & 20.2222 \\ -0.3111, & -0.1778, & 20.2222, & 401.1482 \end{bmatrix}$$

and the controller matrix has now the form

$$k^T = [0.8, 0.6, -0.5, -0.6667]$$

It is again useful to compare the resultant matrices P and k^T with the results of the preceding example.

Using the extended state equation we now calculate:

k	0	1	2	3	4
$x_1(k)$	0	0.6667	2.1667	2.9167	4.0417
$x_2(k)$	0	1.3333	-0.6667	-0.4167	-1.2917
$\omega_1(k)$	0	1	2	3	4
$\omega_2(k)$	1	1	1	1	1
$u(k)$	0.6667	-0.1667	0.3333	0.0833	0.2084

where

$$y(k) = x_1(k) \quad \text{and} \quad r(k) = \omega_1(k)$$

Example 16.6: Consider again the plant of Example 16.1 and calculate the optimum control according to the cost function (16.177) with $\gamma_1 = 1$, assuming that the command variable has the form of a unit velocity step, i.e. $r(t) = t$. The initial vector is $\psi^T(0) = [0, 0, 0, 1]$. The sampling period $T = 1$.

CONTROLLER DESIGN BASED ON QUADRATIC COST FUNCTIONS 375

Solution: The matrices V, W, A^*, b^*, c^{*T} are identical with the corresponding matrices of Example 16.5.

The weighting matrices are determined according to eqns. (16.182) and (16.185):

$$Q_1 = \begin{bmatrix} 1, & 0, & -1, & 0 \\ 0, & 0, & 0, & 0 \\ -1, & 0, & 1, & 0 \\ 0, & 0, & 0, & 0 \end{bmatrix}$$

$$Q_2 = \begin{bmatrix} -0.5 \\ -1 \\ 0 \\ 1 \end{bmatrix} [-0.5, -1, 0, 1] = \begin{bmatrix} 0.25, & 0.5, & 0, & -0.5 \\ 0.5, & 1, & 0, & -1 \\ 0, & 0, & 0, & 0 \\ -0.5, & -1, & 0, & 1 \end{bmatrix}$$

$$S = \begin{bmatrix} -0.5 \\ -1 \\ 0 \\ 1 \end{bmatrix}, \quad R = 1$$

Using relations (16.4) and (16.7), we eliminate from the cost function (16.178) the mixed term containing the matrix S. We obtain

$$\hat{Q} = Q - SR^{-1}S^T = Q_1 + Q_2 - Q_2 = Q_1$$
$$\hat{R} = R$$
$$\hat{A} = A + bR^{-1}S^T = \begin{bmatrix} 1, & 0, & 0, & 1 \\ -1.5, & -2, & 0, & 2 \\ 0, & 0, & 1, & 1 \\ 0, & 0, & 0, & 1 \end{bmatrix}, \quad \hat{b} = b$$

The solution of the Riccati equation determined by the matrices $\hat{A}, \hat{b}, \hat{Q}, \hat{R}$ is

$$P_{17,18} = \begin{bmatrix} 2.6633, & 0.9100, & -2.2083, & -0.7584 \\ 0.9100, & 0.7922, & -0.5139, & -0.6602 \\ -2.2083, & -0.5139, & 3.3333, & 25.3036 \\ -0.7584, & -0.6602, & 25.3036, & 449.3072 \end{bmatrix}$$

and, finally, the controller matrix according to (16.21) has the form

$$k^T = [0.0708, -0.4764, -0.3090, 0.3970]$$

Using the extended state equation we find

k	0	1	2	3	4
$x_1(k)$	0	0.6030	2.0468	2.9186	4.0185
$x_2(k)$	0	1.206	−0.4288	−0.4690	−1.2399
$\omega_1(k)$	0	1	2	3	4
$\omega_2(k)$	1	1	1	1	1
$u(k)$	−0.3970	0.4438	−0.1282	0.0999	−0.0362

where $y(k) = x_1(k)$ and $r(k) = \omega_1(k)$. Note that in this case the state vector of the extended system has the following components:

$$\psi(k) = \begin{bmatrix} x_1(k) \\ x_2(k) \\ \omega_1(k) \\ \omega_2(k) \end{bmatrix} = \begin{bmatrix} y(k) \\ -1.5y(k) + y(k+1) - u(k) \\ r(k) \\ r(k+1) - r(k) \end{bmatrix}$$

The first two components follow from definitions (6.19) and (6.20) and the last two components are discrete versions of eqns. (15.66).

Note 16.3: In Examples 16.5 and 16.6, where the command variable $r(t)$ is permanently varying, the solution P of the Riccati equation does not stabilize itself at a constant matrix although the control loop is time-invariant. That part of the matrix P which corresponds to a block of the matrix W is permanently varying in each iteration cycle. On the other hand, the controller matrix k^T or, with multi-input/multi-output loops, the matrix K, stabilizes itself in time-invariant loops at a constant matrix and it could be proved that k^T or K does not depend on the time-variant portion of the matrix P. For these reasons, in such cases the iterative solution of the Riccati equation cannot be terminated according to the required accuracy of the elements of P but according to the required accuracy of the elements of k^T or K.

Note 16.4: The synthesis of multi-input/multi-output systems may be carried out in complete analogy with the synthesis of single-input/single-output systems.

Example 16.7: Consider the same double-input/double-output system as in Example 11.3. Determine a controller matrix K such that the sum of square errors of both output variables from the relevant command variable be minimum. Consider that the first command variable is a unit position step and the second command variable a unit velocity step. Assume equal weights of both errors. Let the initial vector be $\psi^T(0) = [0, 0, 0, 1, 0, 1]$ and the sampling period $T = 1$.

Solution: The example may be solved mechanically by virtue of the previously given procedures and relations. We are to determine

$$J = \min \sum_{k=0}^{\infty} e^T(k) \, e(k) = \min \sum_{k=0}^{\infty} \psi^T(k) \, Q \, \psi(k)$$

where

$$e^T(k) = [e_1(k), e_2(k)]$$
$$e_1(k) = r_1(k) - y_1(k) = r_1(k) - x_1(k)$$
$$e_2(k) = r_2(k) - y_2(k) = r_2(k) - x_3(k)$$
$$\psi^T(k) = [x_1(k), x_2(k), x_3(k), r_1(k), r_2(k), \Delta r_2(k)]$$

CONTROLLER DESIGN BASED ON QUADRATIC COST FUNCTIONS

The matrices of the state description are

$$A^* = \begin{bmatrix} A, & 0, & 0 \\ 0, & W_1, & 0 \\ 0, & 0, & W_2 \end{bmatrix} = \begin{bmatrix} 0.6, & 0.7, & 0, & 0, & 0, & 0 \\ 0, & 0.75, & 0, & 0, & 0, & 0 \\ 0, & 0, & 0.8, & 0, & 0, & 0 \\ 0, & 0, & 0, & 1, & 0, & 0 \\ 0, & 0, & 0, & 0, & 1, & 1 \\ 0, & 0, & 0, & 0, & 0, & 1 \end{bmatrix}$$

$$B^* = \begin{bmatrix} B \\ 0 \end{bmatrix} = \begin{bmatrix} 0.8, & 0.4 \\ 0, & 0.9 \\ 0.9, & 0.9 \\ 0, & 0 \\ 0, & 0 \\ 0, & 0 \end{bmatrix}, \quad C^{*T} = \begin{bmatrix} C^T, & 0 \end{bmatrix} = \begin{bmatrix} 1, & 0, & 0, & 0, & 0, & 0 \\ 0, & 0, & 1, & 0, & 0, & 0 \end{bmatrix}$$

The matrix Q of the cost function J is determined by (16.179) through (16.182)

$$Q = Q_1 + Q_2 = q_1 q_1^T + q_2 q_2^T$$

where, in accordance with the introduced errors $e_1(k)$ and $e_2(k)$,

$$q_1^T = [-1, 0, 0, 1, 0, 0]$$
$$q_2^T = [0, 0, -1, 0, 1, 0]$$

so that

$$Q = \begin{bmatrix} 1, & 0, & 0, & -1, & 0, & 0 \\ 0, & 0, & 0, & 0, & 0, & 0 \\ 0, & 0, & 1, & 0, & -1, & 0 \\ -1, & 0, & 0, & 1, & 0, & 0 \\ 0, & 0, & -1, & 0, & 1, & 0 \\ 0, & 0, & 0, & 0, & 0, & 0 \end{bmatrix}$$

The solution of the Riccati equation (16.34) corresponding to the matrices A^*, B^*, C^{*T}, Q and $R = 0$ enables us to determine by (16.32) the controller matrix

$$K = \begin{bmatrix} 0.1256, & -0.6171, & 0.7401, & 0.4823, & -1.2325, & -3.0012 \\ 0.2302, & 1.2298, & -0.2729, & -1.2543, & 0.7280, & 2.9545 \end{bmatrix}$$

The matrix of dynamics of the closed control loop is, according to (16.33),

$$A^{**} = \begin{bmatrix} 0.4074, & 0.7017, & -0.4829, & 0.1159, & 0.6948, & 1.2191 \\ -0.2072, & -0.3569, & 0.2456, & 1.1289, & -0.6552, & -2.6591 \\ -0.3202, & -0.5515, & 0.3795, & 0.6948, & 0.4541, & 0.0419 \\ 0, & 0, & 0, & 1, & 0, & 0 \\ 0, & 0, & 0, & 0, & 1, & 1 \\ 0, & 0, & 0, & 0, & 0, & 1 \end{bmatrix}$$

and, finally, the values of the state vector for $k = 0, 1, \ldots, 5$ are

k	0	1	2	3	4	5
x_1	0	1.3350	1.1442	1.0622	1.0269	1.0118
x_2	0	−1.5302	−1.7350	−1.9951	−2.2790	−2.5731
x_3	0	0.7367	1.8868	2.9514	3.9792	4.9913
r_1	1	1	1	1	1	1
r_2	0	1	2	3	4	5
Δr_2	1	1	1	1	1	1

The above table shows that both output variables $y_1 = x_1$ and $y_2 = x_3$ converge very well towards the values of the command variables r_1 and r_2, respectively.

The resultant minimum value of the cost function could be calculated by means of eqn. (16.25) using the initial vector $\psi(0)$.

16.7 Application of the estimator of non-measurable state variables

So far we have always assumed in Chap. 16 that all components of the state vector are measurable and that the values of these components at the instants $k = 0, 1, 2, \ldots$ may be used for the determination of controlling variables by means of the general relation (16.20). As we know, in practical cases only the input and output variables of the controlled plant are measurable, as a rule. Therefore, if we select such a representation of the plant of the order n that p of its output variables y_i, $i = 1, 2, \ldots, p$, $p < n$, are also the components of the state vector x, then the estimate of the remaining $n - p$ components of the vector x must in some way be calculated. For this purpose we may use the estimator of the reduced order described in Par. 14.1.2 and used in Sec. 15.6 for the control and in Sec. 15.7 for the compensation of disturbing variables.

The previously given relations for the application of estimators hold without change also for such cases where the controller matrix is determined by means of quadratic cost functions. The procedure is to calculate first, according to the selected cost function, the controller matrix assuming all components of the state vector to be measurable and to complete the control loop by an estimator of non-measurable components of the plant state vectors. This procedure will now be shown in detail by an example.

Example 16.8: Consider again the plant given in Example 16.1, determined by the matrices

$$A = \begin{bmatrix} 1.5, & 1 \\ -0.5, & 0 \end{bmatrix}, \quad b = \begin{bmatrix} 1 \\ 2 \end{bmatrix}, \quad c^T = [1, 0]$$

whose state vector $x^T = [x_1, x_2]$ has the non-measurable component x_2. According to the quadratic cost function

$$J = \sum_{k=0}^{\infty} x^T(k) \, Q \, x(k), \quad Q = \begin{bmatrix} 1 & 0 \\ 0 & 0 \end{bmatrix}$$

the controller matrix was determined to be $k^T = [0.8, 0.6]$. Design a control loop with an estimator of the state variable x_2. The input variable of the loop is a command variable in the form of a unit step applied to the controller input at the instant $k = 0^+$. Assume that the initial vector for the numerical solution is $x^T(0) = [0, 0]$ and that the error estimate of the non-measurable component of the state vector is $\Delta v(0) = -0.5$.

Solution: First we rearrange the state variables according to (14.9) and calculate the matrices P, Q, R, S, B_3 of (14.12). We obtain

$$A^* = \begin{bmatrix} A_{11} & A_{12} \\ A_{21} & A_{22} \end{bmatrix} = \begin{bmatrix} 0 & -0.5 \\ 1 & 1.5 \end{bmatrix}, \quad b^* = \begin{bmatrix} B_1 \\ B_2 \end{bmatrix} = \begin{bmatrix} 2 \\ 1 \end{bmatrix}$$

$$c^{*T} = [C_1, C_2] = [0, 1]$$

Since $C_1 = 0$, we have

$$P = A_{11} = 0, \quad Q = A_{12} C_2^{-1} = -0.5$$

$$R = C_2 A_{21} = 1, \quad S = C_2 A_{22} C_2^{-1} = 1.5, \quad B_3 = C_2 B_2 = 1$$

For the control design with the estimator of the state variable x_2 we use the general expression (15.172) where we determine the matrix H, for instance, according to (15.150). For example, with the selected root value $z_1 = 0.2$ of the characteristic polynomial of the estimator we determine $H = -0.2$, and the estimator equation according to (14.17),

$$\hat{v}(k+1) = A_E \, \hat{v}(k) + H_E \, y(k) + B_E \, u(k)$$

where

$$A_E = -HR = 0.2$$

$$H_E = -HRH + Q - HS = -0.24$$

$$B_E = B_1 - HB_3 = 2.2$$

In eqn. (15.172) it remains to determine $r(k)$ where, according to (15.178),

$$r(k) = V w(k)$$

The transformation matrix V is calculated so as to obtain a zero control error in the equilibrium state. For the equilibrium state it holds that

$$\hat{v}(k+1) = \hat{v}(k) \quad \text{for} \quad k \to \infty$$
$$x(k+1) = x(k) \quad \text{for} \quad k \to \infty$$
$$y(k) \quad = x_1(k) = w(k)$$

From the first equality it follows that

$$\hat{v}(k) = \frac{H_E}{1 - A_E} y(k) + \frac{B_E}{1 - A_E} u(k) =$$
$$= -0.3 y(k) + 2.75 u(k) =$$
$$= M\, y(k) + N\, u(k)$$

and from the second equality we obtain

$$(E - A^*)\, x(k) = b^*\, u(k)$$

$$\begin{bmatrix} 1, & 0.5 \\ -1, & -0.5 \end{bmatrix} \begin{bmatrix} x_2(k) \\ x_1(k) \end{bmatrix} = \begin{bmatrix} 2 \\ 1 \end{bmatrix} u(k)$$

$$u(k) = 0 \quad \text{for} \quad k \to \infty$$
$$x_2(k) = -0.5 \quad \text{for} \quad k \to \infty$$

According to (15.159)

$$r(k) = u(k) + k^T\, \hat{x}(k)$$

where, using (15.153),

$$\hat{x}(k) = \begin{bmatrix} \hat{x}_2(k) \\ y(k) \end{bmatrix} = \begin{bmatrix} \hat{v}(k) + H\, y(k) \\ y(k) \end{bmatrix} = \begin{bmatrix} E, & H \\ 0, & E \end{bmatrix} \begin{bmatrix} \hat{v}(k) \\ y(k) \end{bmatrix}$$

$$\hat{x}(k) = L \begin{bmatrix} \hat{v}(k) \\ y(k) \end{bmatrix}$$

Making use of the preceding partial results we have

$$V w(k) = k^T L \begin{bmatrix} M \\ E \end{bmatrix} w(k)$$

so that

$$V = [0.6,\ 0.8] \begin{bmatrix} 1, & -0.2 \\ 0, & 1 \end{bmatrix} \begin{bmatrix} -0.3 \\ 1 \end{bmatrix} = 0.5$$

It is now a simple matter to write eqn. (15.172) where, for the example being solved, the disturbing variables are zero and the command variable $w(k) = 1$, $k = 0^+$, 1, 2, ...

$$\begin{bmatrix} x_2(k+1) \\ x_1(k+1) \\ \Delta v(k+1) \end{bmatrix} = \begin{bmatrix} -1.2, & -2.1, & 1.2 \\ 0.4, & 0.7, & 0.6 \\ 0, & 0, & -0.2 \end{bmatrix} \begin{bmatrix} x_2(k) \\ x_1(k) \\ \Delta v(k) \end{bmatrix} + \begin{bmatrix} 1 \\ 0.5 \\ 0 \end{bmatrix} w(k)$$

where $\Delta v(k) = v(k) - \hat{v}(k)$.

With the given initial vector and the given error estimate of the component $x_2(0)$ we calculate

k	0	1	2	3	4	5	6	etc.
$x_2(k)$	0	0.4	0.22	−1.094	−0.1562	−0.6793	−0.4118	
$x_1(k)$	0	0.2	0.86	1.178	0.8894	1.0606	0.9708	
$\Delta v(k)$	−0.5	0.1	−0.02	0.004	0.0008	0.0002	0	

In this example, the control process changes the component $x_1(k)$ from the value $x_1(0) = 0$ to the value $x_1(k) = 1$ for $k \to \infty$, and the estimate $\hat{x}_2(k)$ from the value $\hat{x}_2(0) = \hat{v}(0) = 0.5$ to the value $\hat{x}_2(k) = x_2(k) = -0.5$ for $k \to \infty$, and, according to (15.169), $\hat{v}(k) = -0.3$ for $k \to \infty$.

The structure of the complete control loop with separate block diagrams for the controller and the estimator may be set up analogously to Example 15.9.

17

Synthesis of Stochastic Systems

17.1 Synthesis in the Kalman sense

The synthesis of control loops in the Kalman sense solves the problem of optimum control by means of the quadratic cost function in such cases where random variables are acting on the plant. This problem is also referred to as the stochastic linear controller problem. The plant is represented in the state space by the equations

$$x(k + 1) = A(k) x(k) + B(k) [u(k) + w(k)]$$
$$y(k) = C(k) x(k) + v(k) \quad (17.1)$$

where $x \in X^n$, $y \in Y^p$, $u \in U^r$ and $w(k)$, $v(k)$ are mutually independent Gaussian random sequences with the properties described in Sec. 14.1. Let the matrices (A, B, C) and their transforms be real.

Since the random variables are acting on the plant, all components of the state vector are distorted and, therefore, their actual values cannot be obtained by an ordinary measurement. This is to say that in this case of stochastic synthesis no complete information is available on the plant state and the state vector must be estimated.

The quadratic cost function (16.1) estimating the quality of control is in the case of the stochastic controlled plant (17.1) a random variable as well so that there is no straightforward way to define its smallest value. However, the quality of control of stochastic controlled plants may be considered by virtue of the mean value defined, for example, by the expression

$$\mathcal{E}J = \mathcal{E}\{\|x(N)\|_{P_N}^2 + \sum_{k=0}^{N-1} [\|C(k) x(k)\|_{Q_k}^2 + \|u(k)\|_{R_k}^2]\}$$

The minimization of this cost function is a difficult task. However, for the quadratic cost function it was proved that estimation of the plant state and determination of the optimum controlling variable values may be calculated separately using a state estimate found in advance for the calculation of the values of the controlling variables.

Control synthesis may, therefore, be performed formally in the same way as in the deterministic case. The only difference is that we apply estimated values of the state variables instead of their actual values. The cost function may, for instance, be of the form

$$J = \|\hat{x}(N \mid N)\|^2_{P_N} + \sum_{k=0}^{N-1} [\|C(k)\,\hat{x}(k \mid k)\|^2_{Q_k} + \|u(k)\|^2_{R_k}] \qquad (17.2)$$

which is identified as the modified quadratic cost function (16.1). In the cost function (17.2) the quadratic form of components of the output vector y is applied instead of the quadratic form of the state vector components because $y = Cx$ for $v = 0$.

The separation of the stochastic linear controller problem into the problem of state estimation and the problem of a linear controller optimal in the sense of the selected quadratic cost function ensures the optimum solution of the stochastic problem as a whole. Thus, the gain matrix of the controller does not depend on the statistical parameters of the problem and the optimum filter for the state estimate does not depend on the matrices of the quadratic control cost function.

The solution of the stochastic linear controller problem in the described manner takes advantage of the so-called *separation principle*. The proof of this important principle will not be repeated here as its various modifications have been published in the literature (see e.g. [69.6, 70.1]). In Sec. 17.2 we shall only mention some of the basic properties of this principle.

The state vector estimate in the statistical and probabilistic sense and particularly the Kalman's state estimate were discussed in detail in Sec. 14.2. The results given in that section apply fully even to the case where the state vector being estimated is used further for the purpose of synthesis.

Since both the state vector estimate and the deterministic control synthesis according to quadratic control cost functions were already described in detail in the appropriate chapters we can mostly confine ourselves to a summary of previous results.

Following the same procedure as in Chap. 16 we obtain for the cost function (17.2) the solution

$$u(k) = -K_{OR}(k)\,\hat{x}(k \mid k) \qquad (17.3)$$

In this equation $\hat{x}(k \mid k)$ denotes the plant state vector estimate in the sense of Theorem 14.1 and K_{OR} is the optimum controller matrix given by

$$K_{OR}(k) = [B^T(k)\,P(k+1)\,B(k) + R(k)]^{-1}\,B^T(k)\,P(k+1)\,A(k) \qquad (17.4)$$

where $P(k+1)$ is defined by the Riccati equation

$$P(k) = A^T(k)\,P(k+1)\,A(k) - $$
$$- A^T(k)\,P(k+1)\,B(k)\,[B^T(k)\,P(k+1)\,B(k) + R(k)]^{-1}\,B^T(k)\,P(k+1)\,A(k) +$$
$$+ C^T(k)\,Q(k)\,C(k) \qquad (17.5)$$

The state vector estimate $\hat{x}(k \mid k)$ may be derived as in Sec. 14.2:

$$\hat{x}(k+1 \mid k+1) = \hat{x}(k+1 \mid k) + K_E(k+1) \left[y(k+1) - C(k+1) \hat{x}(k+1 \mid k) \right] \tag{17.6}$$

where

$$\hat{x}(k+1 \mid k) = A(k) \hat{x}(k \mid k) + B(k) u(k) \tag{17.7}$$

In eqn. (17.6) $K_E(k+1)$ is the estimator matrix given by

$$K_E(k+1) =$$
$$= P(k+1 \mid k) C^T(k+1) \left[C(k+1) P(k+1 \mid k) C^T(k+1) + R(k+1) \right]^{-1} \tag{17.8}$$

where

$$P(k+1 \mid k) = A(k) P(k \mid k) A^T(k) + B(k) Q(k) B^T(k) \tag{17.9}$$

and

$$P(k+1 \mid k+1) = \left[E - K_E(k+1) C(k+1) \right] P(k+1 \mid k) \tag{17.10}$$

Relations (17.8) through (17.10) represent a generalization of the appropriate relations of Theorem 14.2 for the time-variant case.

Using eqns. (17.1), (17.3), (17.6) and (17.7) we may now construct the block diagram of the complete control loop (Fig. 17.1) consisting of the controlled plant, the state vector estimator and of the feedback controller K_{OR}.

From the resultant relations and from the block diagram in Fig. 17.1 it is apparent that the controller matrix K_{OR}, optimal in the defined sense, does not depend on the statistical parameters of the problem and, on the contrary, the optimum state esti-

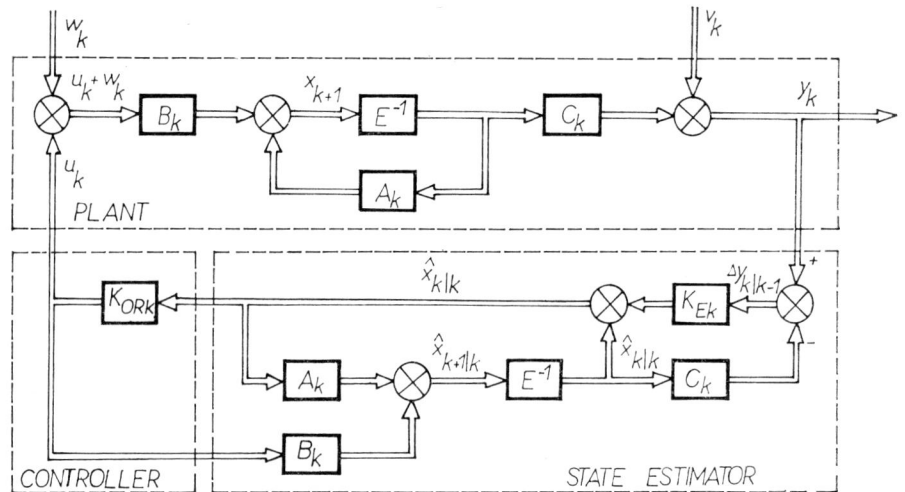

Fig. 17.1. Control loop with Kalman filter.

SYNTHESIS OF STOCHASTIC SYSTEMS

mator does not depend on the controller matrix K_{OR}. Therefore, the instructions given separately for the state estimate in Sec. 14.2 and for the calculation of the controller gain matrix in Chap. 16 apply to the numerical solution. Let us only remind that $\hat{x}(N \mid N)$ need not be determined because the controlling variable is acting last at the instant $k = N - 1$. Besides, from the estimate $\hat{x}(0 \mid 0) = \mathscr{E} x(0)$ it follows that $u(0) = -K_{OR}(0) \hat{x}(0 \mid 0) = 0$ if $x(0)$ has a zero mean value.

Example 17.1: Consider the controlled plant

$$x(k+1) = \begin{bmatrix} 1.5, & 1 \\ -0.5, & 0 \end{bmatrix} x(k) + \begin{bmatrix} 1 \\ 0.6 \end{bmatrix} [u(k) + w(k)]$$

$$y(k) = [1, 0] x(k) + v(k)$$

Calculate:

(a) The optimum controller

$$u(k) = -k^T x(k) + V r(k)$$

minimizing the cost function (16.175), i.e.

$$J = \sum_{k=0}^{\infty} [\psi^T(k) Q \psi(k) + R u^2(k)]$$

where

$$\psi^T(k) = [x_1(k), x_2(k), r(k)]$$

$$Q = \begin{bmatrix} 1, & 0, & -1 \\ 0, & 0, & 0 \\ -1, & 0, & 1 \end{bmatrix} \text{ and } R = 1$$

The command variable $r(k)$ may take the values of ± 1. The sampling period $T = 1$.

(b) The stochastic state vector estimator of the plant disturbed by the variables $w(k)$ and $v(k)$ being independent Gaussian noises with the properties given by (14.21) through (14.26) and with the dispersion $\sigma_w^2 = 0.01$ and $\sigma_v^2 = 0.0025$, i.e. with the standard deviation $\sigma_w = 0.1$ and $\sigma_v = 0.05$. The initial values are $x(0) = 0$, $\hat{x}(0 \mid 0) = 0$ and $P(0 \mid 0) = 0$. Calculate the gain function $K_E(k)$, $k = 0, 1, 2, \ldots$, and determine the time behaviour of the state variables $x_1(k)$ and $x_2(k)$ for a closed control loop with the controller determined in (a) when the command variable $r(0) = 0$ and $r(k)$, $k > 0$, assumes alternately the values of ± 1 considering that the change occurs at randomly selected instants kT. Compare the time behaviours with the estimates $\hat{x}_1$ and $\hat{x}_2$ and with the deterministic solution as formulated in (c).

(c) The deterministic estimator of the reduced order for estimating the state variable $x_2(k)$, assuming that the disturbing variables $w(k) = v(k) = 0$. Let the root of the characteristic equation of the estimator be $z_1 = 0.2$. Calculate the time behaviour of the state variables $x_1(k)$ and $x_2(k)$ for a closed control loop with the controller given in the present example, considering that the command variable $r(k)$ varies in the same way as in (b).

Solution:

(a) When calculating the controller we may make use of the fact that the upper left block of the matrix P of the dimension $(n; n)$, where $n = 2$ is the plant order, is invariant with respect to the elements in the third column and in the third row of the matrix Q. The solution of the Riccati equation (16.30) may be carried out with the simpler matrix

$$Q = \begin{bmatrix} 1, & 0 \\ 0, & 0 \end{bmatrix}$$

and with the matrix $R = 1$; using the general relation (16.21) we may then determine

$$k^T = [0.7886, \ 0.6119]$$

For a single-input/single-output control loop the matrix V in the controller equation has the dimension $(1; 1)$, i.e. it is a constant which may be determined from the condition of zero steady-state control error. For a value of k approaching to infinity it holds that $x(k+1) = x(k)$ and $x_1(k) = y(k) = r(k)$ and, therefore,

$$x(k) = (A - bk^T) x(k) + bV r(k)$$

From this condition we calculate $V = 0.4826$ and the steady-state value $x_2 = -0.5$. With this result the controller equation is

$$u(k) = [-0.7886, \ -0.6119, \ 0.4826] [x_1(k), \ x_2(k), \ r(k)]^T$$

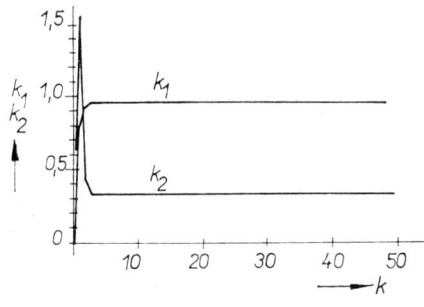

Fig. 17.2. Variation of estimator gain matrix elements.

Note that the constant V does not depend on the value of the command variable $r(k)$.

(b) The state vector estimate is calculated by means of the recursive relations (17.6) through (17.10) and the estimated state vector is used for the calculation of the controlling variable

$$u(k) = -k^T \hat{x}(k) + V r(k)$$

When calculating the state vector estimate we also obtain the values of the estimator gain matrix

$$K_E(k) = [k_1(k), k_2(k)]$$

$k = 0, 1, 2, \ldots$. The changes in both elements of the gain matrix $K_E(k)$ are shown in Fig. 17.2 where the individual values are connected by lines. It is apparent that $K_E(k)$ converges, after a few sampling periods, to constant values of the elements.

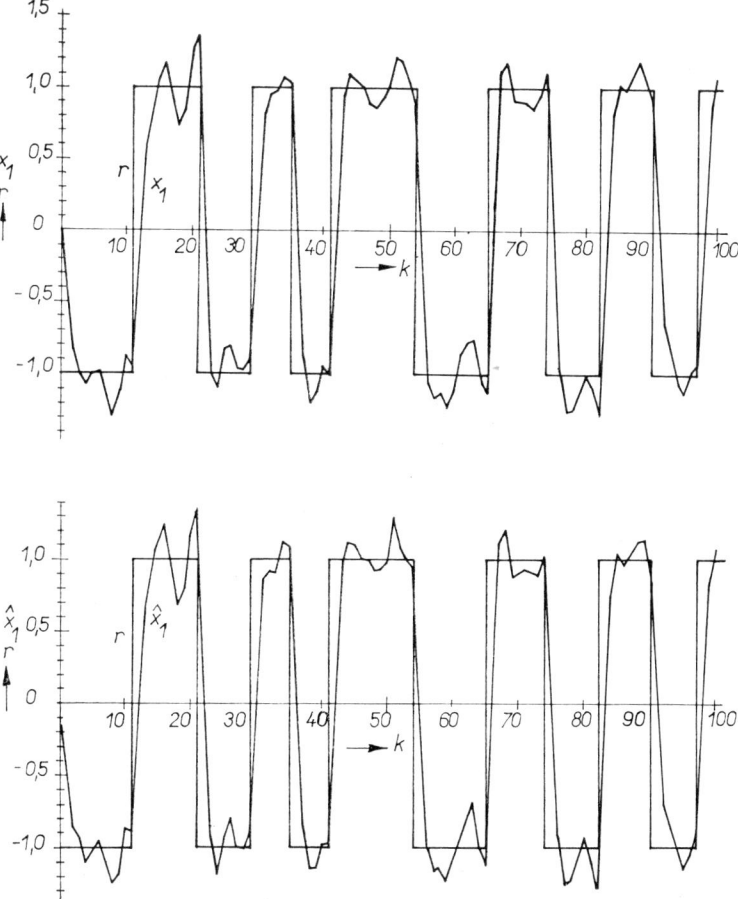

Fig. 17.3. Transient behaviour of the state variables $x_1(k)$ and $\hat{x}_1(k)$.

The time behaviour of the state variables and their estimates calculated according to the formulation of the problem is shown in Figs. 17.3 and 17.4. As can be seen, the agreement between the actual and estimated values of the state variables is very good. If $r(k) = \pm 1$, then $\mathscr{E}[x_2] = \mathscr{E}[\hat{x}_2] = \pm r(k)/2$. The error dispersion of the state estimates, calculated from a transient behaviour within 100 steps, is $\sigma_1^2 = 0.2550 \cdot 10^{-2}$ and $\sigma_2^2 = 0.3769 \cdot 10^{-2}$, where the subscripts 1 and 2 denote correspondence to the variables $\hat{x}_1$ and $\hat{x}_2$, respectively. The individual values of the variables calculated for the instants kT, $k = 0, 1, 2, \ldots$, are connected by lines in Figs. 17.3 through 17.6.

(c) The estimator of the reduced order is determined according to Par. 14.1.2. Using the given values we calculate successively

$$P = 0, \quad S = 1.5, \quad A_E = 0.2$$
$$Q = -0.5, \quad B_3 = 1, \quad H_E = -0.24$$
$$R = 1, \quad H = -0.2, \quad B_E = 0.8$$

With these values the estimator equation (14.18) is

$$\hat{v}(k+1) = 0.2\hat{v}(k) - 0.24y(k) + 0.8u(k)$$

The state variable $\hat{x}_2(k)$ being estimated may be calculated using the first of eqns. (15.153), that is

$$\hat{x}_2(k) = \hat{v}(k) - 0.2y(k)$$

The time behaviour of the state variables $x_1(k)$ and $x_2(k)$, $k = 0, 1, 2, \ldots$, of the

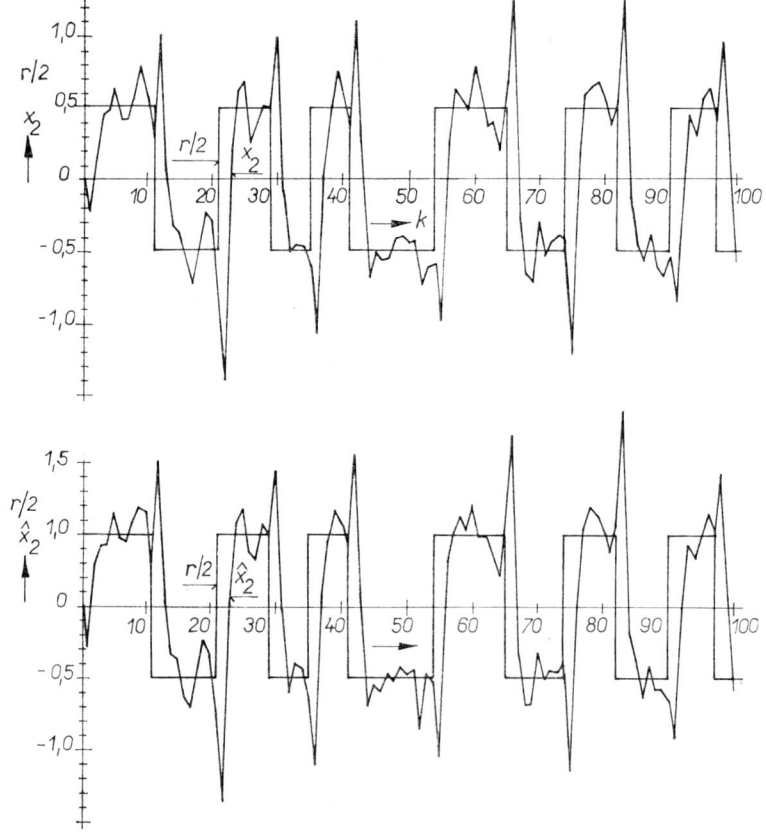

Fig. 17.4. Transient behaviour of the state variables $x_2(k)$ and $\hat{x}_2(k)$.

SYNTHESIS OF STOCHASTIC SYSTEMS

plant in a closed control loop with a controller is shown in Figs. 17.5 and 17.6. Here the controlled variable $y(k) = x_1(k)$ and the command variable $r(k)$ was varying according to the problem formulation. It will be seen that both state variables converge well to the steady values. The simplified block diagram of the control loop is shown in Fig. 17.7.

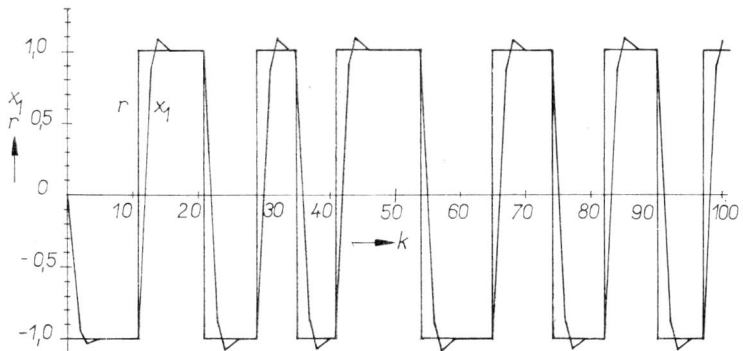

Fig. 17.5. Transient behaviour of the variables $y(k) = x_1(k)$ in the control loop with a deterministic estimator.

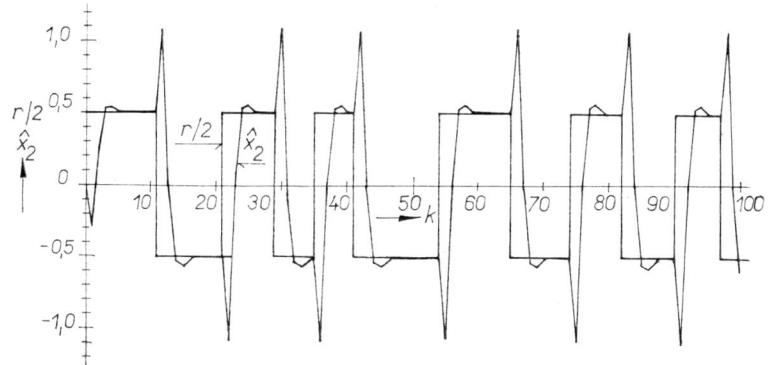

Fig. 17.6. Transient behaviour of the variable $x_2(k)$ in a control loop with a deterministic estimator.

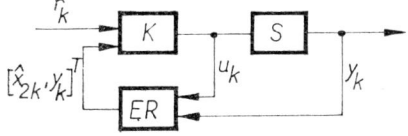

Fig. 17.7. Block diagram representation of the control loop with a deterministic estimator.

In conclusion it should be noted that the deterministic as well as the stochastic case of solution is very good. Particularly the estimate of non-measurable state variables agrees very well with reality. It cannot be excluded, however, that still smaller deviations of the controlled variable from the desired values could be achieved if a different controller were designed.

17.2 The separation principle

The possibility of a separate determination of the optimum state vector estimate and the optimum controller was first pointed out by Kalman and Koepke [58.3]. The proof was independently arrived at by Joseph [61.1] and Gunckel [63.3]. The extension for more general problems was worked out by Striebel [65.4]. In the literature this possibility is usually referred to as the separation principle.

Consider the plant equation (17.1), the controller equation (17.3) and eqns. (17.6) and (17.7) for the state vector estimation.

Let us define the error of the state estimate by

$$\Delta x(k+1 \mid k) = x(k+1) - \hat{x}(k+1 \mid k) =$$
$$= A(k) x(k) + B(k) [u(k) + w(k)] - A(k) \hat{x}(k \mid k) - B(k) u(k) =$$
$$= A(k) x(k) + B(k) w(k) - A(k) \{\hat{x}(k \mid k-1) +$$
$$+ K_E(k) [y(k) - C(k) \hat{x}(k \mid k-1)]\} \tag{17.11}$$

where we may substitute for $y(k)$ from eqn. (17.1). Equation (17.11) may be further modified. It holds that

$$\Delta x(k+1 \mid k) = A(k) x(k) + B(k) w(k) - A(k) \hat{x}(k \mid k-1) -$$
$$- A(k) K_E(k) C(k) x(k) - A(k) K_E(k) v(k) +$$
$$+ A(k) K_E(k) C(k) \hat{x}(k \mid k-1) =$$
$$= [A(k) - K_{OE}(k) C(k)] \Delta x(k \mid k-1) + B(k) w(k) - K_{OE}(k) v(k) \tag{17.12}$$

In eqn. (17.12) we have denoted the optimum filter for the state estimation according to the relation

$$K_{OE}(k) = A(k) K_E(k) \tag{17.13}$$

The state vector at the instant $k+1$ is

$$x(k+1) = A(k) x(k) - B(k) K_{OR}(k) \hat{x}(k \mid k) + B(k) w(k) \tag{17.14}$$

Substituting for $\hat{x}(k \mid k)$ from eqn. (17.6) we obtain

$$x(k+1) = A(k) x(k) - B(k) K_{OR}(k) \{\hat{x}(k \mid k-1) +$$
$$+ K_E(k) [C(k) x(k) + v(k) - C(k) \hat{x}(k \mid k-1)]\} + B(k) w(k) \tag{17.15}$$

Extending the right-hand side of eqn. (17.15) by $\pm B(k) K_{OR}(k) x(k)$ we have

$$x(k+1) = [A(k) - B(k) K_{OR}(k)] x(k) +$$
$$+ B(k) K_{OR}(k) \Delta x(k \mid k-1) -$$
$$- B(k) K_{OR}(k) K_E(k) C(k) \Delta x(k \mid k-1) -$$
$$- B(k) K_{OR}(k) K_E(k) v(k) + B(k) w(k) \qquad (17.16)$$

or

$$x(k+1) = [A(k) - B(k) K_{OR}(k)] x(k) +$$
$$+ B(k) K_{OR}(k) [E - K_E(k) C(k)] \Delta x(k \mid k-1) -$$
$$- B(k) K_{OR}(k) K_E(k) v(k) + B(k) w(k) \qquad (17.17)$$

where, according to (17.10) it holds that

$$E - K_E(k) C(k) = P(k \mid k) P^{-1}(k \mid k-1) \qquad (17.18)$$

Equations (17.12) and (17.17) may now be combined into a simple system described by the state equations

$$\begin{bmatrix} x(k+1) \\ \Delta x(k+1 \mid k) \end{bmatrix} = \begin{bmatrix} A(k) - B(k) K_{OR}(k) & B(k) K_{OR}(k) P(k \mid k) P^{-1}(k \mid k-1) \\ 0 & A(k) - K_{OE}(k) C(k) \end{bmatrix} \cdot$$
$$\cdot \begin{bmatrix} x(k) \\ \Delta x(k \mid k-1) \end{bmatrix} + \begin{bmatrix} B(k) \\ B(k) \end{bmatrix} w(k) +$$
$$+ \begin{bmatrix} -B(k) K_{OR}(k) K_E(k) \\ K_{OE}(k) \end{bmatrix} v(k) \qquad (17.19)$$

The output equation of the complete system has the form

$$y(k) = [C(k) \mid 0] \begin{bmatrix} x(k) \\ \Delta x(k \mid k-1) \end{bmatrix} + v(k) \qquad (17.20)$$

In eqn. (17.19) the matrix of dynamics of the closed control loop is

$$A^*(k) = A(k) - B(k) K_{OR}(k) \qquad (17.21)$$

and the matrix of dynamics of the closed loop of the estimator

$$\hat{A}^*(k) = A(k) - K_{OE}(k) C(k) \qquad (17.22)$$

From eqn. (17.19), it is evident that the dynamics of the plant with a feedback controller and the dynamics of the estimator feedback loop do not influence each other since the eigenvalues of the matrix $A^*(k)$ do not depend on the estimator matrix $K_{OE}(k)$ and, conversely, the eigenvalues of the matrix $\hat{A}^*(k)$ do not depend on the controller matrix $K_{OR}(k)$. In the synthesis of the control loop it is, therefore, possible to solve the problem of optimum state estimate and the problem of optimum controller separately.

Comparing the resultant relations (17.19) and (17.20) with a similar solution expressed by (15.163) in Sec. 15.4 we see that the separation of the control and of the state estimation is identically proved. The differences in the resultant equations follow from the different formulation of the problems and of the corresponding mathematical models.

From the above results follow the theorems:

THEOREM 17.1: *Consider a controlled plant, state estimator and controller, expressed by the equations*

$$x(k + 1) = A(k) x(k) + B(k) [u(k) + w(k)]$$

$$y(k) = C(k) x(k) + v(k)$$

$$\hat{x}(k + 1 \mid k + 1) = \hat{x}(k + 1 \mid k) + K_E(k + 1) [y(k + 1) - C(k + 1) \hat{x}(k + 1 \mid k)]$$

$$\hat{x}(k + 1 \mid k) = A(k) \hat{x}(k \mid k) + B(k) u(k)$$

$$u(k) = -K_{OR}(k) \hat{x}(k \mid k) \tag{17.23}$$

Then the characteristic polynomial of the complete system is

$$\chi(z) = \chi[A^*(k)] \chi[\hat{A}^*(k)]$$

where

$$\chi[A^*(k)] = \det [zE - A(k) + B(k) K_{OR}(k)]$$

and

$$\chi[\hat{A}^*(k)] = \det [zE - A(k) + K_{OR}(k) C(k)]$$

Since, in general, it does not matter in which way the filter $K_{OE}(k)$ for the state vector estimation is determined and which cost function is used for the controller design, the only condition being that the estimation loop and the control loop be stable, we may state the following theorem:

THEOREM 17.2: *Consider a fully controllable and fully reconstructable plant, a control matrix $K_{OR}(k)$ ensuring the stable characteristic polynomial $\chi[A(k) - B(k) K_{OR}(k)]$ and an estimation matrix $K_{OE}(k)$ ensuring the stable characteristic polynomial $\chi[A(k) - K_{OR}(k) C(k)]$. Then the complete system is described by eqn. (17.19) and is stable. The dynamic behaviour of the complete system may be expressed as the dynamics of control and the dynamics of state estimation.*

17.3 Principle of duality

The optimum plant state estimation in the sense of Kalman's filtration and the optimum control according to the quadratic cost function are related by mathematical and physical relationships which were first pointed out by R. E. Kalman [60.4] who denominated them as the *principle of duality*. These relationships result directly from a comparison of the relations for the control and for the optimum state estimation. The relations needed will now be recapitulated.

If the equation of plant dynamics has the form (17.1) and the control cost function the form (17.2), then, provided that the state x is known, the optimum control, i.e. the sequence of values of the vector of controlling variables $u(0), u(1), \ldots, u(N-1)$ is defined by the relations

$$u(k) = -K_{OR}(k)\, x(k) \tag{17.24}$$

where

$$K_{OR}(k) = [B^T(k)\, P(k+1)\, B(k) + R(k)]^{-1} B^T(k)\, P(k+1)\, A(k) \tag{17.25}$$

$$\begin{aligned} P(k) &= A^T(k)\, P(k+1)\, A(k) - K_{OR}^T(k)\, B^T(k)\, P(k+1)\, A(k) + \\ &\quad + C^T(k)\, Q(k)\, C(k) \end{aligned} \tag{17.26}$$

where the relation (17.26) was obtained by substituting $K_{OR}(k)$ of eqn. (17.4) into eqn. (17.5).

The matrix of dynamics of the closed control loop is

$$A^*(k) = A(k) - B(k)\, K_{OR}(k) \tag{17.27}$$

The plant state estimate (17.1) with a minimum error dispersion is given by

$$K_{OE}(k) = A(k)\, P(k \mid k-1)\, C^T(k)\, [C(k)\, P(k \mid k-1)\, C^T(k) + R(k)]^{-1} \tag{17.28}$$

$$\begin{aligned} P(k+1 \mid k) &= A(k)\, P(k \mid k-1)\, A^T(k) - K_{OE}(k)\, C(k)\, P(k \mid k-1)\, A^T(k) + \\ &\quad + B(k)\, Q(k)\, B^T(k) \end{aligned} \tag{17.29}$$

where the relation (17.29) is obtained by substituting for $P(k \mid k)$ in eqn. (17.9) according to eqn. (17.10).

The matrix of dynamics of the closed loop for the plant state estimation is

$$\hat{A}^*(k) = A(k) - K_{OE}(k)\, C(k) \tag{17.30}$$

Comparing the results relating to the optimum control and to the optimum plant state estimate we find that they are dual in the sense expressed by Theorem 17.3, the so-called *theorem of duality*.

THEOREM 17.3: *The problem of optimum state estimation and the problem of optimum control, according to the definitions accepted in Sec. 14.2 and in the present section, are dual in the sense that by replacing each of the matrices* $X(k) = X(k_0 + \tau)$, $\tau = 0, 1, 2, \ldots$, *in the state estimation model, i.e. in eqns.* (17.28) *through* (17.30), *by the corresponding matrix* $Y^T(k) = Y^T(N - \tau)$ *from the control model, we obtain the set of equations* (17.25) *through* (17.27) *and vice versa. Herein the time sequence of the arguments of the matrices P must be preserved.*

Theorem 17.3 can be proved by peforming the defined substitutions. The matrices and vectors corresponding to each other may be tabulated as follows:

State estimation	Control
$x(k)$ — unobservable state of a random process (stochastic plant)	$x(k)$ — observable state of a controlled plant
$y(k)$ — observed random output variable of a plant	$u(k)$ — controlling variable, i.e. useful input variable of a plant
k_0 — initial instant of observation	N — last instant of observation
$k = k_0 + \tau$ — sampling instant, $\tau = 0, 1, 2, \ldots$	$k = N - \tau$ — sampling instant, $\tau = 0, 1, 2, \ldots$
$A(k_0 + \tau)$ — matrix of plant dynamics	$A^T(N - \tau)$ — matrix of plant dynamics
$P(k_0 + \tau + 1 \mid k_0 + \tau)$ — covariance matrix of optimized estimate error	$P(N - \tau)$ — matrix of quadratic form of the cost function
$K_{OE}(k_0 + \tau)$ — weighting matrix of observed outputs (matrix of optimum estimator)	$K_{OR}^T(N - \tau)$ — weighting matrix of state vector (matrix of optimum controller)
$C(k_0 + \tau)$ — weighting matrix of the components of a state vector acting on observed output	$B^T(N - \tau)$ — weighting matrix of the components of an input vector acting on a state vector
$Q(k_0 + \tau)$ — covariance matrix of a random plant input w	$Q(N - \tau)$ — matrix of the quadratic form defining the cost function of quality of control
$R(k_0 + \tau)$ — covariance matrix of a random variable v acting on the plant output	$R(N - \tau)$ — matrix of a quadratic form defining the cost function of quality of control

In conclusion it may be stated that both the problem of optimum state estimation and the problem of optimum control lead to the same form of the Riccati equation. In other words, eqns. (17.26) and (17.29) are identical in the sense of Theorem 17.3 (theorem of duality).

The physical significance of Theorem 17.3 lies in the duality of the matrices K_{OE} and C in the case of state estimation and of the matrices B and K_{OR} in the case of control, respectively, as it clearly follows from a comparison of Figs. 17.8 and 17.9 derived from Fig. 17.1.

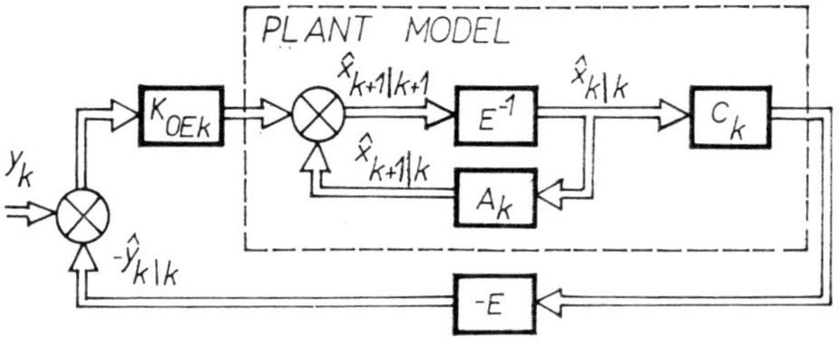

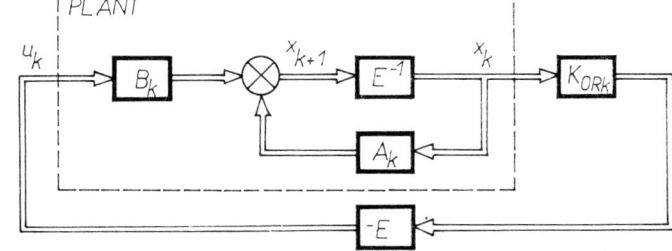

Fig. 17.8. Estimation loop dual to the control loop.

Fig. 17.9. Control loop dual to the estimation loop.

It is evident that the state estimation system (Fig. 17.8) is described by the equations

$$\hat{x}(k+1 \mid k+1) = A(k)\,\hat{x}(k \mid k) + K_{OE}(k)\,\Delta y(k \mid k)$$
$$\hat{y}(k \mid k) = C(k)\,\hat{x}(k \mid k) \tag{17.31}$$

and the control system (Fig. 17.9) by the equations

$$x(k+1) = A(k)\,x(k) + B(k)\,u(k)$$
$$u(k) = -K_{OR}(k)\,x(k) \tag{17.32}$$

Finally it should be noted that in the classical control theory the relationships between the state estimation and the control of a plant remained quite concealed while in the state space they can be fully revealed and used for a deeper study of the theory of state estimation and control.

On the other hand, it cannot be denied that the simple formulation of Theorem 17.3 could only be arrived at by applying a certain form of the cost functions, by a convenient form of derivation and by ensuring formal identity of the denotation of matrices, variables and their arguments. For these reasons the principle of duality must be comprehended, above all things, as a tool of theory, which may be applied only indirectly in the solution of practical problems.

Appendices

Appendix A:

Matrix inversion lemma

When rearranging matrix expressions we frequently meet with the matrix

$$M = (A \pm BCD)^{-1} \tag{A1}$$

where A and C are non-singular matrices. Often there is also $D = B^T$. The inversion of the matrix (A1) may be expressed as follows:

$$(A \pm BCD)^{-1} = A^{-1} - A^{-1}B(DA^{-1}B \pm C^{-1})^{-1} DA^{-1} \tag{A2}$$

The proof can be carried out by the substitution

$$(A \pm BCD)\left[A^{-1} - A^{-1}B(DA^{-1}B \pm C^{-1})^{-1} DA^{-1}\right] =$$
$$= E \pm BCDA^{-1} - B(DA^{-1}B \pm C^{-1})^{-1} DA^{-1} \mp$$
$$\mp BCDA^{-1}B(DA^{-1}B \pm C^{-1})^{-1} DA^{-1} =$$
$$= E \pm BCDA^{-1} - B(E \pm CDA^{-1}B)(DA^{-1}B \pm C^{-1})^{-1} DA^{-1} =$$
$$= E \pm BCDA^{-1} - BC(C^{-1} \pm DA^{-1}B)(DA^{-1}B \pm C^{-1})^{-1} DA^{-1} =$$
$$= E \pm BCDA^{-1} \mp BCDA^{-1} = E$$

Q. E. D.

If the matrix C is not non-singular, the matrix (A1) may be modified into the form

$$(A \pm BCD)^{-1} = A^{-1} - A^{-1}BC(C + CDA^{-1}BC)^{-1} CDA^{-1} \tag{A3}$$

Appendix B:

Test of the linear dependence of vectors

In many problems involving an analysis of the properties of systems represented in the state space, one must answer the question whether the vectors $x_1, x_2, \ldots, x_n$ are linearly dependent or not. In solving this problem we may take advantage of the Schmidt orthogonalization [63.7].

The procedure consists in the conversion of the vectors $x_1, x_2, \ldots, x_n$ into the orthogonal vectors $g_1, g_2, \ldots, g_n$. If, in testing the linear dependence, we start with the vector with the lowest subscript and proceed successively to the vectors with a higher subscript, and if we find that, say, the vector $g_k = 0$, $k > 1$, it means that the vector x_k is linearly dependent on the preceding vectors $x_1, x_2, \ldots, x_{k-1}$.

Thus, if we start with the vector x_1, we put

$$g_1 = x_1 \tag{B1}$$

The vector g_2 is sought as the linear combination of the vectors g_1 and x_2 in the form

$$g_2 = x_2 + k_{21} g_1 \tag{B2}$$

The constant k_{21} is determined so that the vectors g_2 and g_1 be mutually normal, i.e. so that their scalar product be zero:

$$g_2^T g_1 = x_2^T g_1 + k_{21} g_1^T g_1 = 0 \tag{B3}$$

$$k_{21} = -\frac{x_2^T g_1}{g_1^T g_1} \tag{B4}$$

Similarly, the vector g_3 is sought as the linear combination of the vectors g_1, g_2 and x_3 in the form

$$g_3 = x_3 + k_{32} g_2 + k_{31} g_1 \tag{B5}$$

and the constants k_{31} and k_{32} are determined so that the vector g_3 be normal to both vectors g_1 and g_2. The corresponding scalar products must, therefore, be zero:

$$g_3^T g_1 = x_3^T g_1 + k_{32} g_2^T g_1 + k_{31} g_1^T g_1 = 0$$
$$g_3^T g_2 = x_3^T g_2 + k_{32} g_2^T g_2 + k_{31} g_1^T g_2 = 0 \tag{B6}$$

From the set of equations (B6) we calculate the constants k_{31} and k_{32} and then, from (B5), we find the vector g_3.

The next vectors g_i, $i = 4, 5, \ldots$, are calculated in a similar manner. If each of the orhogonal vectors g_i, $i = 1, 2, \ldots$, is normalized by its length in the Euclidean space, we obtain the set of orthonormal vectors $e_1, e_2, \ldots, e_n$. It holds that

$$e_1 = \frac{x_1}{\|x_1\|} = \frac{g_1}{\|g_1\|}, \quad \text{where} \quad \|g_1\| = (g_1^T g_1)^{1/2} \tag{B7}$$

$$e_2 = \frac{g_2}{\|g_2\|} = \frac{x_2 - (x_2^T e_1) e_1}{\|x_2 - (x_2^T e_1) e_1\|} \tag{B8}$$

and, in general,

$$e_k = \frac{g_k}{\|g_k\|} = \frac{x_k - \sum_{i=1}^{k-1} (x_k^T e_i) e_i}{\left\|x_k - \sum_{i=1}^{k-1} (x_k^T e_i) e_i\right\|} \tag{B9}$$

This procedure is referred to as the Gram-Schmidt orthonormalization. When testing linear vector dependence we can usually manage with vector orthogonalization.

Appendix C:

Definiteness and semidefiniteness

The quadratic form $x^T A x$, where A is a real symmetrical matrix (or the Hermitian form $x^* A x$, where A is the Hermitian matrix and x^* is the complex conjugate and transposed vector corresponding to the vector x), is

(a) *positive definite* if

$$x^T A x > 0 \quad (\text{or } x^* A x > 0) \text{ for an arbitrary } x \neq 0$$
$$x^T A x = 0 \quad (\text{or } x^* A x = 0) \text{ for } x = 0$$

(b) *positive semidefinite* if

$$x^T A x \geq 0 \quad (\text{or } x^* A x \geq 0) \text{ for an arbitrary } x \neq 0$$
$$x^T A x = 0 \quad (\text{or } x^* A x = 0) \text{ for } x = 0$$

(c) *negative definite* if

$$x^T A x < 0 \quad (\text{or } x^* A x < 0) \text{ for an arbitrary } x \neq 0$$
$$x^T A x = 0 \quad (\text{or } x^* A x = 0) \text{ for } x = 0$$

(d) *negative semidefinite* if

$$x^T A x \leq 0 \quad (\text{or } x^* A x = 0) \text{ for an arbitrary } x \neq 0$$
$$x^T A x = 0 \quad (\text{or } x^* A x = 0) \text{ for } x = 0$$

If $x^T A x$ (or $x^* A x$) assumes positive and negative values we speak of $x^T A x$ (or $x^* A x$) as being *indefinite*.

The necessary and sufficient conditions for the quadratic form $x^T A x$ (or $x^* A x$) to be positive definite, negative definite, positive semidefinite or negative semidefinite, may be expressed by the following Sylvester's theorems:

THEOREM C1: The quadratic form $x^T A x$, where A is a symmetric real matrix of the dimension $(n; n)$, is positive definite if and only if det A and all successive main subdeterminants are positive, that is

$$a_{11} > 0, \quad \begin{vmatrix} a_{11}, a_{12} \\ a_{21}, a_{22} \end{vmatrix} > 0, \quad \begin{vmatrix} a_{11}, a_{12}, a_{13} \\ a_{21}, a_{22}, a_{23} \\ a_{31}, a_{32}, a_{33} \end{vmatrix} > 0, \ldots, \det A > 0$$

$$(a_{ij} = a_{ji})$$

THEOREM C2: The quadratic form $x^T A x$, where A is a symmetric real matrix of the dimension $(n; n)$, is negative definite if and only if det $A > 0$ for an even n and det $A < 0$ for an odd n and the successive main subdeterminants of the even order are positive and those of the odd order are negative, that is

$$a_{11} < 0, \quad \begin{vmatrix} a_{11}, a_{12} \\ a_{21}, a_{22} \end{vmatrix} > 0, \quad \begin{vmatrix} a_{11}, a_{12}, a_{13} \\ a_{21}, a_{22}, a_{23} \\ a_{31}, a_{32}, a_{33} \end{vmatrix} < 0$$

$$\det A > 0 \quad \text{for an even } n$$

$$\det A < 0 \quad \text{for an odd } n$$

$$(a_{ij} = a_{ji})$$

THEOREM C3: The quadratic form $x^T A x$, where A is a symmetric real matrix of the dimension $(n; n)$, is positive semidefinite if and only if all main subdeterminants of the matrix A are non-negative, that is

$$A \begin{pmatrix} i_1, i_2, \ldots, i_p \\ k_1, k_2, \ldots, k_p \end{pmatrix} \geq 0, \quad k_j = i_j, \quad j = 1, 2, \ldots, p$$

$$1 \leq i_1 < i_2 \ldots < i_p \leq n, \quad p = 1, 2, \ldots, n, \quad a_{ij} = a_{ji}$$

where i_j and k_j are the row and column indexes of the elements of the matrix A, respectively,

$$A \begin{pmatrix} i_1, i_2, \ldots, i_p \\ k_1, k_2, \ldots, k_p \end{pmatrix} = \begin{vmatrix} a_{i_1 k_1}, a_{i_1 k_2}, \ldots, a_{i_1 k_p} \\ a_{i_2 k_1}, a_{i_2 k_2}, \ldots, a_{i_2 k_p} \\ \ldots \ldots \ldots \ldots \ldots \ldots \\ a_{i_p k_1}, a_{i_p k_2}, \ldots, a_{i_p k_p} \end{vmatrix}$$

THEOREM C4: The quadratic form $x^T A x$, where A is a symmetric real matrix of the dimension $(n; n)$, is negative semidefinite if and only if for all main subdeterminants of the matrix A the following inequalities hold:

$$(-1)^p \, A \begin{pmatrix} i_1, i_2, \ldots, i_p \\ k_1, k_2, \ldots, k_p \end{pmatrix} \geq 0, \quad k_j = i_j, \quad j = 1, 2, \ldots, p$$

$$1 \leq i_1 < i_2 \ldots < i_p \leq n, \quad p = 1, 2, \ldots, n, \quad a_{ij} = a_{ji}$$

where $A(:::)$ is defined in Theorem C3.

Note C1: Theorems C1 through C4 apply as well to the Hermitian form if $x^T A x$ is replaced by the form $x^* A x$, where A is a Hermitian matrix with symmetric complex conjugate elements, i.e. $a_{ij} = \bar{a}_{ji}$.

Appendix D:

Numerical solution of an overdetermined set of linear algebraic equations

Very many problems encountered in control theory, identification and other fields require to solve an overdetermined set of linear algebraic equations. Such a solution satisfies the individual given equations only with a certain error. If we minimize these errors we may consider the resultant solution to be optimum in the chosen sense. For instance, the regression model (13.57) of a single-input/single-output plant has the solution (13.59) satisfying the minimum of the sum of square errors (13.58). This is the case of an overdetermined set of equations, for we may assume that $K \gg \gamma$, where γ is the number of sought parameters of the vector ϑ.

In general, in a multi-input/multi-output case, we are solving a set of equations given by (13.104), i.e.

$$E_e^* = [Y - Z\Theta] R^{-1/2} \tag{D1}$$

where the individual matrices E_e^*, Y, Z, Θ, R have the dimensions $(K; p)$, $(K; p)$, $(K; v)$, $(v; p)$, $(p; p)$, respectively; K is the number of linearly independent equations, p the number of linearly independent solution sets, v the number of sought parameters in one solution set. The matrix E_e^* is the random error matrix, Y and Z are the matrices of the data enabling us to determine the sought parameters in the matrix Θ and R is the noise covariance matrix of the individual components of the vector $y^T(k) = [y_1(k), y_2(k), \ldots, y_p(k)]$ which, for $k = 1, 2, \ldots, K$, gives the matrix $Y(K, p)$.

In the single-input/single-output case, i.e. for $p = 1$, where the matrix Θ has only one column denoted by ϑ, the sought parameters are usually determined by minimization of the sum of square errors. The corresponding generalized cost function for the multi-input/multi-output case, where $p > 1$, is the square of the error matrix norm, $\|E_e^*\|^2$.

For a numerical solution we first rewrite eqn. (D1) in the form

$$E_e^* = D\tilde{\Theta} R^{-1/2} \tag{D2}$$

where $D = [Z, Y]$ and $\tilde{\Theta} = [-\Theta^T, E]^T$ of the dimension $(K, v + p)$ and $(v + p; p)$, respectively. Since the matrix norm does not change by orthogonal transformation, we select such a transformation matrix T that the matrix D will be transformed to the upper triangular matrix $D_{\mathbf{v}}(v + p; v + p)$, i.e.

$$\|E_e^*\|^2 = \|D\tilde{\Theta}R^{-1/2}\|^2 =$$
$$= \|T_K D\tilde{\Theta}R^{-1/2}\|^2 =$$
$$= \|D_{\mathbf{v}}\tilde{\Theta}R^{-1/2}\|^2 \quad \text{for} \quad T_K^T T_K = E \tag{D3}$$

The indicated transformation to the upper triangular matrix $D_{\mathbf{v}}$ is performed stepwise by means of the transformation matrices T_m of the dimension $(v + p + 1; v + p + 1)$. The resultant transformation matrix

$$T = \prod_m T_m \tag{D4}$$

where T_m are the so-called *matrices of elementary rotations* [64.1] of the form

$$T_m = \begin{bmatrix} 1 & & & & & & & & \\ & \ddots & & & & & & & \\ & & 1 & & & & & & \\ & & & c_m & \cdots & s_m & & & \cdots i \\ & & & & 1 & & & & \\ & & & & & \ddots & & & \\ & & & & & & 1 & & \\ & & & (-s_m) & \cdots & c_m & & & \cdots j \\ & & & & & & & 1 & \\ & & & & & & & & \ddots \\ & & & & & & & & & 1 \end{bmatrix} \tag{D5}$$

where the elements not indicated are zero. In order that the matrix T_m be orthogonal, the condition

$$c_m^2 + s_m^2 = 1 \tag{D6}$$

must be satisfied, where one of the coefficients may be chosen. If only a part of the matrix D, corresponding to its $(v + p + 1)$ rows, is multiplied by the matrix T_m, then, denoting the elements of the resulting matrix by d_{ij} and those of the matrix $T_m D$ by $\tilde{d}_{ij}$, we have

$$\begin{aligned}
\tilde{d}_{kl} &= d_{kl} & \text{for} \quad k \neq i \text{ and } k \neq j \\
\tilde{d}_{il} &= c_m d_{il} + s_m d_{jl} & \text{for} \quad k = i \\
\tilde{d}_{jl} &= -s_m d_{il} + c_m d_{jl} & \text{for} \quad k = j
\end{aligned} \quad \Bigg\} \quad l = 1, 2, \ldots, v + p \tag{D7}$$

Therefore, one of the coefficients in eqn. (D6) may be selected in such a way that the element $\tilde{d}_{j\mu} = 0$, $\mu = 1, \ldots, j - 1$. Repeating the above transformation we arrive at the upper triangular matrix of the dimension $(v + p; v + p)$. The elements of the row $(v + p + 1)$ are all zero. Putting at this place the row $(v + p + 2)$ of the original matrix D we may again nullify all elements of this row by means of the transformations described above and correct the preceding triangular matrix. The matrix D_{∇} is obtained after exhausting all K rows of the matrix D. For the nullification of the row $v + p + l$, $v + p < l \leq K$, obviously $(v + p)$ transformation matrices T_m will be needed in (D4).

It should be noted that it is sufficient to store in the computer memory only the upper triangular matrix while the new rows to be transformed into this triangular matrix may be only the newly entered numerical data. This way of proceeding is very convenient particularly when new data are obtained successively from a running experiment and the evaluation is executed simultaneously.

The algorithm may be designed in such a way that there is always $j = v + p + 1$ in the matrix (D5) and only the subscript $i = 1, 2, \ldots, v + p$ is changing. Then the algorithm may start with a zero matrix D_{∇} and the transformation matrix T_m may be denoted as T_i, $i = 1, 2, \ldots, (v + p)$.

In order that the elements of D_{∇} do not increase with increasing value of K it is useful to reduce them continuously according to the relation

$$F = \frac{1}{\sqrt{K}} D_{\nabla} \tag{D8}$$

where F was denoted in eqn. (13.109) as the *information matrix*. If we partition this matrix into blocks according to the dimensions v and p

$$F = \begin{bmatrix} F_{\Theta\Theta} & F_{\Theta R} \\ \hline 0 & F_{RR} \end{bmatrix} \tag{D9}$$

then $F_{\Theta\Theta}(v; v)$ and $F_{RR}(p; p)$ are again triangular matrices; $F_{\Theta R}(v; p)$ is a rectangular matrix.

Using eqn. (D9), the norm (D3) may now be written as follows:

$$\|\tilde{E}_e^*\|^2 = \|F\tilde{\Theta}R^{-1/2}\|^2$$

$$\left\| \begin{matrix} \tilde{E}_{e1}^* \\ \tilde{E}_{e2}^* \end{matrix} \right\|^2 = \left\| \begin{bmatrix} F_{\Theta\Theta}, & F_{\Theta R} \\ 0, & F_{RR} \end{bmatrix} \begin{bmatrix} -\Theta \\ E \end{bmatrix} R^{-1/2} \right\|^2 \tag{D10}$$

From eqn. (D10) follow the two equations

$$\tilde{E}_{e1}^* = (F_{\Theta R} - F_{\Theta\Theta}\Theta) R^{-1/2} \tag{D11}$$

$$\tilde{E}_{e2}^* = F_{RR} R^{-1/2} \tag{D12}$$

If we put in (D10)
$$F_{\Theta R} - F_{\Theta\Theta}\Theta = 0 \tag{D13}$$
then $\tilde{E}_{e1}^* = 0$ as well. Equation (D13) makes it possible to determine the matrix of the sought parameters because the matrix $F_{\Theta\Theta}$ is non-singular:
$$\hat{\Theta} = F_{\Theta\Theta}^{-1} F_{\Theta R} \tag{D14}$$
The calculation is simple since $F_{\Theta\Theta}$ as well as $F_{\Theta\Theta}^{-1}$ are upper triangular matrices.

The limit of the matrix E_{e2}^* as $K \to \infty$ is a unit matrix (see note to (13.114)) so that from eqn. (D12) it follows that
$$F_{RR} = \hat{R}^{1/2} \tag{D15}$$
which is the estimate of the *right Cholesky root of the covariance matrix* $R(p; p)$
$$\hat{R} = F_{RR}^T F_{RR} = \hat{R}^{T/2} \hat{R}^{1/2} \tag{D16}$$

The above results (D14) and (D16) correspond to the maximum likelihood estimate of the matrix Θ if the random errors in the matrix E_e^* have a normal probability distribution and are mutually independent.

When calculating the matrix Θ by the least squares method we assume that the random errors in the matrix E_e, although being also mutually independent, have an equal dispersion σ^2. The covariance matrix R is, therefore,
$$R = \sigma^2 E \tag{D17}$$
and
$$R^{1/2} = \sigma E \tag{D18}$$

In certain problems, the solution Θ of an overdetermined set of linear algebraic equations represents the parameters of the mathematical model as it is the case, for instance, in identification problems. In such cases it may be useful to assign the lowest weight to the data with the smallest subscript, i.e. to the oldest ones, in order that the newest data with the greatest weight may be applied in the calculation. Such a requirement may be satisfied, for example, by *exponential forgetting*, i.e. by multipying each row of the matrices E_e^*, Y, Z in eqn. (D1) by the weighting coefficient φ^{K-k}, $0 < \varphi \leq 1$, $k = 1, 2, ..., K$. In eqn. (D8) $\sqrt{\varkappa}$ must then be substituted for $\sqrt{K}$, where
$$\varkappa = \sum_{i=0}^{K-1} (\varphi^2)^i \tag{D19}$$

With increasing value of K, the value of $\varkappa$ may be calculated recursively, for it holds that
$$\varkappa_{K+1} = \sum_{i=0}^{K} (\varphi^2)^i = 1 + \varphi^2 \sum_{i=0}^{K-1} (\varphi^2)^i \tag{D20}$$
so that
$$\varkappa_{K+1} = 1 + \varphi^2 \varkappa_K \tag{D21}$$

In conclusion it should be noted that the above solution procedure may, according to the suggestion of V. Peterka [73.18], be modified by calculating the so-called *regression matrix* $G = F^{-1}$ defined as the *left Cholesky root of the matrix* $\varkappa[D^T D]^{-1}$, where $D = [Z, Y]$. Introducing, as in (D9), the blocks $G_{QQ}(v; v)$, $G_{QS}(v; p)$, $G_{SS}(p; p)$ we may express the matrices $\hat{\Theta}$, $\hat{R}^{-1/2}$ by the relations

$$\hat{\Theta} = -G_{QS} G_{SS}^{-1} \qquad \text{(D22)}$$

$$\hat{R}^{-1/2} = G_{SS}^{-1/2} \qquad \text{(D23)}$$

The advantage of this modification lies particularly in the fact that in single-input/single-output cases, where $p = 1$, G_{SS} is only a number and that, according to (D22), the matrices need not be inverted at all. For more details see the work quoted.

References

References in the text are indicated by the last two numerals of the year of publication followed by a point and one or two numerals according to the alphabetical order in that year, e.g. [69.4].

1809   Chapter

1. GAUSS, K. F.: Theoria motus corporum coelestium. F. Perthes and I. H. Besser, Paris, 1809.
   The Theory of the Motion of the Heavenly Bodies Moving about the Sun in Conic Sections. Reprint, Dover Publications, Inc. New York 1963.   13, 14

1877

1. ROUTH, E. J.: Stability of a Given State of Motion; London 1877.   12

1892

1. LYAPUNOV, A. M.: The General Problem of Stability of Motion. (Ляпунов А. М.: Общая задача об устойчивости движения). Ph.D. Thesis. Kharkov, 1892, pp. 250. Reprinted in French: Problème général de la stabilité du mouvement. In Ann. of Mathematics Studies, No. 17, Princeton University Press, 1907, 1949 and Academic Press, New York, 1966. Reprinted in Russian: Гостехиздат 1950.   12, 16

1912

1. FISHER, R. A.: On an Absolute Criterion for Fitting Frequency Curves. Mess. of Math. **41** (1912), 155.   13

1917, 1918

1. SCHUR, I.: Über Potenzreihen, die im Inneren des Einheitskreises beschränkt sind. J. für Math., **147** (1917), pp. 205—232; **148** (1918), pp. 122—145.   12

1947

1. FRAZER, R. A., DUNCAN, W. J., COLLAR, A. R.: Elementary Matrices and Some Applications to Dynamics and Differential Equations. University Press, Cambridge, 1947.   12

1955

1. CODDINGTON, E. A. and LEVINSON, N.: Theory of Ordinary Differential Equations. McGraw-Hill Book Company, Inc., New York, 1955. monograph

1957

1. BELLMAN, R.: Dynamic Programming. Princeton University Press, Princeton, N. Y., 1957. monograph

1958

1. ANDERSON, T. W.: An Introduction to Multivariate Statistical Analysis. John Wiley and Sons, Inc., New York, Chapman and Hall, Ltd., 1958. 13
2. HAHN, W.: Über die Anwendung der Methode von Ljapunov auf Differenzengleichungen. Math. Ann. **136** (1958), pp. 430—441. 12
3. KALMAN, R. E., KOEPCKE, R. W.: Optimal Synthesis of Linear Sampling Control Systems Using Generalized Performance Indexes. Trans. ASME **80 D** (1958), pp. 1820—1826. 17

1959

1. KALMAN, R. E., BERTRAM, J. E.: A Unified Approach to the Theory of Sampling Systems. Journ. of the Franklin Inst. 1959, pp. 405—436. 6, 7
2. KALMAN, R. E., BERTRAM, J. E.: General Synthesis Procedure for Computer Control of Single and Multi-Loop Linear Systems. Trans. AIEE **77** II (1959), pp. 602—609. 6, 7

1960

1. BERTRAM, J. E., KALMAN, R. E.: Control Systems Analysis and Design via the Second Method of Ljapunov. Trans. ASME, J. Basic Eng. **82 D** (1960), pp. 371—400. 12
2. DURBIN, J.: Estimation of Parameters in Time-Series Regression Models. J. Royal Stat. Soc., Ser. B, **22** (1960), pp. 139—153. 1
3. FADDEYEV, D. K., FADDEYEVA, V. N.: Computing Methods of Linear Algebra. (Фаддеев Д. К., Фаддеева В. Н.: Вычислительные методы линейной алгебры. Физматгиз, Москва 1960.) 7 monograph
4. KALMAN, R. E.: A New Approach to Linear Filtering and Prediction Problems. Trans. ASME, J. Basic Eng. **82 D** (1960), pp. 35—45. 14, 17
5. KALMAN, R. E.: On the General Theory of Control Systems. Proc., IFAC Congress, Moscow 1960, pp. 481—492. 16, 17

1961

1. JOSEPH, P. D., TOU, J. T.: On Linear Control Theory. Trans. AIEE **80** II Applications and Industry (1961), pp. 193—196. 17
2. KALMAN, R. E., BUCY, R. S.: New Results in Linear Filtering and Prediction Theory. Trans. ASME, J. Basic Eng. **83D** (1961), pp. 95—108. 14, 16, 17
3. PONTRYAGIN, L. S., BOLTYANSKIJ, V. G., GAMKRELIDZE, R. V., MISHCHENKO, E. F.: Mathematical Theory of Optimal Processes. (Понтрягин Л. С., Болтянский В. Г., Гамкрелидзе Р. В., Мищенко Е. Ф.: Математическая теория оптимальных процессов. Физматгиз, Москва 1961.) monograph

REFERENCES

1962

1. ZADEH, L. A.: An Introduction to State-Space Techniques. Proc. 1962 Joint Automatic Control Conference, AIEE 1962, New York. 2
2. ZADEH, L. A.: From Circuit Theory to System Theory. Proc. of the IRE, 1962, Part II, pp. 856—865. 2

1963

1. DIDUK, G. A.: On the Problem of Investigating Automatic Systems by the Method of Matrix Transformations. (Дидук Г. А.: К вопросу об исследовании автоматических систем методом матричных преобразований. Известия Академии Наук СССР, Техническая кибернетика **6** (1963), pp. 89—92.) 12
2. GILBERT, E. G.: Controllability and Observability in Multivariable Control Systems. SIAM J. on Control (1963), pp. 128—151. 9
3. GUNCKEL, T. L., FRANKLIN, G. F.: A General Solution for Linear Sampled Data Control. Trans. ASME, J. Basic Eng. **85** D (1963), pp. 197—203. 17
4. KALMAN, R. E.: Mathematical Description of Linear Dynamical Systems. SIAM J. on Control (1963), pp. 152—192. 3, 5, 9
5. KALMAN, R. E.: New Methods in Wiener Filtering Theory. Proceedings of the first symposium on Engineering Applications of Random Function Theory and Probability. John Wiley and Sons, Inc., New York 1963, pp. 270—388. 14, 17
6. MAYNE, D. Q.: Optimal Non-Stationary Estimation of the Parameters of a Linear System with Gaussian Inputs. Journ. Elect. Cont., **14** (1963), pp. 101—112. 13
7. REKTORYS, K.: Přehled užité matematiky. SNTL, Praha 1963. Also in English: Survey of Applicable Mathematics. Iliffe Books, London 1969. monograph
8. WING, J., DESOER, C. A.: The Multiple-Input Minimum-Time Regulator Problem. IEEE Trans. on Automatic Control, AC-8 (1963), pp. 125—136. 15
9. ZADEH, L. A., DESOER, C. A.: Linear System Theory — the State Space Approach. McGraw-Hill, New York 1963. monograph

1964

1. HO, Y. C., LEE, R. C. K.: A Bayesian Approach to Problems in Stochastic Estimation and Control. IEEE Trans. on Automatic Control, AC-9 (1964), pp. 333—339. 14
2. LEE, R. C. K.: Optimal Estimation, Identification and Control. M. I. T. Press, Cambridge, Mass., 28, 1964. monograph

1965

1. BROCKETT, R. W.: Poles, Zeros and Feedback: State Space Interpretation. IEEE Trans. on Automatic Control AC-10 (1965), pp. 129—135. 16
2. DEUTSCH, R.: Estimation Theory. Prentice-Hall, Inc., Englewood Cliffs, N. J., USA, 1965. 13, 14
3. STREJC, V. et al.: Syntéza regulačních obvodů s číslicovým počítačem (Design of Computer Controlled Systems). ČSAV, Praha 1965. Also in German: Synthese von Regelungssystemen mit Prozessrechner. Akademie-Verlag, Berlin 1967. 12, monograph
4. STRIEBEL, C.: Sufficient Statistics in the Optimum of Stochastic Systems. J. Math. Anal. Appl. **12** (1965), pp. 576—592. 17

1966

1. ACKERMANN, J.: Beschreibungsfunktionen für die Analyse und Synthese von nichtlinearen Abtast-Regelkreisen. Regelungstechnik **14** (1966), pp. 497—544.
2. GANTMACHER, F. R.: Theory of Matrices (Гантмахер Ф. Р.: Теория матриц. Изд. Наука, Москва). First edition 1953—54, second edition 1966. Also in English: Theory of Matrices. Chelsea, New York 1960. monograph
3. HO, B. L., KALMAN, R. E.: Effective Construction of Linear Statevariable Models from Input/Output Functions. Regelungstechnik **14** (1966), pp. 545 to 548. 11
4. CHINDAMBARA, M. R., JOHNSON, C. D., RANE, D. S., TUEL, W. C.: On the Transformation to Phase-Variable Canonical Form. IEEE Trans. on Automatic Control, AC-11 (1966), pp. 607—610. 10
5. LUENBERGER, D. G.: Observers for Multivariable Systems. IEEE Trans. on Automatic Control, AC-11 (1966), pp. 190—197. 15

1967

1. ANDERSON, B. D. O., LUENBERGER, D. G.: Design of Multivariable Feedback Systems. Proc. IEE **114** (1967), pp. 395—399. 15
2. AOKI, M.: Optimization of Stochastic Systems. Academic Press, New York, London 1967. monograph
3. KUČERA, V.: Syntéza diskrétního regulačního obvodu podle kritéria konečného počtu kroků (Discrete Control Loop Synthesis According to the Criterion of Finite Number of Steps) Ph.D. Thesis, ÚTIA ČSAV, Praha 1967. 15
4. LUENBERGER, D. G.: Canonical Forms for Multivariable Systems. IEEE Trans. on Automatic Control, AC-12 (1967), pp. 290—293. 15

1968

1. ACKERMANN, J.: Zeitoptimale Mehrfach-Abtastregelsysteme. Proc., IFAC Symposium: Multivariable Control Systems. VDI/VDE, Düsseldorf 1968, Vol. 1.5. 15
2. ÅSTRÖM, K. J.: Lectures on the Identification Problem — The Least Squares Method. Lund Institute of Technology, Division of automatic control, Report 6806, 1968. 13
3. BUCY, R. S.: Canonical Forms for Multivariable Systems. IEEE Trans. on Automatic Control, AC-13 (1968), pp. 567—569. 11
4. CHEN, C. T., DESOER, C. A.: A Proof of Controllability of Jordan Form of State Equations. IEEE Trans. on Automatic Control, AC-13 (1968), pp. 195—196. 9, 10
5. JEŽEK, J.: Algoritmus pro výpočet diskrétního přenosu lineární dynamické soustavy (Algorithm for the Calculation of the Discrete Transfer Function of a Linear Dynamic Plant). Kybernetika **4** (1968), pp. 246—259. 7
6. KITAMURA, S., HIRAI, K., NISHIMURA, M.: Stability of Nonlinear Multivariable Control Systems with Time Lags by Lyapunov's Direct Method. Proc., IFAC Symposium: Multivariable Control Systems. VDI/VDE, Düsseldorf 1968, Vol. 1.2. 12
7. LUDYK, G.: Zeitoptimale Abtastsysteme mit mehreren beschränkten Stellgrössen. Proc., IFAC Symposium: Multivariable Control Systems. VDI/VDE, Düsseldorf 1968, Vol. 1.5. 15

8. LORIQUET, P.: Optimisation d'un système linéaire discrèt avec critère quadratique. Proc., IFAC Symposium: Multivariable Control Systems. VDI/VDE, Düsseldorf 1968, Vol. 1.6. 16
9. SAGE, A. P.: Optimum Systems Control. Prentice-Hall, Inc., Englewood Cliffs, N. J., USA, 1968. monograph
10. WONHAM, W. M.: On a Matrix Riccati Equation of Stochastic Control. SIAM J. on Control 6 (1968), pp. 681—698. 9, 17

1969

1. ACKERMANN, J.: Diskussionsbeitrag zur Arbeit von O. FÖLLINGER "Synthese von Mehrfachregelungen mit endlicher Einstellzeit". Regelungstechnik 17 (1969), pp. 170—174. 11, 15
2. BRYSON, A. E., YU-CHI-HO: Applied Optimal Control. Blaisdell Publishing Company, Waltham, Mass., Toronto, London 1969. monograph
3. JEŽEK, J.: Převody přenosové funkce lineární soustavy ze spojitého tvaru na diskrétní a obráceně (Conversions of the Transfer Function of a Linear System from the Continuous to the Discrete Form and Vice Versa). Ph.D. Thesis, ÚTIA ČSAV, Praha 1969. 7
4. KALMAN, R. E., FALB, P. L., ARBID, M. A.: Topics in Mathematical System Theory. McGraw-Hill, New York 1969. monograph
5. KUČERA, V.: Stabilita diskrétních regulačních obvodů (Stability of Discrete Control Systems). Research Report ÚTIA ČSAV No. 261, Praha 1969. 12
6. MEDITCH, J. S.: Stochastic Optimal Linear Estimation and Control. McGraw-Hill, New York 1969. monograph
7. MEHRA, R. K.: On the Identification of Variances and Adaptive Kalman Filtering. Proc. JACC 1969. Science Press, Inc., Ephrata, Penna. 1969, pp. 494 to 505. 14, 17
8. SAIN, M. K., MASSEY, J. L.: Invertibility of Linear Time-Invariant Dynamical Systems. IEEE Trans. on Automatic Control, AC-14 (1969), pp. 141—149. 16
9. SILVERMAN, L. M.: Inversion of Multivariable Linear Systems. Proc. JACC 1969. Science Press, Inc., Ephrata, Penna., 1969, pp. 853—859. 15
10. SILVERMAN, L. M.: Inversion of Multivariable Linear Systems. IEEE Trans. on Automatic Control AC-15 (1969), pp. 270—276. 16
11. WOLOVICH, W. A., FALB, P. L.: On the Structure of Multivariable Systems. Proc. JACC 1969. Science Press, Inc., Ephrata, Penna., 1969, pp. 860—868. 11

1970

1. ÅSTRÖM, K. J.: Introduction to Stochastic Control Theory. Academic Press, New York, London 1970. monograph
2. BUCY, R. S., ACKERMANN, J.: Über die Anzahl der Parameter von Mehrgrössensystemen. Regelungstechnik 18 (1970), pp. 451—452. 11
3. CADZOW, J. A., MARTENS, H. R.: Discrete-Time and Computer Control Systems. Prentice-Hall, Inc., Englewood Cliffs, N. J., USA, 1970. monograph
4. KUČERA, V.: Optimální řízení lineárních mnohoparametrových soustav podle kritéria konečného počtu kroků (Optimum Control of Linear Multiparameter Systems According to the Criterion of Finite Number of Steps). Research Report ÚTIA ČSAV No. 305, Praha 1970. 15
5. LOCATELLI, A., RINALDI, S.: Controllability versus Sensitivity in Linear Discrete Systems. IEEE Trans. on Automatic Control AC-15 (1970), pp. 254 to 255. 9

6. Man, F. T.: Comments on "A Theorem on the Lyapunov Matrix Equation". IEEE Trans. on Automatic Control AC-15 (1970), pp. 279—280. 12
7. Mehra, R. K.: On the Identification of Variances and Adaptive Kalman Filtering. IEEE Trans. on Automatic Control AC-15 (1970), pp. 175—184. 14
8. Rosenbrock, H. H.: State-Space Theory and Multivariable Theory. Nelson, Melbourne 1970. monograph
9. Tether, A. J.: Construction of Minimal Linear State-Variable Models from Finite Input-Output Data. IEEE Trans. on Automatic Control AC-15 (1970), pp. 427—436. 13
10. Tse, E., Athans, M.: Optimal Minimal Order Observer-Estimators for Discrete Linear Time-Varying Systems. IEEE Trans. on Automatic Control AC-15 (1970), pp. 416—426. 14

1971

1. Ackermann, J.: Die minimale Ein-Ausgangs-Beschreibung von Mehrgrössensystemen und ihre Bestimmung aus Ein-Ausgangs-Messungen. Regelungstechnik 19 (1971), pp. 203—206. 11
2. Ackermann, J., Bucy, R. S.: Canonical Minimal Realization of a Matrix of Impulse Response Sequences. Information and Control 19 (1971), pp. 224 to 231. 11
3. Anderson, B. D. O.: Stability Properties of Kalman-Bucy Filters. J. Franklin Inst. 291 (1971), pp. 137—144. 14
4. Budin, M. A.: Minimal Realization of Discrete Linear Systems from Input-Output Observations. IEEE Trans. on Automatic Control AC-16 (1971), pp. 395—401. 13
5. Cooper, C. A., Nahi, N. E.: An Optimal Stochastic Control Problem with Observation Cost. IEEE Trans. on Automatic Control AC-16 (1971), pp. 185—188. 14, 17
6. Cheneveaux, B.: Synthesis of a Multivariable Control System via Multilevel Techniques. Proc., IFAC Symposium: Multivariable Technical Control Systems. North — Holland Publishing Company, Amsterdam 1971, Vol. 2, Ref. 3.2. 16
7. Freud, E.: Zeitvariable Mehrgrössensysteme. Springer, Berlin 1971.
8. Jury, E. I.: "Inners" Approach to Some Problems of System Theory. IEEE Trans. on Automatic Control AC-16 (1971), pp. 233—240. 12
9. Kaminski, P. G., Bryson, A. E., Schmidt, S. F.: Discrete Square Root Filtering: A Survey of Current Techniques. IEEE Trans. on Automatic Control, AC-16 (1971), pp. 727—736. 16
10. Kant, D., Winkler, D.: Numerische Lösung schlecht konditionierter linearer Gleichungssysteme auf dem Prozessrechner. Regelungstechnik 19 (1971), pp. 145—149, 211—214. D, 13, 16
11. Kučera, V.: The Structure and Properties of Time-Optimal Discrete Linear Control. IEEE Trans. on Automatic Control AC-16 (1971), pp. 375—377. 15
12. La Cava, M., Nicosia, S.: A New Way to the Solution of the Quadratic Optimization Problem for Discrete Multivariable Systems. Proc., IFAC Symposium: Multivariable Technical Control Systems. North-Holland Publishing Company, Amsterdam, 1971, Vol. 2, Ref. 2.2.6. 16
13. Laurent, F., Gentina, J. C., Staroswiecki, M., Borne, F.: On the Operating Time Optimalization of Incompletely Observable Sampled Data Systems with a Plurality of Monitoring. Proc., IFAC Symposium: Multivariable Technical

Control Systems. North-Holland Publishing Company, Amsterdam 1971, Vol. 2, Ref. 2.4.2. 15
14. LUENBERGER, D. G.: An Introduction to Observers. IEEE Trans. on Automatic Control AC-16 (1971), pp. 596—602. 15
15. MEHRA, R. K.: On-Line Identification of Linear Dynamic Systems with Applications to Kalman Filtering. IEEE Trans. on Automatic Control AC-16 (1971), pp. 12—21. 14
16. MEIER III, L., LARSON, R. E., TETHER, A. J.: Dynamic Programming for Stochastic Control of Discrete Systems. IEEE Trans. on Automatic Control AC-16 (1971), pp. 767—775. 17
17. MENDEL, J. M.: Computational Requirements for a Discrete Kalman Filter. IEEE Trans. on Automatic Control AC-16 (1971), pp. 748—758. 14
18. RAPPAPORT, D., SILVERMAN, L. M.: Structure and Stability of Discrete-Time Optimal Systems. IEEE Trans. on Automatic Control AC-16 (1971), pp. 227 to 233.
19. ROJTENBERG, Ya. N.: Automatic Control. (Ройтенберг Я. Н.: Автоматическое управление. Изд. Наука, Москва 1971.) monograph
20. SAGE, A. P., MELSA, J. L.: Estimation Theory with Applications to Communications and Control. McGraw-Hill, New York 1971. monograph
21. SAGE, A. P., MELSA, J. L.: System Identification. Academic Press, New York, London 1971, Vol. 80. 13, 14
22. SINGER, R. A., SEA, R. G.: Increasing the Computational Efficiency of Discrete Kalman Filters. IEEE Trans. on Automatic Control AC-16 (1971), pp. 254 to 257.
23. TARN, T. J., RAO, S. K., ZABORSZKY, J.: Singular Control of Linear-Discrete Systems. IEEE Trans. on Automatic Control AC-16 (1971), pp. 401—410. 15
24. TSE, E.: On the Optimal Control of Stochastic Linear Systems. IEEE Trans. on Automatic Control AC-16 (1971), pp. 776—785. 17
25. TUZAR, A., ŠINDELÁŘ, J.: Úvod do teorie stability regulovaných soustav (Introduction to the Theory of Stability of Controlled Systems). Inset of the journal Kybernetika **6** (1970), and **7** (1971). 12
26. TUZAR, A.: Stabilita a optimální stabilizace nestacionárních impulsních soustav (Stability and Optimum Stabilization of Time-Variant Impulse Systems). Research Report ÚTIA ČSAV No. 367, Praha 1971. 12
27. YÜKSEL, Y. O., BONGIORNO, J. J.: Observers for Linear Multivariable Systems with Applications. IEEE Trans. on Automatic Control AC-16 (1971), pp. 603 to 612. 15
28. ZUBOV, V. I.: Lectures on Control Theory. State University Press, Leningrad 1971. (Зубов В. И.: Лекции по теории управления. Изд. Ленинградского Государственного университета, 1971.) monograph

1972

1. ACKERMANN, J.: Abtastregelung. Springer-Verlag, Berlin, Heidelberg, New York, 1972. monograph
2. ACKERMANN, J.: Der Entwurf linearer Regelungssysteme im Zustandsraum. Regelungstechnik **20** (1972), pp. 297—300. 15
3. HAVIRA, R. M., LEWIS, J. B.: Computation of Quantized Controls Using Differential Dynamic Programming. IEEE Trans. on Automatic Control AC-17 (1972), pp. 191—196. 16

4. JURY, E. I., AHN, S. M.: Symmetric and Innerwise Matrices for the Root-Clustering and Root-Distribution of a Polynomial. J. Franklin Inst. **293** (1972), pp. 433—450. — 12
5. KUČERA, V.: The Discrete Riccati Equation of Optimal Control. Kybernetika **8** (1972), pp. 430—447. — 9, 16
6. KUČERA, V.: State Space Approach to Discrete Optimal Control. Kybernetika **8** (1972), pp. 233—251. — 15, 16
7. KUČERA, V.: A Contribution to Matrix Quadratic Equations. IEEE Trans. on Automatic Control, AC-17 (1972), pp. 344—347. — 16
8. KUČERA, V.: On Non-Negative Definite Solutions to Matrix Quadratic Equations. Proc., 5th IFAC World Congress, Vol. 4, Paris 1972. — 16
9. KWAKERNAAK, H., SIVAN, R.: Linear Optimal Control Systems. John Wiley and Sons, Inc., New York 1972. — monograph
10. NAHI, N. E., SCHAEFER, B. M.: Decision-Directed Adaptive Recursive Estimators: Divergence Prevention. IEEE Trans. on Automatic Control AC-17 (1972), pp. 61—68. — 14
11. NOMURA, T., NAKAMURA, K.: Discrete-Time Regulator of an Infinite-Dimensional System. J. Franklin Inst. **293** (1972), pp. 229—241. — 17
12. OESTERHELT, G.: Optimale Kontrolle endlicher diskreter Prozesse mit konstantem Regler. Regelungstechnik **20** (1972), pp. 522—524. — 15
13. PREUSCHE, G.: Verallgemeinerung des Entwurfs auf endliche Einstellzeit. Regelungstechnik **20** (1972), pp. 477—480, 518—522. — 15
14. RAMASWAMI, B., RAMAR, K.: A New Method of Solving Optimal Servo Problems. IEEE Trans. on Automatic Control AC-17 (1972), pp. 131—135. — 11
15. SARIDIS, G. N., LOBBIA, R. N.: Parameter Identification and Control of Linear Discrete-Time Systems. IEEE Trans. on Automatic Control AC-17 (1972), pp. 52—60. — 13, 15
16. STREJC, V.: State Space Synthesis of Discrete Linear Systems. Kybernetika **8** (1972), pp. 83—113. — 16
17. TSE, E., ANTON, J. J.: On the Identifiability of Parameters. IEEE Trans. on Automatic Control AC-17 (1972), pp. 637—646. — 9, 13

1973

1. ALEXANDROV, A. G.: Design of Discrete Control Systems with Given Properties. (Александров А. Г.: Построение дискретных систем управления с заданными свойствами. Автоматика и телемеханика (1973)); No. 9, pp. 57—66. — 15
2. ANTONOV, V. G., SHEPELYAVYI, A. I.: Optimal Control in a Finite Time Interval for Discrete Systems in the Problem of Minimization of a Nonhomogeneous Quadratic Functional. (Антонов В. Г., Шепелявый А. И.: Оптимальное управление на конечном интервале времени для дискретных систем в задаче минимизации неоднородного квадратичного функционала. Автоматика и телемеханика (1973)); No. 4, pp. 43—50. — 15
3. BOLTYANSKYI, V. G.: Optimum Control of Discrete Systems. (Болтянский В. Г.: Оптимальное управление дискретными системами. Изд. Наука, Москва 1973.) — monograph
4. BOOZER, D. D., MCDANIEL, Jr. W. L.: Linear Stochastic System Identification Using Correlation Techniques. J. Franklin Inst. **296** (1973), pp. 59—69. — 13

5. CAREW, B., BÉLANGER, P. R.: Identification of Optimum Filter Steady-State Gain for Systems with Unknown Noise Covariances. IEEE Trans. on Automatic Control AC-18 (1973), pp. 582—587. 14
6. CSÁKI, F.: Die Zustandsraummethode in der Regelungstechnik. Akadémiai kiadó, Budapest 1973. 2, 3, 5, 6, 7
7. DREYFUS, S. E., KAN, Y. C.: A General Dynamic Programming Solution of Discrete-Time Linear Optimal Control Problems. IEEE Trans. on Automatic Control AC-18 (1973), pp. 286—289. 15
8. GEVERS, M. R., KAILATH, T.: An Innovations Approach to Least-Squares Estimation — Part IV: Discrete-Time Innovations Representations and Recursive Estimation. IEEE Trans. on Automatic Control AC-18 (1973), pp. 588—600. 14
9. GRUJIČ, L. T., ŠILJAK, D. D.: On Stability of Discrete Composite Systems. IEEE Trans. on Automatic Control AC-18 (1973), pp. 522—524. 9
10. YANUSHEVSKYI, P. T.: Theory of Linear Optimal Multivariable Systems. (Янушевский П. Т.: Теория линейных оптимальных многосвязных систем. Изд. Наука, Москва 1973). monograph
11. JURY, E. I.: Inner Algorithm Test for Controllability and Observability. IEEE Trans. on Automatic Control AC-18 (1973), pp. 682—683. 9
12. KUČERA, V.: A Review of the Matrix Riccati Equation. Kybernetika 9 (1973), pp. 42—61. 16
13. KUNTSEVICH, V. M.: On the Synthesis of Optimum Discrete Control Systems with Permanently Acting Disturbing Variables. (Кунцевич В. М.: О синтезе оптимальных дискретных систем управления при постоянно действующих возмущениях. Автоматика и телемеханика (1973)); No. 4, pp. 60—69. 15
14. KUNTSEVICH, V. M., LYCHAK, M. M.: Synthesis of Optimal and Suboptimal Discrete Control Systems of Deterministic and Stochastic Plants by Means of Lyapunov Functions. (Кунцевич В. М., Лычак М. М.: Синтез оптимальных и субоптимальных дискретных систем управления детерминированными и стохастическими объектами с помощью функций Ляпунова. Автоматика и телемеханика (1973)); No. 1, pp. 70—78. 17
15. MULLIS, C. T.: On the Controllability of Discrete Linear Systems with Output Feedback. IEEE Trans. on Automatic Control AC-18 (1973), pp. 608—615. 9
16. PAYNE, H. J., SILVERMAN, L. M.: On the Discrete Time Algebraic Riccati Equation. IEEE Trans. on Automatic Control AC-18 (1973), pp. 226—234. 15
17. PERELMUTER, V. M.: Suboptimal Algorithm for Estimating the Parameters and State of Dynamic Systems. (Перельмутер В. М.: Субоптимальный алгоритм оценки параметров и состояния динамических систем. Автоматика и телемеханика (1973)); No. 12, pp. 52—59. 13, 14
18. PETERKA, V.: Průběžné odhadování parametrů vícerozměrových regresních modelů (Continuous Estimation of the Parameters of Multidimensional Regression Models). Research Report ÚTIA ČSAV No. 541, Praha 1973. 13
19. SOEDA, T., YOSHIMURA, T.: A Practical Filter for Systems with Unknown Parameters. ASME Journal of Dynamic Systems, Measurement and Control 95 D (1973), pp. 396—401. 14
20. ŠTECHA, J., KOZÁČIKOVÁ, A., KOZÁČIK, J., LIDICKÝ, J.: Optimal Control of a Linear Discrete System. Kybernetika 9 (1973), 5, pp. 374—388. 16
21. TSE, E.: Observer-Estimators for Discrete-Time Systems. IEEE Trans. on Automatic Control AC-18 (1973), pp. 10—16. 14

22. VORCHIK, B. G., FETISOV, V. N., SHTEYNBERG, SH. E.: Identification of a Stochastic Closed Control System. (Ворчик Б. Г., Фетисов В. Н., Штейнберг Ш. Е.: Идентификация стохастической замкнутой системы. Автоматика и телемеханика (1973)); No. 7, pp. 41—52. 13

23. WENNER, G.: Bestimmung der Zustandsraum-Beschreibung linearer zeitinvarianter Mehrgrössensysteme aus ihrer Eingangs-Ausgangs-Beschreibung. Regelungstechnik 21 (1973), pp. 294—297. 13

1974

1. ANDERSON, B. D. O., JURY, E. I.: On the Reduced Hermite and Reduced Schur-Cohn Matrix Relationships. Int. J. Control 19 (1974), pp. 877—890. 12
2. BARNETT, S.: Application of the Routh Array to Stability of Discrete-Time Linear Systems. Int. J. Control 19 (1974), pp. 47—55. 12
3. CADZOW, J. A.: Minimum-Amplitude Control of Linear Discrete Systems. Int. J. Control 19 (1974), pp. 765—780. 15
4. CASTI, J.: Necessary and Sufficient Conditions in the Minimal Control Field Problem for Linear Systems. Research Report IIASA, RM-74-10, 1974. 15
5. CASTI, J.: Minimal Control Fields and Pole-Shifting by Linear Feedback. Research Report IIASA, RM-74-9, 1974. 15
6. CROSSLEY, T. R., PORTER, B.: Dead-Beat Control of Sampled-Data Systems with Bounded Input. Int. J. Control 19 (1974), pp. 869—876. 16
7. EYKHOFF, P.: System Identification. John Wiley and Sons, London 1974. 13
8. FLOWER, J. O., GUPTA, R. K.: Optimal Control Considerations of Diesel Engine Discrete Models. Int. J. Control 19 (1974), pp. 1057—1068. 14, 15
9. FORMALSKII, A. M.: Controllability and Stability of a System with Limited Sources. (Формальский А. М.: Управляемость и устойчивость систем с ограниченными ресурсами. Изд. Наука, Москва 1974.) 12
10. GELB, A.: Applied Optimal Estimation. M. I. T. Press, Cambridge 1974. 13
11. GRAY, CH. E.: Minimum Energy Control of Systems Using N-State Controls. IEEE Trans. on Automatic Control AC-19 (1974), pp. 367—373. 16
12. HEALEY, M., AL-BAHRANI, H. M. S.: A Stochastic Optimal Discrete Implicit Model-Following Control Law. Int. J. Control 19 (1974), pp. 789—795. 17
13. ISERMANN, R., BAUR, U., KURZ, H.: Identifikation dynamischer Prozesse mittels Korrelation und Parameterschätzung. Regelungstechnik 22 (1974), pp. 235—242.
14. LIFSHIC, N. A., VINOGRADOV, V. N., GOLUBYEV, G. A.: Correlation Theory of the Optimal Control of Multivariable Processes. (Лифшиц Н. А., Виноградов В. Н., Голубев Г. А.: Корреляционная теория оптимального управления многомерными процессами. Сов. радио, Москва 1974.)
15. KATKOVNIK, V. YA., KULCHITSKYI, O. YU.: Identification of Linear Dynamic Systems with a Random Disturbing Variable. (Катковник В. Я., Кульчицкий О. Ю.: Идентификация линейных динамических систем со случайным возмущением. Кибернетика и вычислительная техника 24 (1974)); pp. 19—22. 13
16. KIM, C. H., LINDORF, D. P.: Input Frequency Requirements for Identification through Liapunov Methods. Int. J. Control 20 (1974), pp. 35—48. 13
17. KREBS, V., THÖM, H.: Parameter-Identifizierung nach der Methode der kleinsten Quadrate — ein Überblick. Regelungstechnik 22 (1974), pp. 1—10.
18. KUNTSEVICH, V. M.: On the Synthesis of the Control of Optimum Discrete Servomechanisms. (Кунцевич В. М.: О синтезе оптимальных дискретных

следящих систем управления. Кибернетика и вычислительная техника **24** (1974)); pp. 3—9.

19. MELIN, C., MICHAILESCO, G., SIRET, J. M.: Partial Decoupling of Multivariable Nonlinear Sampled-Data Control Systems and the Asymptotic Stability Analysis. Int. J. Control **19** (1974), pp. 977—988.   12, 15
20. MORF, M., SIDHU, G. S., KAILATH, T.: Some New Algorithms for Recursive Estimation in Constant, Linear, Discrete-Time Systems. IEEE Trans. on Automatic Control AC-19 (1974), pp. 315—323.
21. REDCHENKO, I. F.: On the Problem of Asymptotic Stability in the Large of One Class of Nonlinear Impulse Systems. (Редченко И. Ф.: К вопросу об асимптотической устойчивости в целом одного класса нелинейных импульсных систем. Кибернетика и вычислительная техника, **24** (1974)); pp. 37—42.   12
22. SAUER, G. J., MELSA, J. L.: Stochastic Control with Continuously Variable Observation Costs for a Class of Discrete Nonlinear Systems. IEEE Trans. on Automatic Control AC-19 (1974), pp. 234—239.   17
23. SHERIF, A., WU, M. Y.: Identification of Linear Dynamical Systems. Int. J. Control **19** (1974), pp. 185—192.   13
24. SIMS, G. S.: An Algorithm for Estimating a Portion of the State Vector. IEEE Trans. on Automatic Control AC-19 (1974), pp. 391—393.   14
25. WOLOVICH, W. A.: Linear Multivariable Systems. Springer-Verlag, New York 1974.   11
26. ZAHR, K. M., SLIVINSKIJ, CH.: State Variable Feedback in Computer-Controlled Multivariable Systems. IEEE Trans. on Automatic Control AC-19 (1974), pp. 404—407.   16

1975

1. AKAIKE, H.: Markovian Representation of Stochastic Processes by Canonical Variables. SIAM J. Control **13** (1975), pp. 162—173.   14
2. AKASHI, H., NOSE, K.: On Certainty Equivalence of Stochastic Optimal Control Problem. Int. J. Control **21** (1975), pp. 857—863.   17
3. BAR-NESS, Y.: Pole Sensitivity of the Quadratic Optimal Regulator. Electronics Letters **12** (1976), pp. 341—343.   15
4. BAR-NESS, Y.: The Ripple Minimization Problems in Sample Data Systems. Int. J. Control **22** (1975), pp. 689—699.   15
5. BAR-NESS, Y., LANGHOLZ, G.: Preservation of Controllability under Sampling. Int. J. Control **22** (1975), pp. 39—47.   9, 10
6. BAR-NESS, Y., LANGHOLZ, G.: Preservation of Controllability under Linear Transformation. Int. J. Systems Sci. **6** (1975), pp. 1089—1092.   9, 10
7. BARNET, S.: Insensitivity of Optimal Linear Discrete-Time Regulators. Int. J. Control **21** (1975), pp. 843—848.   16
8. BLAIR, W. P., SWORDER, D. D.: Feedback Control of a Class of Linear Discrete Systems with Jump Parameters and Quadratic Cost Criteria. Int. J. Control **21** (1975), pp. 833—841.   16
9. FURUTA, K., PAQUET, J. G.: Determination of Matrix Transfer Function in the Form of Matrix Fraction from Input-Output Observations. IEEE Trans. on Automatic Control AC-20 (1975), pp. 392—396.   13
10. GLOVER, K.: Some Geometrical Properties of Linear Systems with Implications in Identification. Proc., Congr. IFAC Boston, 1975.   13
11. GUIDARZI, R.: Canonical Structures in the Identification of Multivariable Systems. Automatica **11** (1975), pp. 361—374.   13

12. HAZEWINKEL, M., KALMAN, R. E.: On Invariants, Canonical Forms and Moduli for Linear, Constant, Finite Dimensional, Dynamical Systems. Proceedings on Mathematical Systems Theory, Udine, Italy, 1975. Springer-Verlag, Berlin, 1976.   10, 11, 15
13. HIRVONEN, J., BLOMBERG, H., YLINEN, R.: An Algebraic Approach to Canonical Forms and Invariants for Linear Time-Invariant Differential and Difference Systems. Int. J. Systems Sci. **6** (1975), pp. 1119—1134.   10, 11, 15
14. LUO, Z., BULLOCK, T. E.: Discrete Kalman Filtering Using a Generalized Companion Form. IEEE Trans. on Automatic Control AC-20 (1975), pp. 227 to 230.   14, 17
15. KWON, W. H., PEARSON, A. E.: On the Stabilization of a Discrete Constant Linear System. IEEE Trans. on Automatic Control AC-20 (1975), pp. 800—801.   9, 15
16. MORF, M., KAILATH, T.: Square-Root Algorithms for Least-Squares Estimation. IEEE Trans. on Automatic Control AC-20 (1975), pp. 487—497.   13
17. NAGATA, A., NISHIMURA, T., IKEDA, M.: Linear Function Observer for Linear Discrete-Time Systems. IEEE Trans. on Automatic Control AC-20 (1975), pp. 401—407.   14
18. SINHA, A. K., SHARMA, J. P.: An On-Line Technique for Weather Forecast. Int. J. Systems Sci. **6** (1975), pp. 681—688.   15, 17
19. STAMM, J. C., PRIEMER, R.: Measurement Designs to Improve the Computational Efficiency of Linear System State Estimators. Int. J. Control **22** (1975), pp. 445—460.   14
20. TANAKA, A.: Parallel Computation in Linear Discrete Filtering. IEEE Trans. on Automatic Control AC-20 (1975), pp. 573—575.   14, 17
21. TANAKA, A.: Treatment of Independent Subsystems in LQG Estimation/Control Problem. Memoirs of Metropolitan College of Technology Tokyo, No. 3 (1975), pp. 19—27.   14, 17
22. TAYLOR, F. J.: Finite Fading Memory Filtering. IEEE Trans. on Systems, Man, and Cybernetics SMC-5 (1975), pp. 134—137.   14, 17
23. TSE, E., WEINERT, H. L.: Structure Determination and Parameter Identification for Multivariable Stochastic Linear Systems. IEEE Trans. on Automatic Control AC-20 (1975), pp. 603—613.   13
24. WARREN, M. E., ECKBERG, A. E.: On the Dimensions of Controllability Subspaces: A Characterization via Polynomial Matrices and Kronecker Invariants. SIAM J. Control **13** (1975), pp. 434—445.   9, 10

1976

1. BAR-NESS, Y.: Sufficient Conditions for the Solution of the Discrete Infinite-Time Linear Regulator. Int. J. Control **24** (1976), pp. 335—343.
2. BRADSHAW, A., PORTER, B.: Design of Linear Multivariable Discrete-Time Tracking Systems for Plants with Inaccessible States. Int. J. Control **24** (1976), pp. 275—281.   15
3. CLARK, J. M. C.: The Consistent Selection of Parametrizations in System Identification. Proc. JACC, West Lafayette, Ind., 1976.   13
4. DEBALQUE, B., GEVERS, M., INSTALLÉ, M.: Combined Identification of the Input-Output and Noise Dynamics of a Closed-Loop Controlled Linear System. Int. J. Control **24** (1976), pp. 345—360.   13, 15
5. DICKINSON, B. W.: The Classification of Linear State Variable Control Laws. SIAM J. Control and Optimization **14** (1976), pp. 467—477.   15, 16

6. DÖRRSCHEIDT, F.: Entwurf auf endliche Einstellzeit bei linearen Regelungssystemen mit veränderlichen Parametern. Regelungstechnik 24 (1976), pp. 89 to 96. ... 15
7. EMRE, E., SILVERMAN, L. M.: Minimal Dynamic Inverses for Linear Systems with Arbitrary Initial States. IEEE Trans. on Automatic Control AC-21 (1976), pp. 766—769. ... 15
8. HALYO, N., CAGLAYAN, A. K.: A Separation Theorem for the Stochastic Sampled-Data LQG Problem. Int. J. Control 23 (1976), pp. 237—244. ... 17
9. IRWIN, G. W., ROBERTS, A. P.: The Luenberger Canonical Form in the State/Parameter Estimation of Linear Systems. Int. J. Control 23 (1976), pp. 851—864. ... 13
10. ISERMANN, R., KNEPPO, P.: Rechnergestützter Entwurf von Regelalgorithmen für Prozessrechner. Regelungstechnik 24 (1976), pp. 189—196. ... 15
11. KÁRNÝ, M.: Pravděpodobnostní identifikace a reprodukující se tvary distribucí (s aplikací na určování řádu dynamické soustavy) [Probabilistic Identification and Self-Reproducing Distribution Forms (with Application to Dynamic Plant Order Determination)]. Ph.D. Thesis at the Inst. for Information Theory and Automation of the Czechoslovak Academy of Sciences, Praha 1976. ... 13
12. KOUPAN, A., MÜLLER, P. C.: Zur numerischen Lösung der Ljapunovschen Matrizengleichung $A^T P + PA = -Q$. Regelungstechnik 24 (1976), 167—169.
13. LAINIOTIS, D. G.: Partitioning: A Unifying Framework for Adaptive Systems, I: Estimation and II: Control. Proceedings of the IEEE 64 (1976), pp. 1126—1143 and 1182—1198. ... 13, 14, 17
14. LINDQUIST, A.: On Integrals in Multi-Output Discrete-Time Kalman-Bucy Filtering. North-Holland, Mathematical Programming Study 5 (1976), pp. 145—168. ... 14, 16, 17
15. LINDQUIST, A.: Linear Least-Squares Prediction Based on Covariance Data from Stationary Processes with Finite-Dimensional Realizations. Proc. Congr. on Operations Research, Stockholm 1976. ... 13, 14, 17
16. LINDQUIST, A.: Some Reduced-Order Non-Riccati Equations for Linear Least-Squares Estimation: The Stationary Single-Output Case. Int. J. Control 24 (1976), pp. 821—842. ... 13
17. LJUBOJEVIC, M.: Hierarchische Verkopplung dynamischer Abtastregler bei Zerlegung in zwei Ebenen. Regelungstechnik 24 (1976), pp. 411—414. ... 15
18. LUDYK, G.: Steuerbarkeit, Erreichbarkeit, Beobachtbarkeit und Rekonstruierbarkeit zeitvarianter linearer zeitdiskreter Systeme. Regelungstechnik 24 (1976), pp. 417—426. ... 9
19. MORRIS, J. M.: The Kalman Filter: A Robust Estimator for Some Classes of Linear Quadratic Problems. IEEE Trans. on Information Theory IT-22 (1976), pp. 526—534. ... 14, 17
20. O'REILLY, J., NEWMANN, M. M.: On the Design of Discrete-Time Optimal Dynamical Controllers Using a Minimal-Order Observer. Int. J. Control 23 (1976), pp. 257—275. ... 14
21. PIECI, G.: Stochastic Realization of Gaussian Processes. Proceedings of the IEEE 64 (1976), pp. 112—122. ... 14
22. RAO, T. S.: A Note on the Bias in the Kalman-Bucy Filter. Int. J. Control 23 (1976), pp. 641—645. ... 14
23. SEBAKHY, O. A.: A Discrete Model Reference Adaptive System Design. Int. J. Control 23 (1976), pp. 799—804. ... 15
24. SILVERMAN, L. M.: Discrete Riccati Equations: Alternative Algorithms, Asymptotic Properties and System Theory Interpretations. Control and Dynamic

Systems. Advances in Theory and Applications, Vol. 12 (1976), Academic Press, Inc., New York.   14, 16, 17

25. TSE, E., BAR-SHALOM, Y.: Actively Adaptive Control for Nonlinear Stochastic Systems. Proceedings of the IEEE **64** (1976), pp. 1172—1181.   17

1977

1. AIDALA, V.: Parameter Estimation via the Kalman Filter. IEEE Trans. on Automatic Control AC-22 (1977), pp. 471—472.   13
2. ANDREEV, YU., N.: Algebraical Methods of State Space in Linear Plant Control Theory. (Андреев Ю. Н.: Алгебраические методы пространства состояний в теории управления линейными объектами. Автоматика и телемеханика **42** (1977)), pp. 5—50.   monograph
3. BAGCHI, A.: Consistent Estimates of Parameters in Noisy Dynamical Systems. Int. J. Control **26** (1977), pp. 883—900.   13
4. BITSORIS, G., BURGAT, C.: Stability Analysis of Complex Discrete Systems with Locally and Globally Stable Subsystems. Int. J. Control **25** (1977), pp. 413—424.   12
5. BOHN, E. V., DE BEER, M. K.: Consistent Parameter Estimation in Multi-Input/Multi-Output Discrete Systems. Automatica **13** (1977), pp. 301—305.   13
6. BREWER, H. W., LEONDES, C. T.: Least Squares Estimation of Nonstationary Covariance Parameters in Linear Systems. Automatica **13** (1977), pp. 265—277.   13, 14
7. CASTI, J. L.: Dynamical Systems and Their Applications. Academic Press, New York, 1977.   monograph
8. CZÁKI, F.: State-Space Methods for Control Systems. Akadémiai kiadó, Budapest, 1977.   monograph
9. CLEMENTS, D. J., ANDERSON, B. D. O.: Linear-Quadratic Discrete-Time Control and Constant Directions. Automatica **13** (1977), pp. 255—264.   15
10. FORRASINI, E., MARCHESINI, G.: Computation of Reachable and Observable Realizations of Spatial Filters. Int. J. Control **25** (1977), pp. 621—635.
11. ISERMANN, R.: Digitale Regelsysteme. Springer-Verlag, Berlin, 1977.   monograph
12. KITAGAWA, G.: An Algorithm for Solving the Matrix Equation $X = FXF^T + S$. Int. J. Control **25** (1977), pp. 745—753.   12, 15
13. KLAMKA, J.: Relative and Absolute Controllability of Discrete Systems with Delays in Control. Int. J. Control **26** (1977), pp. 65—74.   9
14. LANGHOLZ, G., BAR-NESS, Y.: On Observability of Sampled-Data Systems. Int. J. Systems Sci. **8** (1977), pp. 697—704.   9
15. LEDEN, B.: Multivariable Dead-Beat Control. Automatica **13** (1977), pp. 185 to 188.   15
16. LIU, R., SUEN, L. C.: Minimal Dimension Realization and Identifiability of Input-Output Sequences. IEEE Trans. on Automatic Control AC-22 (1977), pp. 227—232.   9
17. LIU, R., SUEN, L. C.: Determination of the Structure of Multivariable Stochastic Linear Systems. Conference on Decision and Control, USA, 1977.   11
18. LJUNG, L.: The Extended Kalman Filter as a Parameter Estimator for Linear Systems. Report Linköping University, ISY-I-0154 (1977).   13
19. LUO, Z.: Transformations between Canonical Forms for Multivariable Linear Constant Systems. IEEE Trans. on Automatic Control AC-22 (1977), pp. 252 to 256.   10, 11
20. MAHMOUD, M. S., VOGT, W. G., MICKLE, M. H.: Multilevel Control and Optimization Using Generalized Gradients Technique. Int. J. Control **25** (1977), pp. 525—543.   15

REFERENCES

21. Mita, T., Arakawa, H.: On Eigenvectors of the Canonical Matrix for Multi-Input Controllable Systems. IEEE Trans. on Automatic Control AC-22 (1977), pp. 262—263. 11
22. Myoken, H., Uchida, Y.: A Minimal Canonical Form Realization for a Multivariable Econometric System. Int. J. Systems Sci. 8 (1977), pp. 801—811. 9, 10
23. Neuman, C. P.: Discrete System Identification and Iterative Solutions of Linear Systems. Int. J. Control 25 (1977), pp. 637—645. 13
24. Rasmy, M. E., Hamza, M. H.: Suboptimal Control of Minimum-Time Linear Discrete Systems. Int. J. Control 25 (1977), pp. 361—373. 15
25. Sagara, S., Wada, K.: On-Line Modified Least-Squares Parameter Estimation of Linear Discrete Dynamic Systems. Int. J. Control 25 (1977), pp. 329—343. 13
26. Sakava, M., Narutaki, R., Suwa, T.: Optimal Control of Linear Systems with Several Cost Functionals through a Multicriteria Simplex Method. Int. J. Control 25 (1977), pp. 901—914. 15
27. Sinha, N. K., Kwong, Y. H.: Recursive Estimation of the Parameters of Linear Multivariable Systems. IFAC Symp. on Multivariable Technological Systems, 1977, Fredericton, N. B., Canada. 13
28. Strejc, V.: Least Squares in Identification Theory. Kybernetika 13 (1977), pp. 83—105. 13
29. Sundareswaran, K. K., McLane, P. J., Bayoumi, M. M.: Observers for Linear Systems with Arbitrary Plant Disturbances. IEEE Trans. on Automatic Control AC-22 (1977), pp. 870—871. 14
30. Vaněček, A.: Control of Initially Unknown Plants. Kybernetika 13 (1977), pp. 371—420. 15
31. Wu, M. Y., Sherif, A.: On Explicit Solution, Stability and Reduction of a Class of Linear Time-Varying Discrete-Time Systems. Int. J. Control 25 (1977), pp. 303—310. 12
32. Young, P., Whitehead, P.: A Recursive Approach to Time-Series Analysis for Multi-Variable Systems. Int. J. Control 25 (1977), pp. 457—482. 13

Subject Index

Algorithm Faddeev's   76, 140, 142

Basis   111, 279, 287, 293
Bayes rule   253
Bayes state estimate   253
Bellman's dynamic programming   13, 248

Causality   16
Central moment   233
Compensation of disturbing variable   323
Condition of
  controllability   99, 100, 101
  identifiability   106
  observability   102, 104, 124, 157
  reachability   99, 101, 113
  reconstructability   102, 104
  stabilisability   102
  transversality   342
Configuration,
  feedback   94
  parallel   92
  series   93
Consistence   16
Control,
  command   317
  law   334
  time optimal   270
Control error   260, 291
Controllability,   97, 100
  complete   98
Controller   259, 260, 263, 264, 294, 310
Convolution   60
Cost function,
  finite number of control steps   270
  finite number of control steps, strong version   279, 284
  finite number of control steps, weak version   279, 290
  pole assignment   263
  quadratic   246, 249, 326, 332, 362, 368, 382
Coupling,
  one way   165
  two way   170
Covariance matrix,   216, 227, 234, 241, 242, 243, 245, 248, 250, 254
  error   227, 228, 243, 245
Crammer rule   76

Decomposition,   44, 61, 107, 268
  canonical   106
Delay   292
Detectability   102, 105
Dispersion   227, 228, 405
Divisor,
  elementary   130, 131
  greatest common   129, 131
Duality   393
Dynamic programming   13, 348

Eigenvalue,   111, 113, 117, 118, 130, 189, 355
  cyclic   360
  distinct   113
  multiple   117, 118, 131, 189
  split   81
Eigenvector,   114, 119, 294, 363, 364
  generalised   119, 294, 364
Equation,
  characteristic   148, 191, 304
  controller   259, 263, 267, 268, 271, 286, 293
  difference   20, 48, 51, 52, 55, 64, 212
  difference, stochastic   218
  differential   21, 25 27, 32, 34

SUBJECT INDEX 423

homogeneous 42, 88
impulse model 219
likelihood 228
Lyapunov 202
nonhomogeneous 43
of dynamics 26, 50
output 26, 50
Riccati 202, 338, 342, 347, 349, 358, 361, 363, 365, 383
state 24, 25, 29, 31, 35, 50, 52, 54, 59, 65, 333
state, block diagram 26, 29, 31, 50, 53
state, continuous time 28, 31, 33, 34
state, continuous time, solution 42
state, discrete time 50, 52, 54, 56
state, discrete time, solution 59, 67, 68

Equivalence,
  observable 40
  strict 41
Error covariance matrix 227, 228, 243, 245
Estimator of
  the order n 237, 298, 310, 317
  the reduced order 241, 302, 311, 317
  the state, deterministic 237, 310, 311, 323, 325, 378
  the state, stochastic 384, 392, 394

Factor,
  input 51
  invariant 129
Faddeev's algorithm 76, 140, 142
Feedback 268, 271
Finite
  number of control steps 270
  sampling time 90
Forgetting 405
Form,
  canonical of controllability 141, 170, 172, 316, 318
  canonical of observability 143, 165, 166, 171, 182
  canonical of reachability 139, 165, 170, 172
  canonical of reconstructability 145, 170, 299
  Frobenius 29, 140, 265
  Jordan 44, 47, 61, 112, 128, 162
Function,
  Hamilton 340
  likelihood 228, 232, 253
  Lyapunov 198, 201, 335

regression 216
staircase 68, 71
transfer 44, 47, 61, 77, 79, 116, 147
transition 16

Hamiltonian 340

Identifiability 105
Index of
  observability 154, 157, 171, 302
  reachability 154, 156, 173, 276, 314, 316
Indicator of reachability 316
Initial conditions 18, 42, 43, 62
Input
  alphabet 18
  concatenation 16
Interpolation 242
Invariants 313
Inversion generalized 363

Jordan
  block 118, 124, 162
  form 44, 61, 112, 128, 162

Kalman
  filter 241, 249
  filtering 13, 225, 241, 382
Kronecker indicator 316

Least squares,
  geometrical interpretation 221
  method 214, 243, 405
  properties 226

Markov process 244
Matrix, 81
  adjoint 78
  characteristic 130, 346
  Cholesky 230, 232, 353, 405
  complex 137
  controllability 272
  controller 259, 263, 265, 276, 278, 290, 294, 295, 313, 318, 320, 336, 394
  covariance 216, 228, 232, 241, 402
  cyclic 119, 130, 132, 134
  diagonal, canonical 129
  elementary rotations 403
  error covariance 227, 228, 234, 243
  equivalent 129
  Euler 344, 355, 357, 363, 365

feedback 268, 270, 316, 324
Frobenius 29, 140, 265
fundamental 43, 88
impulse response 60, 150
information 231, 404
input 26, 50
inversed system 291
inversion lemma 248, 256, 347, 396
Jordan 44, 47, 63, 65, 113, 130, 134, 137, 356
negative definite 399
negative semidefinite 399
nilpotent 271
observability 112, 144, 146, 168, 171, 177, 179, 182, 209, 298
of dynamics 26, 29, 50
output 26, 29, 50
positive definite 399
positive semidefinite 399
quasidiagonal 132
reachability 112, 124, 140, 141, 173, 209, 269, 272, 292, 313
similar 111, 130
split 81
square root 230, 232
Vandermonde 114
weighting 60, 245, 333
Maximum likelihood, 228, 252
  properties 233
Maximum principle, Pontryagin's 13, 340
Method of
  least squares 214, 243, 405
  perturbations 340
Minimum description 181
Model, 215
  autoregression 217
  generalised 216
  ordinary 216
  regression, finite memory 215
  regression, infinite memory 217

Noise 206, 252, 254, 385
Number of control steps 270, 279, 284, 290, 295

Observability 102, 104, 106, 124, 157
  minimum 159
Operator, 38
  homogeneous 39
  linear 39

  matrix 39
  product 38
  summation 38
  vector 39
Order,
  controlled plant, system 25, 166, 169, 171, 173, 178, 181, 208, 210, 222, 302
  relative 291, 362
Orthogonal projection 221
  relative 291, 362
Orthogonal projection 221
Orthonormalisation 398
Output, 15
  representation 16

Parameter identification, estimation, 205, 228 233
  deterministic 206, 207
  least squares 214
  maximum likelihood 228
  stochastic 206
  unbiased 226, 228, 234
Plant,
  multi-input/multi-output controlled, continuous (multivariate, multidimensional) 34
  multi-input/multi-output controlled, decomposition 269
  multi-input/multi-output controlled, discrete 55, 209, 213, 237, 276, 376
  single-input/single-output controlled, continuous (univariate, one-dimensional) 27
  single-input/single-output controlled, discrete 48, 208, 212, 220, 368
Pole assignment 263
Polynomial,
  characteristic 81, 111, 113, 130, 167, 169, 190, 263, 268, 269, 301, 302, 392
  characteristic of the estimator 298
  characteristic, reduction 196
  minimum 76, 81, 124
Prediction 224, 242, 250
Process Markov 244

Random sequence 241, 382
Reachability, 97, 98, 100, 106, 124, 155
  complete 98
  minimum 157
Realization,
  minimum 107, 207
  of the impulse response 107, 207

SUBJECT INDEX

Reconstructability 102, 104
Reduction of characteristic polynomial 196
Regression model,
  linear, finite memory 215
  linear, infinite memory 217
Response,
  impulse 60, 140, 145, 147, 207, 208
  unit-step 207
Routh-Shur stability test 192, 195

Sampling, 20, 86
  asynchronous 87, 150
  interval 86, 87, 149
  period 20, 48, 271
  time, finite 90
  time (instant) 68, 87
Segment,
  input 15, 16, 19
  output 16, 19
Separability 259, 309, 311, 383, 390
Separation principle 383
Space,
  input 16, 18, 17
  output 17
  state 15, 18
Spectral factorization 357
Stabilisability 97, 101
Stability, 186, 392
  in the large 188
  region 187
  uniform 187
Stability test,
  characteristic polynomial 190
  fundamental matrix 189, 191
  impulse response 194
  Lyapunov 187, 188, 197, 201, 334
  Routh-Shur 192, 195
  system input and output 193
  trace of the fundamental matrix 190
Stacionarity 40
State, 15
  assymptotically stable 187
  basis 111, 279, 287, 293
  equilibrium 186, 188, 334
  equivalent 19
  initial 18
  steady 279
  system, controlled plant 15, 16, 19, 237, 241, 253, 259
State estimate, 241

  deterministic 237
  in the Bayes sense 253
  least squares 243
  maximum likelihood 252
  predicted 243, 245
  probabilistic 253
  statistical 241
Sum of
  squares 220
  square errors 368
Superposition 44
Synthesis,
  deterministic 259
  in Kalman's sense 382
  stochastic 382
System, 15
  analysis 206
  assymptotically stable 187, 189, 190, 191, 293
  autonomous 187
  continuous 18
  cyclic 124, 299
  deterministic 17, 19
  dimension 25
  discrete 18
  dynamic 15
  free 106, 186
  identification 206
  input 15, 16, 18
  model 15, 21, 23, 205, 215, 226
  multi-input/multi-output 34, 55, 154, 209, 213, 237, 276, 376
  neutrally stable 187
  on the boundary of stability 187
  order 25, 166, 169, 171, 173, 178, 181, 208, 210, 222
  order, relative 291
  output 15, 17, 18, 362
  physically realisable 16
  reversible 98
  single-input/single-output 27, 48, 208, 212, 220, 368
  stable 188
  stochastic 17
  time invariant 186, 338, 343, 346, 351, 354

Theorem,
  Cayley-Hamilton 192, 267
  Lyapunov 197, 334
Transformation, 20

Laplace 47, 75
linear 110
vector 38
Z 65
Transition 87, 270, 279

Value,
  error 233
  mean 216, 226
  shifted 82
Variable,
  admissible input 287, 293, 294
  command 146, 261, 279, 291, 318, 369
  command, admissible 283, 284
  command, exponential 282
  command, polynomial 281
  command, sinus 281
  controlled 25, 55, 149
  controlling (control, forcing) 55, 233, 270
  disturbing 32, 54, 206, 216, 324, 385
  input 18, 206
  output 17, 25, 206
  random 33, 215, 216, 226, 233
  state 15, 17, 28, 29, 33, 49, 51, 56, 64, 206, 332
Vector,
  dependence 99, 397
  input 17
  of the state variables 17, 28, 34, 46, 56
  of the state variables, extended 152, 285
  orthogonal 398
  orthonormal 398
  output 17
  random 216

Wiener-Kolmogorov filtering 357

Z-transformation 65